TESTIMONIALS

"Ade's book provides a rare opportunity to learn and to have a sober reflection on the many steps already taken by the Nigerian state in her space journey. Drawing from his wealth of experience, Ade Abiodun examines what could have been done differently and makes profound suggestions on how to chart a progressive and sustainable space programme in Nigeria."

—Peter O. Adeniyi,
Emeritus Professor of Geography and Geomatics, University of Lagos, Nigeria
& Pioneer President of African Association
of Remote Sensing of the Environment

"In this book, Adigun Ade Abiodun presents a candid and somewhat troubling assessment of Nigeria's current state of space affairs. He also proffers ideas to take the country in a very different direction. His thorough and detailed set of recommendations on space priorities, backed by a robust science and technology foundation, should guide a successful space engagement both in Nigeria and in other developing countries. A very worthwhile prognosis from a person who has immersed himself, for decades, in the subject, both in Nigeria and within the international community."

Tom Alfoldi,
Retired Geoscience Specialist at Canada Centre for Remote Sensing,
and Renowned Nature Photographer, Ottawa, Canada

"Adigun Ade ABIODUN's book, *Nigeria's Space Journey—Understanding its Past, Reshaping its Future*, is insightful and prescriptive. As the former United Nations Expert on Space Applications and former adviser on space science and technology to the President of Nigeria, Abiodun gained a unique and compelling perspective which informed his vision for Nigeria's space future. The book lucidly demonstrates, with rich examples, how space can be harnessed to make a difference in the lives of Nigerians and in Nigeria's contributions to the solution of global problems. It is an eye-opener, an indispensable guide for building a viable national space programme, and a must read book."

Yewande Afonja,
Co-Founder of STATCO Publishers, Nigeria and Co-Editor
of Family meditations over the Internet

NIGERIA'S SPACE JOURNEY

NIGERIA'S SPACE JOURNEY

UNDERSTANDING ITS PAST
RESHAPING ITS FUTURE

ADIGUN ADE ABIODUN

African Space Foundation
www.africanspacefoundation.org

Printed in the United States of America

Library of Congress Control Number: 2016920275

ISBN Paperback: 978-0-9983321-0-9

Interior Design: Ghislain Viau
Cover Design: Samuel Obadina

*To the Youth of Nigeria, its future generations, and their compatriots in
Africa and in the emerging economies of the world, in my belief that as He did and
continues to do for me, God will guide them, even when they cannot see the path ahead.*

*I hope that this book will inspire them to strive to make a difference
and be agents of change, wherever they are in this world.*

CONTENTS

Nigeria's Space Priorities

Contributions to the
Resolution of Global Concerns

Moving Ahead

FOREWORD

Ade Abiodun is a dear friend and mentor and I am honoured to write the Foreword to this book.

The book is part personal memoire, part historical account and, above all, a call to action. One of the challenges that Ade has faced in writing this book has been to decide to whom it is addressed. Is it for political leaders or officials? Is it just for Nigerians or for Africans or for the developing world, more generally? Is it for professionals or students? In the end, his book speaks to all of these groups and indeed to anybody who seeks a better understanding of Africa's space journey.

Dr. Abiodun describes Nigeria's aspirations to participate in global space development and to establish a space programme of its own. He speaks with authority, passion and deep understanding of the Nigerian political and social landscape that has played such an important role in shaping the directions that the space programme has taken. He chronicles successes and disappointments, excitements and frustrations and he does so in the context of the broader development agenda for the nations and peoples of Africa.

He is passionate about science and technology and especially about the potential for space technologies to help Africa and Africans in two ways: to inform public policy decision-making; and to give the peoples of Africa goals to which they can aspire and in which they can take pride.

He has an unshakable conviction that the peoples of Africa can be lifted from the shackles of the various challenges that bedevil development in the continent. Attaining this goal demands that the governments of Africa shall invest systematically and strategically in educational programmes that emphasise science and technology in which facts speak for themselves, and lead to more rigour and accountability in decision-making.

He is very well-aware of the difficulties and the magnitude of this challenge but he is an optimist at heart. And he combines his optimism with a profound belief that, ultimately, leaders and their people will act wisely and do the right thing.

I know of no other Nigerian, or African indeed, who could have written a book such as this. I highly recommend it to every African and all the citizens of the developing world as well as those interested in Africa's success in space.

Brett Biddington
Principal, Biddington Research Pty Ltd, Canberra, Australia (space and cyber-space policy, security and industry development)
Admitted as a member of the Order of Australia for services to the Australian space sector in 2012

Canberra, Australia,
December 15, 2016

ACKNOWLEDGEMENTS

It took a loving village, called Araromi ABIODUN, to raise this author. I also offer my heartfelt gratitude to those outside the village, particularly Prof. Lubos Perek of the Checz Republic and former Chief of the United Nations Outer Space Division (UN-OSAD), since renamed the United Nations Office for Outer Space Affairs (UN-OOSA), and Ambassadors Dapo Fafowora and George Dove-Edwin, both of Nigeria, for leading me in the path that contributed to the writing of this book.

I also don my hat to all the 24 members of the Nigerian Experts Consultative Group (NECG) and the staff of the office of the Senior Special Assistant (to the President) on Space Science and Technology (SSAP/SST), whose names appear in Chapter V of this book, and who volunteered their time and talent, and worked tirelessly with me, to develop the report we submitted in September 2000, to President Olusegun Obasanjo, titled: "Nigerian Space Programme: *A Blue Print for Scientific and Technological Development.*" The report served as the foundation document for the first Nigeria's *National Space Policy and Programme* that was adopted by the Federal Executive Council on July 4, 2001. Over the years, many friends and compatriots, including the youth of Nigeria and Africa, particularly the *Youth Forum of the African Leadership Conference on Space Science and Technology,* have also urged me to put on paper a compelling message on space science and technology for the leadership and the youth of Nigeria and Africa. I thank all of them for their encouragement, and I hope that this book inspires them to strive to make a difference and be agents of change in our world. Etim Offiong stands out for his insightful review of a number of chapters, and for his skillful rendition of many of the figures and tables in the book.

You should meet a number of my friends and colleagues who volunteered their time and talent, as readers, and who helped me critique different parts of this book. I sincerely value their outstanding work, their kindness and encouragement—offered to me freely and selflessly. However, I take full responsibility for all its contents. Among the readers are Prof. Peter Olu Adeniyi (Nigeria), Mrs. Yewande Afonja (Nigeria), Tom Alfoldi (Canada), Prof. Ekundayo Balogun (Nigeria), Ms. Frances Brown (United Kingdom), Ms. Carol Rose (USA), Dr. Edward Omotoso (Nigeria/USA), Dennis Stone (NASA-USA) and Brett Biddington (Australia). I am indebted to all of them for their review of the various drafts of my manuscript. Brett, who is among those leading the on-going development of Australia's national space programme, gave his support and encouragement from the inception of this book. He read and critiqued the entire manuscript; and with a great heart, he kept my toes on the burner until I got the work done. He also wrote the Foreword. I am also very grateful to Samuel Oladimeji Obadina of Feather Consult, Ile-Ife, Nigeria, for his very skillful development of the graphic art design for the book cover, and for his contribution to the rendering of several of the images in the book.

Finally, my family has been supportive in many ways. I salute my dear cousin, Dr. Olajide Bababode Odubiyi, his wife, Marion, and their daughter, Sade. I offer them my heartfelt thanks for their guidance and day-to-day encouragement to ensure that this book comes out the way it should. I depended on their wise suggestions and counsel and I am very grateful for their steadfast support. I am also indebted to my dear partner and spouse, and occasional editor, Judith Burgess Abiodun, for her love and patience, and for her belief in me and in the message of the book; and also to her grandniece, Serene Shirley, who reminded me of the importance of *cliff hangers*. Lastly, it brought me great joy that I had the support of my dear daughters, Ololade and Olukemi, their children—Joshua and Shavyla, and their spouses—Emanuel and Daryl—respectively. They showered me with love and encouragement during their regular check-up on Daddy/GrandPa and on the progress of the book.

INTRODUCTORY REMARKS

Our ancestors would be dazzled, and certainly be overwhelmed, to see how far we have come today—from a society that once relied on talking drums and the gong to send messages across villages, to one with access to satellite communication technologies and assets to attain the same goal, not only locally but also globally. What a revolution! This particular revolution has a number of inherent challenges, including those associated with the development of a road map—the national space policy and programme—and the building of a vibrant space institution.

These and other unforeseen challenges notwithstanding, a nation's space programme should attract and engage the right calibre of people that can advance human knowledge of outer space and use that knowledge, not only to improve the wellbeing of the citizenry but also to contribute to the solution of regional and global concerns. Doing so will also require a lesson or two from the earlier space explorers; the latter realised and understood that investment commitments in the space enterprise should be built on a solid science and technology (S&T) foundation and that its growth takes time before the investment can produce a meaningful and bountiful harvest.

In this book, I have focused on Nigeria as a case study and on how some of these challenges have played out in its space programme. It is my hope that these challenges and the manner in which they were addressed would serve as lessons, not only for Nigeria and its people, particularly its decision makers, but also for other countries in similar situations. In writing this book, I drew from (i) my three plus years (March 2000 through June 2003) of service as the Senior Special Assistant (to the President of Nigeria) on Space Science and Technology (SSAP-SST), under President Olusegun Obasanjo's administration, (ii) my many interactions, with the global space community, as the United Nations Expert on Space Applications

(November 1981–September 1999), (iii) my role as the Chairman of the United Nations Committee on the Peaceful Uses of Outer Space—COPUOS—(June 2004–June 2006), and (iv) my role as Adviser to the Nigerian delegation to COPUOS (2000–2004 and 2007–2014).

Since my retirement from the services of the United Nations, I have also had the privilege of participating in and contributing to a number of unique space-related events—some of them are cited here.[1] These interactions and others, including my role as the SSAP-SST, have expanded my horizon about what Nigeria's future activities in space should be. This book thus offers me an opportunity to share with Nigeria and the developing world, my views and *The Thinking in other Lands*, on how a number of countries are setting out and pursuing their respective space agenda for their own future, to drive home the lessons therein for us, and to offer my own proposals for the way forward. The latter includes the rebuilding of the foundation of science and technology in Nigeria, a major pre-requisite for its attainment of its space aspirations. The book is divided into six main segments, namely: *Space in Human Lives; Space Efforts in Nigeria; Reawakening of Nigeria; Nigeria's Space Priorities; Contributions to the Resolution of Global Concern; and Moving Ahead.*

Space in Human Lives

The space revolution that has captured the attention of the global community can perhaps be best understood and appreciated by reflecting on Socrates' challenge to humanity to:

".....rise above the surface of the Earth, to the top of the atmosphere and beyond, in order to understand the world in which we live —Socrates.[2]

---

1 (i) The Workshop on "Space Vistas: The Promise and Challenge of Responsive Space," International Space University, Strasbourg, France, January 26 –29, 2004; (ii) International Workshop on *Humans and Space in the Next 1,000 Years,* Foundation for the Future, Bellevue, State of Washington, USA, June 23-26, 2006; (iii), High Level Panel on "*Global Exploration Strategy—The Framework for Coordination,* 50th Session of COPUOS, Vienna, Austria, June 6, 2007; (iv) Nurturing the development of space technology—Symposium to strengthen the partnership with industry, Scientific and Technical sub-Committee of COPUOS, Vienna, Austria, February 8, 2010; and (v) International Perspectives on Space Sustainability from Africa, Asia, and Latin America, Secure World Foundation, Washington, DC, USA, May 1, 2012.

2 Plato's Dialogue, *Phaedo*, around 109e (Plato. Plato in Twelve Volumes, Vol. 1 translated by Harold North Fowler; Introduction by W.R.M. Lamb. Cambridge, MA, Harvard University Press; London, William Heinemann Ltd. 1966.)

Gaining a better understanding of the world we all live in would have to await the1957-58 International Geophysical Year (IGY). That was when the Union of Soviet Socialist Republics (USSR) successfully launched *Sputnik-1*, the first human artificial satellite, on October 4, 1957, followed by the subsequent launch of *Explorer-1*, by the United States of America (USA) on January 31, 1958. These two key IGY-related achievements marked the beginning of the space race between the USSR and the USA as well as the beginning of the space age that we know today. To douse the fear of global domination by these two new space powers, the United Nations established, in 1959, its Committee on the Peaceful Uses of Outer Space (COPUOS), an action that brought some global calm. With the fall of the Berlin Wall in 1989 and the subsequent breakup of the USSR, the era of space race eventually ended yielding place to the new era of space enterprise. The latter culminated in the joint development, by 2011, of the International Space Station (ISS) by five partners. Between 1957and today, spacecraft from most of these partners and others have landed on most of the planets in our Solar System. Today, Mars, which, on average, is 140 million miles (225 million km) from Earth, has become the target of the next phase of manned space challenge because, in its evolution and formation, it is comparable to Earth.

Over time, satellites that once served as instruments of war became dual technologies that began to deliver economic benefits to humankind, including the meeting of our basic everyday needs. A few African countries, and others in the developing world, as well as a number of private entities, soon became aware of the apparent opportunities that space offers and the need to be active space players in order to gain the rewards of participation. Such active participation has shed the isolation of many nations and their peoples; it has enabled nations and their citizens to acquire and share knowledge in various areas of human endeavour; and it has made a difference in human lives in many societies.

India led the developing world of the 1970s by being the first to throw its hat into the space ring to test the realities of the space applications possibilities referred to earlier. In 1974, it borrowed an American satellite, ATS 6, for one year and used it to determine whether the satellite could deliver a variety of education programmes that would improve the wellbeing of Indians, particularly those in the rural communities. The project, known as Indian Satellite Instruction Television Experiment (SITE), was an instant success, and it set the stage for the continuing space success India has enjoyed ever since. Other countries soon initiated their own space programmes. They established relevant research institutions, undertook research and development programmes, designed, built, and tested a number of space experiments and in later years, successfully designed and built their first satellites—all of which were launched for them by either the USSR or the United States. In most cases, these goals were attained because of the singular contributions of the research establishments in each country's universities and the funding support provided to them by their respective national governments.

Space Efforts in Nigeria between 1957 and Today

Chronicled in this section of this book are the events that gave Nigeria its initial space awareness and interactions. For example, the contributions of Nigerian scientists to the study of the ionosphere, in the late 1950s to the early 1960s, a period that coincided with the IGY, introduced Nigeria to the peripheries of outer space, and may have influenced the request to Nigeria, by the USA, to provide local support for America's initial manned space missions. That support and other factors most likely set the stage for the1963 first ever live satellite telephone conversation between America's then youngest President, and Nigeria's first Prime Minister. Less than eight months before this historical phone call, a 40 pound meteorite (rock) from Mars, named Zagami, landed in Zagami, Katsina State, Nigeria. But what eventually happened to *the Zagami rock*, a national treasure and very important object that is worth more than gold or diamond, in the open market, such as E-Bay, today? The book shows an illustrated diary of *Zagami's* known fate since its landing on Earth and its subsequent contact with man; a portion of it is still in Nigeria, today. That notwithstanding, the current Nigerian administration should seek a full account of *Zagami's* fate from the Nigerian Geological Survey, since renamed the Nigerian Geological Survey Agency, the Nigerian government entity that became the custodian of the Martian rock a day after it landed in 1962.

The type of national focus that guided the Indian initiative and its SITE project cited above eluded Nigeria's early space efforts during its long military era of 1966-1979 and 1983-1999. In a parliamentary democracy, issues are discussed and majority or consensus positions often guide the next step(s) of the government. But that is a luxury that a military government, albeit, a unitary one, does not indulge in. Nigeria was no exception. That was when a number of federal government ministries, unilaterally and unchallenged, made their own space-related commitments and purchases or hi-jacked projects that were outside their mandates and for which they had no competence to execute. One could have imagined the outcome of such actions—wrong choice of technologies; acquisition of technologies that were at the teething stage; and abandoned projects. The projects made no national impact and were financial sinkholes for the country. At that time, the national pre-occupation with what a given technology can do for the country and its people overshadowed the critical importance of why and how a given technology works the way it does. That syndrome lingered on for a long time and stuck itself into Nigeria's psyche, as each ministry dreamt up its own brand of money-gobbling project. At the close of the 20th century, the Federal Ministry of Science and Technology (FMST) came up with its own proposal, a *Nigerian Satellite Project*, which won the approval of the Federal Executive Council (FEC) on March 5, 1999; at that same sitting, the FEC also established the National Space Research and Development Agency (NASRDA) as the entity to carry the nation's space mantle. With the Nigerian Satellite Project on track, Nigeria was able to show-case itself and creditably participated, as a budding African space power, in the 1999 Third United Nations Conference on the Peaceful Uses of Outer Space (UNISPACE III).

Although it would have been prudent to think and plan for a national space programme which would have embodied university-led research and development efforts, instead, the nation embarked on a Nigerian Satellite Project without a space policy or programme. Against sound advice from several quarters, the Nigerian government was internally persuaded, with *unjustifiable justifications*, to go ahead. The nation did, and it subsequently bought a satellite, *NigeriaSat-1*, from abroad; it was also launched abroad in September 2003. But a critical issue was left unaddressed in the process—the crumbling state of the nation's science and technology, an indispensable foundation that is needed to power and sustain its space aspiration. Indeed, when the FMST decided that it was time for Nigeria to be space-bound, it followed the same approach taken by the other government ministries that embarked on unilateral space-related projects before it; it side-lined the nation's universities. Thus, the Nigerian Satellite Project did not benefit from the available local rich scientific talents needed to guide the nation's space effort at its teething stage.

But no country has become space capable or space faring by first buying or purchasing satellites. A nation will earn that badge of honour by utilizing its S&T capabilities to indigenously develop and build its space tools and ancillaries that can be used to positively impact the lives of its people, and to enhance its economic development, on the ground. That is yet to happen in Nigeria.

Part of the problem has been the very limited understanding and appreciation of the role of science and technology (S&T) in the development process, particularly, since the military incursion into the political life of the nation. Our leaders and others in many developing countries bought into the idea that we should not re-invent the wheel, but that we should leap-frog, jump-start and fast-track our development efforts. Instead of focusing on technology development, we became pre-occupied with technology transfer and the resultant turn-key projects which we continue to heavily pay for, in exchange for foreign manufactured equipment, goods and services. However, by their very nature, turn-key projects do not contribute to intellectual stimulation. Indeed, entrenched embracement of turn-key projects as tools of national development has prolonged the chronic under-funding of S&T and the absence of S&T culture in the nation. Subsequently, it plagued the nation's S&T capability and crippled its overall growth and development. I have used a number of globally accepted standard indices to illustrate and compare, through several figures, the S&T situation in Nigeria with those in a number of key African and developing countries.

You will also note the variety of reasons why the activities of both NASRDA and NigComSat Ltd have had no appreciable impact on Nigeria's development and on the wellbeing of Nigerians; their inability to perform as designed have also, on occasions, denied the nation's leadership the needed input in critical decision making. It is thus not surprising that foreign satellite operators have filled the void and their space assets are supplying the Earth observation and the satellite communications needs of the country and its people. Certainly, there is a lesson in all of these, particularly for *those who have Nigeria's interests at heart, because "those who fail to learn from history are bound to repeat its mistakes.*[3]" The actions and inactions highlighted above and what should be the nation's next steps dictated the title of this book—**Nigeria's Space Journey:** *Understanding its Past, Reshaping its Future.*

Reawakening of Nigeria

The nation's science and technology foundation that was decimated by the civil war and by subsequent national inactions must first be rebuilt. Doing so successfully will need a different mindset, one that requires that we recollect what, at one time, were best in/about Nigeria—Nigeria's productive institutions and those capable Nigerians who used their intellect for the good of the nation. How do we emulate them today? According to an adage, "*It is not the strongest that survives, nor the most intelligent, but the one most responsive to change.*" Nigeria and most African countries are today enamored by/with innovation. But innovation has been around for a long

---

3 Santayana, George (1905). *The Life of Reason*, Volume 1, 1905, Originally published by Charles Scribner's Sons in 1905.

time; it is the process of translating an idea or invention into goods or services that creates value; it can also mean a change in a business policy and practice to deliver better products or services. Innovation transformed the former colonial powers into industrialised nations through their proven capacity to translate scientific knowledge into economic productivity as they added value to the raw materials they imported from their former colonies; a process that is still in progress today. Nigeria must equally go through such a transformation.

Change also comes in many forms. For example, Nigeria signalled a positive change in its future direction through the collective political decision of its citizens, in March 2015; the leadership brought into power by that 2015 election is now the custodian of the nation's space journey. How that journey is re-defined and re-designed, and how its programmes are executed will dictate its end result. Because the goal of that journey is to make a positive change in the lives of Nigerians, the nation must understand that science and technology is the overarching tool for Nigeria's future and that the health of its space activities is inseparable from the health of science and technology in Nigeria. To be sound and effective, the new Nigerian science and technology foundation must be built, as is the practice in progressive countries that are planning for the future, on (i) research and development at the federal and state levels and not on the multiplicity of academic institutions that hand out, annually to their graduating students, paper certificates which may not fetch any meaningful employment, nor on additional layers of bureaucracy, such as ministries of science and technology at the State level; and (ii) on the recognition and enforcement of transparency, accountability and meritocracy in the nation. God already endowed Nigeria, abundantly, in human resources and in material endowments. Our task is to change how we engage these human talents, at the public and private sector levels, to ably manage and harness these rich endowments for the good of all.

Nigeria's Space Priorities

Nigeria must harness space to meet the basic daily needs of the Nigerian people, within both its urban and rural communities. Space is a proven safety and security tool; it is also a recognised tool for securing and maintaining peace. Nigeria's space priorities should include an enrichment of our knowledge of the world in which we live and our place in it. In this category belong our greater understanding and effective use of our space-related natural endowments, particularly, the equator. Implementing each of these priorities requires informed decisions on the best nature and level of investment that would impact the nation's development agenda. Nigeria must also note that not every darling-technology of today will survive the next seven to ten years; some of them have become vulnerable to a variety of problems thus necessitating the search for new ideas and more endurable tools, with Nigeria making its own equitable contributions. Thus a national decision on each recommended priority should be predicated by an appropriate study that takes into consideration its sustainability.

Contributions to the Resolution of Global Concern

The world is anxiously awaiting Nigeria to productively use its human resources and endowments to secure its passport to its own future as a knowledge society and a developed country that is capable of meeting, not only its own needs, but also one that can contribute, assuredly, to the solution of major issues of regional and global concern. The latter include global warming and its attendant climate change and other consequences. As a fossil-fuel producing country, Nigeria is part of the problem; hence, it must be part of the solution. As agreed to at the 2015 Paris Climate Summit, Nigeria should recognise that the abatement of global warming is a shared global responsibility that can only be addressed through its concrete climate research programmes, its commitment to knowledge development and sharing, and its compliance with targeted carbon emission reduction goals. Similarly, in all living memory, humans and other living things have been sitting targets for a hit (man-made or anthropogenic) from space, at any time, and at any place; Nigeria and other equatorial countries went through such a space-related induced panic in April 2003. Thus, through a number of dedicated programmes, Nigeria should also become an active partner in the global space debris and Near Earth Objects (NEOs) tracking and mitigation efforts.

Moving Ahead

Moving ahead on this space journey requires the right compass, the nation's commitment to a number of action plans, and specific mechanisms for achieving the nation's space goals and priorities. A key mechanism is a broad-based and space-knowledgeable group of Nigerian experts that are at home and in the Diaspora. Such a group, hereby named *The Think-Tank,* should be constituted by and accountable to the National Space Council (NSC); it shall assist in enhancing the effectiveness of the Council and shall provide technical oversight for the execution of the nation's re-defined and re-designed space programme. *The Think-Tank* shall, *inter-alia*, ensure that all the stakeholders are part of a united national space effort which shall be reviewed at an annual *National Space Dialogue*. I have also recommended that a select few of Nigeria's decision makers and leading members of the nation's scientific and engineering communities should undertake a genuine study and learning mission to a number of key countries in order to gain a better understanding and appreciation of what it means to be an active player in S&T and in the space enterprise of today and tomorrow.

Finally, the effectiveness of the S&T and space investments by Nigeria and other space aspiring countries in Africa and in the developing world will dictate the extent to which they are able to attract international collaboration and partnerships. Today, the reality is that such partnerships are becoming the norm; they are manifested in the formation of cross-continental multi-national organisations—made-up of diverse groups of nations. Each member State has a stake in the organization which is driven by mutual interests and whose primary objective is cooperation on security, technology, economic development, and issues of global concern. Such partnerships have enabled several of the space-faring and the space-capable nations, as well as the nations in-transition of this world, to take bold steps that have enabled them to overcome their own challenges; they are moving ahead, and are reaping the rewards of their astute space investments. Nigeria must also first overcome its own internal challenges, many of which are elaborated upon throughout this book. Thereafter, it would need to resolutely commit itself to the building of a sound S&T foundation, the backbone of a redefined and redesigned national space programme that will enable it to attain its aspirations. Our actions, as a nation and a people, will tell.

Adigun Ade ABIODUN

SPACE IN HUMAN LIVES

Chapter I

HUMAN EXPLORATION OF OUTER SPACE—WHY AND HOW?

From time immemorial, humans around the globe and in their varying degrees of sophistication have always looked into the heavens for clues for their next steps as they journeyed through planet Earth. The outer space environment has always engaged our untiring attention. For example, on any given day whenever the sky is clear and blue, "Wow! What a beautiful day!" is the commentary on the lips of most city dwellers. However, should the sky be overcast and laden with impending rain showers, these same city dwellers would snipe at Mother Nature and complain that the rains would get them wet and impede their movements around the city. The opposite is the reaction of rural farmers whose livelihood depends on rain-fed agriculture; they would express their joy and thanks to God for the prospects of a rainfall that would water their crops and enhance their harvests. For the stargazers and the astronomers, a clear night sky brings excitement because it offers them some unique opportunities to observe and study the stars and other celestial bodies, using the tools at their disposal.

As each day goes by, we humans are fascinated by the outer space environment and particularly brim with excitement when, at sunrise and at sunset, we see the spectacular colours displayed in the horizon, and just marvel at these wonders of nature. These brilliant colours are brought about as planet Earth rotates on its own axis every 24 hours while orbiting the Sun. In the process, the rays of the Sun filter through the sky to bathe the Earth with its natural light. It is this light that is most resplendent at dawn and at twilight. Our appreciation of these colours, of course, assumes that we take an occasional break from our daily chores to gaze at the sky and experience, with joy, these wonders of nature.

Recorded history tells us that from time to time, visionaries and inquiring minds challenge their societies to seek their place in the Sun. Socrates, the great Greek philosopher, was such a thinker. He was credited to have envisioned an inter-relationship between outer space and the Earth and its human society, not only for the Greeks, but also for all humankind. He went on to challenge humanity when he declared that:

'Man must rise above the Earth to the top of the atmosphere and beyond, for only thus will he fully understand the world in which he lives' ...Socrates.[1]

For several centuries thereafter, many serious observers of outer space in different corners of the globe took incremental steps to respond to Socrates' challenge.[2] By the year 1950, newly available tools—such as rockets, radar, cosmic ray recorders, spectroscopes, and radiosonde balloons had opened the upper atmosphere to detailed exploration, while newly developed electronic computers facilitated the analysis of large data sets. But the most dramatic of the new technologies available to the International Geophysical Year (IGY) was the rocket; it rendered a coordinated worldwide study of the Earth's systems possible. With these new tools, the International Committee of Scientific Unions (ICSU) outlined, in 1952, a new approach to respond to the challenge.[3] The availability of these tools also ignited and encouraged

---

1 Plato's Dialogue, *Phaedo*, around 109e (Plato. Plato in Twelve Volumes, Vol. 1 translated by Harold North Fowler; Introduction by W.R.M. Lamb. Cambridge, MA, Harvard University Press; London, William Heinemann Ltd. 1966.)

2 Early pioneers that gave meaning to Socrates' vision included: Nicolaus Copernicus, 1473–1543 (Poland), Galileo Galilei, 1564–1642 (Italy) Konstantin Tsiolkovsky, 1857–1935 (Russia), Hermann Oberth, 1894—1989 (Romania/Germany) and Robert Goddard, 1882-1945 (USA).

3 ICSU, with its headquarters in Paris, France, was founded in 1931. It consists of various national academies of science to promote international scientific activities in the different branches of science and its applications for the benefit of humanity. Over the years, it has addressed specific global issues through the establishment of Interdisciplinary Bodies, and of Joint Initiatives in partnership with other organisations. Amongst its latest programmes are the World Climate Research Programme (WCRP); DIVERSITAS: An International Programme of Biodiversity Science; and the International Human Dimensions Programme on Global Environmental Change (IHDP). While maintaining the same acronym, ICSU, today, the organisation is known as the International Council for Science. It has 31 International Scientific Union Members and 122 National Scientific bodies/members that represent 142 countries. In addition, ICSU has 22 International Scientific Associates. http://www.icsu.org/about-icsu/about-us (Accessed, March 2, 2016).

a competitive spirit that subsequently translated Socrates' vision into reality and led ICSU to declare the 18-month period, July 1, 1957 to December 31, 1958, the International Geophysical Year (IGY). ICSU timed the IGY to coincide with the high point of the eleven-year cycle of sunspot activity.[4]

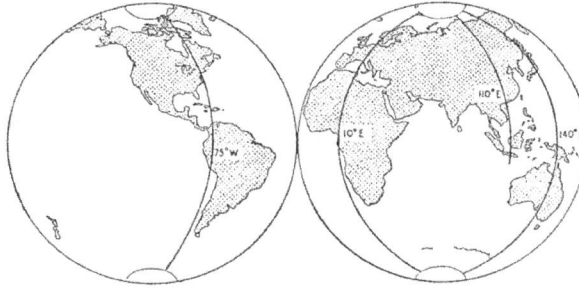

Figure 1:1 The Three Chief Meridians (i.e., Pole-to-Pole geographic lines) where IGY research stations took measurements, plus a fourth for additional coverage of the Soviet Union and Southeast Asia *(Credit: Ron Fraser)*[5]

Over 60,000 scientists from sixty-seven countries and more than 4,000 research stations participated in the IGY. They concentrated their efforts on the Polar Regions, along the Equator, and along several geographic lines, joining the North Pole to the South Pole, and universally referred to as *meridians, as shown in Figure 1:1.* One such line, *at longitude 10⁰E at the Equator*, went through Europe and Africa; another, *at 75⁰W*, went through the Americas; and the third, *at 140⁰E*, went through East Asia and Australia; the fourth meridian, *at 110⁰E*, was added to provide additional coverage for the then Soviet Union and South East Asia.[6] Among the IGY projects, in eleven (11) Earth sciences disciplines, were aurora and airglow, cosmic rays, geomagnetism, glaciology, gravity, the physics of the ionosphere, precision mapping

---

4 A sunspot is a region on the Sun's surface (photosphere) that is marked by a lower temperature than its surroundings and has intense magnetic activity. Its field strength, which is thousands of times stronger than the Earth's magnetic field, inhibits convection, forming areas of reduced surface temperature. Scientists timed the IGY to coincide with an expected peak of sunspot activity and several eclipses. An 18-month long IGY would allow scientists to sample all of these solar events, as well as conduct other timely research. Accordingly, the ICSU designated July 1, 1957 through December 31, 1958, as the IGY.

5 Fraser, Ron (1961). *Once Around the Sun: The International Geophysical Year*, The MacMillan Company, New York.

6 Ibid

(i.e., longitude and latitude determinations), meteorology, oceanography, seismology and solar activity.

As the IGY progressed, data flowed into the World Data Centres that ICSU had established for the IGY projects as well as into the duplicate centres that it later established in order to avoid a loss of data that could result from any unknown global event(s) such as wars and natural disasters.[7] Perhaps the most significant of the rewards from this global effort was the ability of scientists from many nations and from different scientific disciplines to find a common ground and to collaborate on new paths to invention. Hitherto, these scientists had worked in isolation. As of that time, IGY was the largest cooperative international scientific endeavour ever undertaken. Soon, the public began to enjoy the impact of these scientific accomplishments in their daily lives, especially as regards the lessening of international tensions and the increasing opportunities to undertake peaceful collaborative research.

For a very short while, it looked as if politics had receded to the background and that it would not interfere much with all these international collaborative efforts. But not for long! On October 4, 1957, everything changed. That was the day the then Soviet Union launched the Earth's first artificial satellite named *Sputnik-1*. Although the launching of *Sputnik-1* was a planned part of the IGY, the Soviet Union launched it using a military intercontinental ballistic missile. This was against the protocol that all members of ICSU had *agreed* upon—to use *only* non-military satellite designs and deployments to collect data during the IGY.

This violation by the Soviet Union, notwithstanding, the world-wide competition among scientists that ICSU had initiated marked a major turning point in human history. Its highpoint was the successful launch of the first human-made object into space and in the process, gave birth to the *space age* we know today. Many of the basic motivations that catapulted humankind into space in 1957 are still the same today, namely: the yearnings for discovery, for knowledge generation and for the development of those skills needed to address a variety of problems here on Earth. Fortunately, the government funding of needed scientific research, the contributions of the private sector and the challenges by accredited international

---

7 Much of the data collected during the 1932 Polar Year and the scientific analysis of the data were lost forever because of the advance of World War II. In order to avoid a repeat of this experience, ICSU established the IGY World Data Centre system. Branches of the World Data Centre System were hosted by the United States, the then Soviet Union, western European countries, Australia and Japan.

organisations (both governmental and non-governmental) continue to fuel the attainment of these objectives.

It is now close to sixty years since the former Soviet Union and the USA began the exploration of outer space. It started off as an international scientific competition between these two nations; however, each of the two participants soon discovered that it could turn the whole exercise into a tool for exploring the opportunities that abound in outer space and that the successful outcome of such an exploration would serve its national interests. This line of reasoning was, of course, in tune with the geopolitical climate of that era. Before long, an aggressive and prolonged competition arose between the two countries and it soon spilled over into the global arena. And the international community soon concluded that the unhealthy rivalry between these two states was equally unhealthy for our world and all its inhabitants. Most nations also saw the emerging space race as a precursor to the domination of the world from outer space. The prevailing view was that should either of these two adversaries gain an upper hand, it could also be bold enough to dictate future developments, not only here on Earth, but also in the space frontier—a dangerous phase that would confirm the African adage that says: *When two elephants fight, it is the grass that suffers.* This belief engendered fear in the corridors of political power across the globe.

To calm all nerves and in the interest of all humankind for which it was founded, the United Nations stepped in and nurtured the subsequent formal establishment, in 1959, of the United Nations Committee on the Exploration and Peaceful Uses of Outer Space (COPUOS). The United Nations General Assembly (GA) resolution that established COPUOS also mandated the Committee to:

- Review, as appropriate, the area of international co-operation;
- Study practical and feasible means for devising programmes in the peaceful uses of outer space to be undertaken under the auspices of the United Nations; and
- Provide encouragement for national research programmes for the study of outer space, and the rendering of all possible assistance and help towards their realization.[8]

While the need for peace in our world led to the establishment of COPUOS, the era that immediately followed the launch of *Sputnik-1* also had another name—the *space race* era. That was the time when the United States and the Soviet Union

---

8 United Nations General Assembly, *Resolution 1472 (XIV)*

dominated early space efforts with very huge and expensive space assets that focused on multiple manned and unmanned missions undertaken with little or no regard for costs. A number of these efforts had military objectives especially those at the rocket development level; but for what purpose? Table 1:1 gives a glimpse of the competitions that dominated the space race era.

Fortunately for the history of space and that of humankind, the spirit of cooperation between the USA and the Soviet Union was first discreetly displayed in a 1969 American movie titled *Marooned*. In this movie, Soviet cosmonauts helped rescue three U.S. astronauts stranded in Earth's orbit."[9] On July 15, 1975, this spirit of cooperation, played out in a movie, was actualized in the first docking of two spacecraft, namely: Apollo, which belonged to the USA, and Soyuz, which belonged to the USSR.

Table 1:1 Competition between the USA and the USSR
during the Space Race Era[10]

Time	Mission	Human/Machine Involved	Country
Apr. 12, 1961	1st Human in Space and in orbit	Yuri Gagarin	USSR
May 5, 1961	1st American in Space	Alan Sheppard	USA
May 25, 1961	Put a man on the moon by the end of the decade	President John F. Kennedy's challenge to the USA	USA
Feb. 20, 1962	1st American in orbit	John Glenn	USA
March 18, 1965	1st man to walk in space, tethered	Alexi Leonov	USSR
June 3, 1965	1st American to walk in space, tethered	Ed White	USA
July 14, 1965	1st pictures of Mars	Mariner 4	USA
Feb. 3, 1966	1st spacecraft to land on the moon	Luna 9	USSR

9 *Roald Sagdeev (2008). United States-Soviet Space Cooperation during the Cold War* http://www.nasa.gov/50th/50th_magazine/coldWarCoOp.html (Accessed, March 7, 2014).

10 http://www.archives.gov/research/alic/reference/space-timeline.html (Accessed, March 7, 2014).

Time	Mission	Human/Machine Involved	Country
June 2, 1966	1st American spacecraft to land on the moon	Surveyor 1	USA
Sept. 15, 1968 launch	1st spacecraft to orbit the moon and return to Earth	Zond 5	USSR
Dec. 21, 1968 launch	1st men to orbit the moon and return to Earth	Frank Borman, William A. Anders and James A. Lovell	USA
July 20, 1969	1st men to land on the moon	Neil Armstrong and "Buzz" Aldrin in Apollo 11	USA
Sept. 12, 1970 launch	1st automatic spacecraft to return soil samples of the moon	Luna 16	USSR
Dec. 15, 1970	1st probe to land on Venus	Venera 7	USSR
Apr. 19, 1971	1st space station launched	Salyut 1	USSR
July 30, 1971	1st time to drive a moon rover on the moon	David Scott and James Irwin in Apollo 15	USA
Nov. 13, 1971	1st spacecraft to orbit Mars	Mariner 9	USA
May 14, 1973	1st American space station launched	Skylab	USA
July 17, 1975	1st docking of two spacecraft in space (Apollo-Soyuz Test Project)	Apollo 18 and Soviet Soyuz 19	USA & USSR

This concrete act of cooperation set the stage for the end of the space race that began between these two powers some eighteen years earlier and marked the beginning of the space enterprise.

However, the potential of space cooperation, at an international level, would need to be tested and the essential capabilities in each country would need to be nurtured to ensure that the cooperating partners could work together and share the knowledge and experience needed for the accomplishment of the desired common goals. The 1979 COSPAS-SARSAT programme provided a test for the possibility of such cooperation.

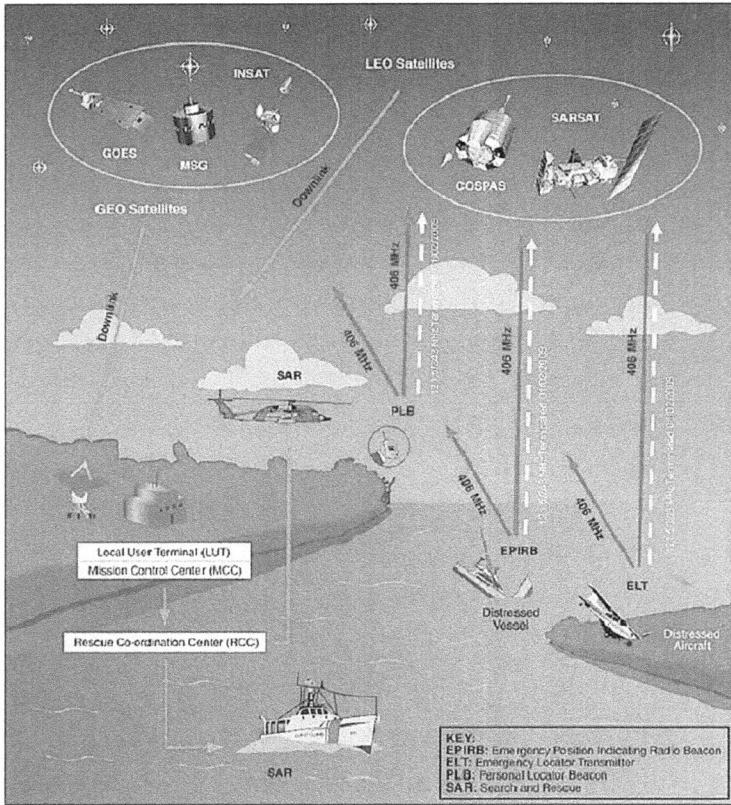

Figure 1:2 The Architecture of COSPAS-SARSAT Operation[11]
(Credit: Cospas-Sarsat)

This programme was based on the coordinated use of the USA-Canadian-French SARSAT and the USSR COSPAS satellites to undertake search and rescue efforts in order to locate airplanes or ships in distress.[12] Figure 1:2 shows the architecture of Cospas-Sarsat cooperation.

With this initial step, space cooperation at a greater level needed some nurturing of the essential capabilities in each participating country. To enhance its capabilities for such an eventuality, the USSR focused its efforts on a continuing refinement of its

---

11 http://www.cospas-sarsat.org; http://ben.br1.net/wp-content/gallery/epirb/cospas-sarsat.png

12 COSPAS-SARSAT is an international satellite-based search and rescue (SAR) distress alert detection and information distribution system established by Canada, France, the United States, and the former Soviet Union. In 1979, five (5) African members States (Algeria, Madagascar, Nigeria, Tunisia and South Africa) currently provide location-related space-based search and rescue services particularly for people and transportation systems in danger, e.g. air crashes, shipwrecks and automobile accidents https://www.cospas-sarsat.int/en/ (Accessed, February 10, 2014).

Salyut space station that paved the way for the *Mir* multiple-modular space station, and on long-term research into the problems of living in space. It launched its first *Mir* module in 1986 and concentrated its efforts on other areas that included the conduct of a variety of astronomical, biological and Earth-resources experiments.

For the United States, a transition period followed between 1973 and 1981, starting with the Apollo programme, to the short-lived Skylab space station, and later to the Space Shuttle. The Shuttle eventually became the space transportation system (STS) that was to serve the Space Station Freedom;[13] the latter was authorized in 1984 to be USA's response to USSR's Mir space station. The crew of the operational shuttle launched numerous satellites, interplanetary probes and the *Hubble Space Telescope (HST)* from the Shuttle's payload-bay and conducted many science experiments while in orbit.

Meanwhile, the fall of the *Berlin Wall* in 1989 re-united the Germans, tempered the world political climate and marked the end of aggressive competition.[14] It also ushered in a new era of cooperation among nations including cooperation in space exploration and utilisation otherwise referred to, in this book, as *space enterprise*. The reduced tension brought an additional dividend for space enterprise—the latter, at one time a taboo, is today embraced by the leading space-faring and space-capable countries of the world. Such a co-operation, proposed by the USA in 1994, subsequently resulted in the co-development, co-construction and the on-going co-use of space assets, such as the *International Space Station (ISS)*, shown in Figure 1:3.

Economic factors soon became the major drivers of space enterprise and the coincidental budget difficulties in the USA and financial difficulties in the USSR made

---

13 *Space Station Freedom* was a NASA project to construct a permanently manned Earth-orbiting space station in the 1980s. Although approved by President Ronald Reagan and announced in the 1984 State of the Union Address, *Freedom* was never constructed or completed as originally developed, and after several cutbacks, the project evolved into the International Space Station program. https://en.wikipedia.org/wiki/Space_Station_Freedom (Accessed, February 10, 2014).

14 The Berlin Wall was a former barrier that surrounded West Berlin; it was a symbol of the Cold War. The East German government built the wall on August 13, 1961 with barbed wire barricade and concrete. The Wall restricted the free movement and interaction of the East Germans with the rest of the world, and in particular, West Germans. But by 1989, East Germany was caught in the wind of political liberalization that roared throughout the Eastern Bloc countries, particularly, its neighbours, Poland and Hungary. On November 9, 1989, the East German government announced to the world that all East German citizens were free to visit West Germany and West Berlin. The fall of the Berlin Wall paved the way for German reunification which was formally concluded on October 3, 1990.

the continuing independent operation of the two competing stations impracticable. Following appropriate political dialogues, *Freedom*, *Mir-2*, and *the European and Japanese modules* were incorporated into a single International Space Station (ISS) in November 1993. The first element of the ISS was launched into space in 1998 and the assembly of the station as we know it today, was completed in 2011 by its five partners, namely Canada, Europe [ESA], Japan, Russia and the United States.

Figure 1: 3 The International Space Station *(Credit: NASA)*

The objective of the ISS is to help improve our everyday life here on Earth. Today, research facilities and capabilities on board the ISS are being used by ISS and non-ISS partners for a variety of research activities, including: physical and material sciences, biology and biotechnology, human research, medical and industrial applications; Earth and space sciences, agricultural experiments on hybrid seeds; technology test beds, robotics, communication and ground control, and automotive and transportation. As shown later in Chapter XII, arrangements are in progress to offer similar research opportunities to the developing countries.

In the period between the 1957 launch of *Sputnik-1* and today (2017), many countries and private entities have become active in space. Specifically, the space enterprise era has also opened windows of opportunity for some developing countries that include Algeria, Chile, Indonesia, Malaysia and Nigeria, to aim for space via the micro-satellite route. Table 1:2 shows the categories of space activities of member states globally. It should be noted, however, that earth-observation, surveillance, intelligence gathering, disaster mitigation and early-warning tasks in the wilderness and jungle areas and along national borders are among many other objectives, which were once the domain of satellites, but are being achieved today without going to the space altitude. This feat is currently made possible by a variety of *unmanned aerial vehicles (UAV)* as shown later in Chapter XII of this book.

Table 1:2 Categories of Member States in the Space Arena[15]

	Space Category	Description
1	**Space-faring Nations**	Countries that have acquired launch capabilities and that continue to successfully develop, build and launch their own satellites, space vehicles & ancillaries.
2	**Space-capable Nations**	Countries that can independently develop and build their own space assets including satellites. A number of these countries opted to forego the development of launch facilities while others are developing theirs.
3	**Space-aspiring Nations**	Countries that have sought external co-operation in the co-development and co-building of their satellites. They may also be actively developing launch capabilities with a demonstrable measure of success.
4	**Space-aiming Nations**	These include (a) those countries that have purchased their own satellites from the open market as turnkey projects; and (b) those countries that may not yet be financially strong enough to embark on satellite technology as a manifestation of their space aspirations. Nevertheless, a number of these same countries are active in astronomical studies, balloon experiments and rocket research. Most countries, in categories (1) to (3) above are far advanced in these fields.
5	**Space Users or Supplicant Nations**	These are countries that have no plans to invest or participate in any space activity but who actively make use of space-based services.

15 Abiodun, Adigun Ade (2004). *The roles of governments, international organisations and the private sector in the promotion of space science and technology*. Keynote Address at the First Asian Space Conference, Chiang Mai, Thailand, November 23-26, 2004.

Today, the old and new entrants into space enterprise are contributing in varying degrees to the growing human knowledge and understanding of the universe and planet Earth, particularly the prevailing status of our life support systems. As of mid-2016, active civilian satellites in space numbered over 1,419, consisting of Earth resources satellites; Oceanographic satellites; Meteorological satellites; Communication satellites; Global positioning satellites (GPS); Search and rescue satellites; and Space and Earth science satellites.[16] These *eyes and ears* of humankind in space are collecting a variety of information that each country depends upon today to meet the daily needs of its people. In addition, all the major planets in the Solar System have now been explored, to varying degrees, by spacecraft launched from the Earth.[17] Through these unmanned missions, humans have been able to get close-up photographs of these planets and their satellites as well as the photographs of smaller bodies that include comets, asteroids, and dust.

As we expand our presence into the solar system, Mars is proving to be an inviting destination for scientific discovery as well as for robotic and human exploration. Its formation and evolution are comparable to those of planet Earth and this fact is helping us learn more about our own planet's history and future. Mars had conditions suitable for life in its past. As humankind continues to expand its activities in the new frontier of space, future explorations of that frontier could uncover evidence of life beyond what we know today and this would answer one of the fundamental mysteries of the cosmos: Does life exist beyond Earth?

Meanwhile, humankind continues to find increasing and new ways that satellite technology can serve as well as meet its daily needs. It would be instructive to know, through an annual relevant publication, the impact of space products and services on the development and wellbeing of the people of each space-aspiring nation. Chapter II that follows examines the different areas where space can make a difference in the development of Nigeria and the developing economies, and in the wellbeing of their people.

* * *

---

16 http://www.ucsusa.org

17 The Solar System, as we know it today, includes the Sun and the nine main planets that orbit it, namely: Mercury, Venus, the Earth and its Moon, Mars, Uranus, Jupiter, Saturn, Neptune and Pluto, including their satellites, as well as smaller bodies such as comets, asteroids, and dust.

Chapter II

THE RELEVANCE OF SPACE TO NIGERIA

Up to the end of the 1990s, the rot and the confusion in the Nigerian communication industry was not only indescribable but also unbelievable. As shown in Figure 2:1, this was the period when telephone subscribers in many parts of the country used coloured tapes to identify their individual telephone lines in a jungle of other telephone lines wrapped around the telephone poles, erected by the Nigerian Telephone and Telegraph Authority (NITEL) in the major cities of the country.

Figure 2:1 Drooping, Wrapped and Hard-to-Distinguish Telephone Lines in the Lagos Metropolis, Nigeria *(Credit: Ade ABIODUN, June 1997)*

Since the year 2000 and now, we are pleasantly relieved to see how Nigerians, at the individual and corporate levels, have been systematically graduating into the space and information age. They have abandoned the obsolete services of the NITEL and are consuming satellite-derived services and products at a rate never imagined possible. As they do this, they automatically become members of the worldwide *Community of Consumers of Space Services and Products*, an association they join simply by patronising these services and products and not because of any formal registration.

Similar to experiences in other lands, Nigeria is becoming more dependent on communication satellite-supported services which are now giving Nigerians the opportunity to break out of their global isolation. In the process, its citizens now readily share information among themselves, their immediate African neighbours and the entire global community—something that was not possible a few decades ago. These satellites are owned by a multiple of overseas satellite operators. And as shown in Chapter VII, they provide Nigerians with satellite communications services which include electronic mail and data transfer; telephone and teleconferencing; tele-education and the sharing of information about academic and non-academic performances; radio and TV programmes, and the entertainment industry. Satellite-based information and services needed for accomplishing human daily tasks are becoming fully integrated into government operations, educational systems and establishments, banks and other financial institutions, the newspaper industry, national security establishments, personal safety, and transportation industries (air, land and sea).

Nigeria also uses its own Earth observation satellites and those of other countries, to provide copious data that the nation can use daily, to monitor and manage its environment and natural resources, including land-use, forestry, water resources and the harvesting and management of fisheries and in the high-seas. There are also meteorological and search and rescue satellites in space; they are not owned by Nigeria, but they respectively provide the nation with continuous data needed to forecast the weather over its territory and to aid it in its search and rescue efforts for those in distress, particularly in time of danger, such as in aircraft crashes. Space-based satellite navigation systems (GNSS),[1] such as the United States *Global Positioning System* (GPS), provide location and time information in all weather conditions, anywhere on or near the Earth where there is an unobstructed line of sight to four or more GPS satellites. The system provides critical capabilities to military, civil,

---

1 Other position, navigation and timing systems, such as GLObal NAvigation Satellite System—GLONASS (Russia), Galileo (Europe), BeiDou Navigation Satellite System—Beidou (China), Indian Regional Navigation Satellite System—IRNSS (India), and Quasi Zenith Satellite *System*—QZSS (Japan) are covered in detail, in Chapter XI.

and commercial users around the world. With the aid of GPS, Nigeria is gradually reducing the loss of human lives on its roads; the system is also facilitating the quick recovery of vehicles which are stolen in Europe and North America and shipped to other parts of the world including Nigeria.

The five examples of space applications given below illustrate how, with requisite knowledge and understanding, satellite capabilities can:

- Enhance the quality of human health.
- Contribute to the sustainability of the nation's fisheries resources.
- Provide improved weather surveillance across the country to meet a variety of the daily needs of Nigerians.
- Provide search and rescue support for those in distress situations, and
- Be used to build a culture of peace in Nigeria and in the rest of Africa.

Healthcare

In line with the universal adage that *Health is wealth*, almost everyone pays much attention to the state of their health. Today, satellite services and spin-off benefits of space exploration are contributing, in practical terms, to the delivery of health-care services in many parts of the world including remote and inaccessible areas including those communities where there is a shortage of health-care personnel. It is worthy of note that a major focus of attention on the International Space Station (ISS) is the development of drugs and vaccines for a variety of human ailments that afflict us here on Earth. To take advantage of such space-related capabilities, many countries are establishing appropriate tele-health/telemedicine policies and related guidelines.[2, 3] Telemedicine is now routinely practised in many parts of the world.

---

2 Telemedicine is the use of telecommunication and information technologies (ICTs) such as satellites, television, video conferencing, internet, mobile phones etc., to provide clinical health care services in such areas as radiology, cardiology, pathology, dermatology, psychiatry, pharmacy, etc. at a distance. In the process, the healthcare provider is able to use ICTs to transmit related information and digital images, such as X-ray images, in the case of a tele-radiology. Telemedicine helps eliminate the barriers of distance and it also improves access to medical services that would often not be consistently available in distant rural communities. It is also used to save lives in critical care and emergency situations, such as in floods, earthquakes, fire, wars etc.

3 Dr. Willem Einthoven, a Dutch doctor and physiologist is the father of telemedicine. In 1903 he invented the first practical electrocardiogram (ECG or EKG) for which he received the Nobel Prize for Medicine in 1924. Because his hospital did not allow him to move patients outside the hospital to his laboratory to test his new device, Dr. Einthoven did tests with the transmission of ECG over telephone lines. In 1906, he came up with a way to transmit the data from the hospital directly to his lab.

Pediatric Advice Service: Doctors-Without-Borders use this service regularly. Located at St. Joseph's Children's Hospital, in Paterson, New Jersey, USA, Medical Missions for Children (MMC) is an institution which has established a long and impressive record of using telemedicine tools to provide secondary pediatric advice from specialist hospitals in the US to medical institutions in developing countries and emerging economies.[4]

Cardiovascular Diseases: In Brazil, for example, by 2006, out of the 5,600 Brazilian cities, 4,300 have been identified as being in need of tele-health services, particularly for the treatment of *Cardiovascular Diseases* which are the major causes of death and illnesses in that country. Because of the high cost of treatment of these diseases, the Brazilian government has made the prevention and treatment of cardiovascular diseases a national priority in public health[5].

Emergency Service for Cardiac Arrest: Students at The Technical University in Delft, The Netherlands, have developed and deployed a drone that can quickly deliver a defibrillator to the emergency scene of a cardiac arrest. This process saves lives and facilitates the early recovery of patients suffering from heart failure, drownings, traumas and respiratory problems. Via a *livestream* video and audio connection, the drone can also provide direct feedback to the emergency services and the persons on site can be instructed on how to treat the patient. The drone, which weighs 4 kg, can fly at around 100 km/h while carrying an additional 4kg load. It is able to locate the patient via the caller's mobile phone signal and to make its way to the emergency scene using GPS.[6]

Monitoring of Diseases and the Spread of Diseases: Health, as an information-intensive sector, also requires extensive data collection, information management and knowledge utilization at all levels and at all times. In addition to aiding tele-health services, satellites are also now indispensable in the monitoring of diseases and disease-spread. Rodents, insects and ants (bugs) are noted not only for being hosts of *Hemorrhagic Fever, Hantavirus Pulmonary Syndrome* and *Dengue fever* and other

---

4 Humanitarian Medicine Event at the European Space Policy Institute, Vienna, Austria, June 14, 2012.

5 Abiodun, Adigun Ade (2006). *Production and exchange of knowledge: Space science and technology perspectives.* Paper invited for presentation by the Government of Brazil and the African Union at the Second Conference of the Intellectuals of Africa and the Diaspora (II CIAD) Salvador, State of Bahia, Brazil, July 12—14, 2006.

6 http://www.tudelft.nl/en/current/latest-news/article/detail/ambulance-drone-tu-delft-vergroot-overlevingskans-bij-hartstilstand-drastisch/ (Accessed, November 5, 2014).

tropical diseases, but also for spreading these same diseases throughout their host communities. Ants and flies also serve as hosts for certain other life-threatening tropical diseases that include *Chagas, Malaria, Dengue* and *Leishmaniasis*. These diseases are prevalent in most warm climates including Nigeria. In such rodent/insect/flies/ants/bug-infested communities, understanding the characteristics of the eco-geographical areas where the diseases develop is a national challenge. Fortunately, space technologies are proving useful in this area.[7]

To control similar diseases and their spread in Argentina, the Argentinian Space Agency (CONAE), in collaboration with the Italian Space Agency and the Cordoba National University in Argentina, has embarked on a mission that addresses the question: "How does a community develop a set of numerical tools, as an early warning system, that is devoted to the surveillance of a population under risk from rodent or insect attacks?"[8] These partners have embarked on a four-pronged approach. The first step is the development of plans of action. The second step is the development of a health information system which integrates the use of communication satellites for data transmission with the use of earth observation satellites for data collection in the affected areas. The third step, called the 'Cartography of Risk Factors,' is the fusion of epidemiological, biological and remote-sensing data to develop the maps of the affected areas. The fourth and final step, known as the 'Space-temporal modeling of epidemics,' models the inter-relationships between the hosts, the vectors, the reservoirs and the ecosystem of the affected areas. Presently, efforts are in progress to build a multi-scale and a multi-factor system based on remote sensing and GIS in order to improve the ability to predict future outbreaks of virological and entomological diseases as well as support Dengue fever control actions in the affected areas.

In the immediate future, projects of this nature will need to take into account the impact of increasing human populations on the population density of these hosts—the rodents, the bugs and the insects. This is necessary because some of these hosts are already adapting to climate change and are migrating to geographical and ecological areas that were at one time alien and/or climatically hostile to them.

---

7 Abiodun, Adigun Ade (2013). *We Must Harness Space for Sustainable Development,* Journal of Space Policy (JSPA_1032), Volume 29, Issue 1, February 2013.

8 Lanfri, Mario Alberto (2012). Space for Sustainable Development—Remote Sensing in Health Applications, Rio+20 Side Panel on "Space for Sustainable Development," Rio +20 Conference, Rio de Janeiro, Brazil. June 19, 2012.

Marine Ecosystems and Fisheries

The Atlantic Coast of West and South-West Africa, from Guinea Conakry to Angola, is regarded as one of the most productive marine and coastal areas of the world, with rich fishery resources, oil and gas reserves, precious minerals, and high coastal tourism potential. As shown in Figure 2:2, the coastline is shared by 16 riparian countries[9] and is an important global reservoir of marine biological diversity. The total economic value of the environmental goods and services provided by the area's coastal and marine resources is estimated to be around US$18 billion annually. The environmental goods and services of these shared resources support the livelihoods of approximately 40% of the region's 350 million peoples.

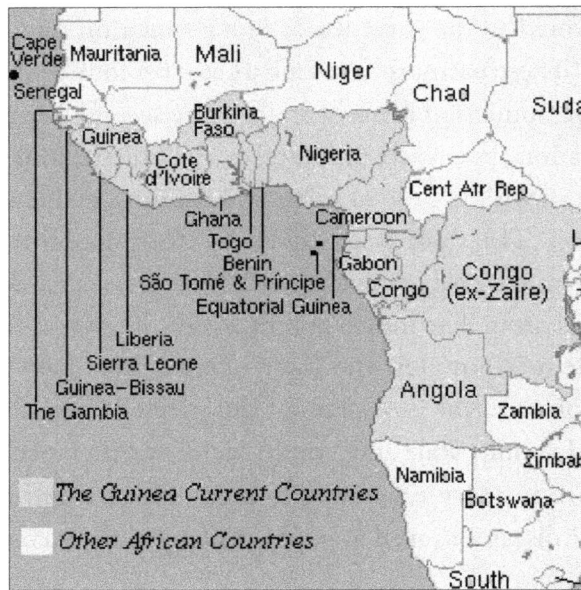

Figure 2:2 Gulf of Guinea Large Marine Ecosystem
(Credit—Global Environmental Facility (GEF))

These people are dependent on fisheries and other coastal and marine resources[10] for their livelihoods. Space tools can offer assistance in the resolution of the

---

9 Angola, Bénin, Cameroon, The Republic of the Congo, Côte d'Ivoire, The Democratic Republic of the Congo, Equatorial Guinea, Gabon, Ghana, Guinea, Guinea-Bissau, Liberia, Nigeria, Sao Tomé and Principe, Sierra Leone and Togo.

10 Tengberg, Anna and Arne Andreasson (2012). Study of the Concept of Large Marine Ecosystems and its Institutional Relevance for Ecosystem-based Management and Development, Swedish Agency for Marine and Water Management, Gothenburg , Sweden, March 2012.

environmental problems of this sub-region, particularly those that are associated with petroleum exploration and transportation in the Niger Delta region of Nigeria, the destruction of West African mangroves, over-fishing within coastal waters and the pollution of the coastlines.

At the global level, most of the 64 coastal and marine ecosystems of the world, known as Large Marine Ecosystems (LME)[11] are suffering from environmental degradation. The degradation is caused by the mismanagement of fisheries resources that include stock depletion, the extinction of mariculture[12], industrial and municipal pollution, contamination by ballast water, coastal erosion and the destruction of upstream spawning areas. The exploration, exploitation and transportation of petroleum and gas resources, particularly in the mangroves and along the coastal waters around the world, have contaminated and degraded these ecosystems which, collectively suffer an average of more than 1,000 oil spills and related fires every year.[13]

In 2008, the Gulf Current Commission (GCC),[14] of which Nigeria is a member, reached an agreement on a long-term Strategic Action Plan to address its own prevailing challenges within the Gulf Current Large Marine Ecosystem (GCLME) shown in Figure 2:2. These challenges include the massive depletion of fisheries resources, unending degradation of habitats (mangroves) and land and ship-based major pollution activities that endanger the sustainability of living and non-living

---

11 See https://www.cbd.int/ecosystems/newsletters/ea-2009-10.htm for more details on Large Marine Ecosystems (LME).

12 *Mariculture* is the farming of aquatic plants and animals, marine organisms (including plants and animals) for food and other products in the open ocean, in an enclosed section of the ocean, or in tanks, ponds or raceways which are filled with seawater.

13 Lichem, Walther (2012). Space-based Data and Sustainable Water and Ecosystem Management, Rio+20 Side Panel on "Space for Sustainable Development," Rio+20 Conference, Rio de Janeiro, Brazil. June 19, 2012.

14 The GCC is the political and policy body established by the sixteen West and South-West African countries to manage and ensure the sustainable development of Gulf Current Large Marine Ecosystem (GCLME). GCLME is that stretch of the East Atlantic Coastlines that is shared by 16 West and South-West African countries—from its northernmost shore in Guinea Bissau to its southernmost shore in Angola. Member States of the GCC include: Angola, Bénin, Cameroon, The Republic of the Congo, Côte d'Ivoire, The Democratic Republic of the Congo, Equatorial Guinea, Gabon, Ghana, Guinea, Guinea-Bissau, Liberia, Nigeria, Sao Tomé and Principe, Sierra Leone and Togo.

resources of the GCLME.[15] Marine and coastal ecosystems in West Africa (this region includes Nigeria) also need the recovery and sustenance of their depleted fisheries, the restoration of degraded habitats, and the reduction of land and ship-based pollution as well as a reversal of both coastal area degradation and living resources depletion. It is true that when it comes to resolving these and other related problems, there would be competing interests and institutions; nevertheless, through cooperation at the regional level and when necessary at international levels, these problems can be resolved with the aid of copious satellite data.

Many of the problems cited above are very prevalent in the Niger Delta region of Nigeria, the nation's fountain of oil wealth. Fortunately, they are easily captured by Earth observation satellites, using instruments such as MER on ESA's *Envisat* satellite, MODIS on the *Aqua* and *Terra* satellites of USA and the IMAGER on board *Insat* of India. Urgently needed by the concerned parties are shared objectives, a degree of coordination of the different governmental, intergovernmental and non-governmental responsibilities, and the willingness to work with shared information and data.[16]

A major natural resource in the GCLME is the abundance and variety of aquaculture including the food-chain that supports and sustains them. Phytoplankton, which is a critical element in the food chain of most fish types, is associated with upwelling systems.[17] As shown in Figure 2:3, coastal upwelling areas are among the most productive regions of the world's oceans—in Africa, these areas include the coastal zones of northwest, west, southwest, and northeast regions.

The practice of giving with the left hand and taking away with the right hand has been sanctioned in West Africa since space became an operational tool of development and Africa and Europe and West Africa became yoked with the Lomé Convention. For a long time, a number of space-faring countries and private companies have engaged and are using several space tools such as the Coastal Zone Color Scanner (CZCS) on board NASA's *Nimbus* satellites and the sensors on board the *Meteosat*

---

15 Strategic Action Programme (2008). A Programme of the Governments of the GCLME Countries *et al*, GCC Secretariat, Accra, Ghana, September 2008.

16 Abiodun, Adigun Ade (2006). *Production and Exchange of Knowledge: Space Science and Technology Perspectives...*African Union Second Conference of the Intellectuals in Africa and the Diaspora, Bahia, Brazil, 2006.

17 *Upwelling* is a process in which deep, cold water in the ocean rises toward the surface. Displaced surface waters are replaced by cold, nutrient-rich water that "wells up" from below. Winds blowing across the ocean surface push water away.

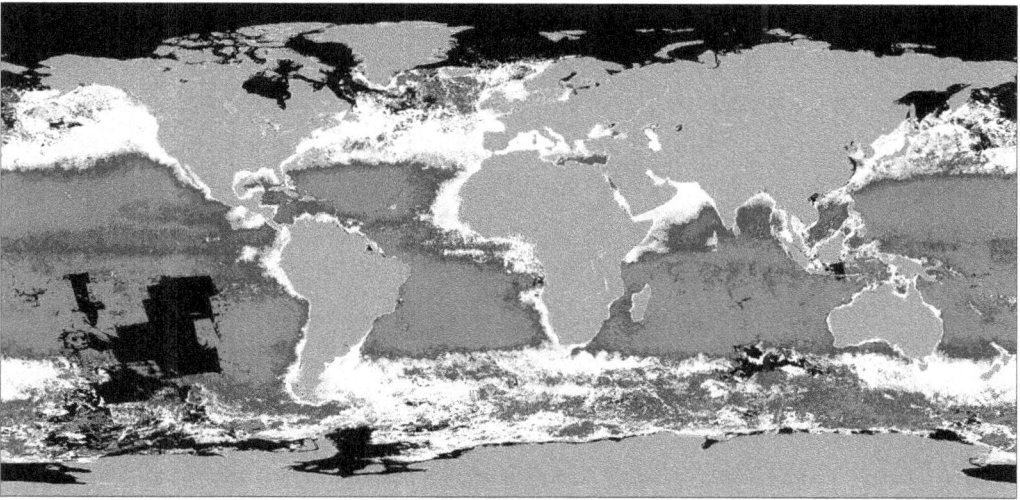

Figure 2:3—Highlighted coastal areas, world-wide, that are often associated with upwelling currents *(Credit—San Francisco State University –http://geosci.sfsu.edu)*

satellites of the European Space Agency and the *GeoEye* and *WorldView* satellites of Digital Globe, to monitor and harvest the fishery resources of these productive shores of Africa. For decades, foreigners (mostly Europeans, Russians, Chinese and Japanese) have been undertaking satellite-aided large-scale fishing operations in West Africa's coastal waters.[18] West Africa's political leaders and decision makers agreed to these foreign fishery-operations by issuing fishing licenses to these mammoth international fishing companies. What has been the end result? The fishermen of Nigeria and other West Africa countries are today mostly idle because of their inability to compete with these international satellite-aided trawlers. Untold damages to the GCLME environment have also become the norm rather than the exception and there is the attendant collapse of the local economy.[19]

Today, affordable satellite-related tools and capabilities exist that would allow Nigeria and other GCC member states to reverse decades of environmental degradation and fisheries depletion within the GCLME as well as ensure the preservation of its fish stocks for today's and future generations. Among the capabilities needed to enforce good fishing practices is the near real-time tracking of the activities of

---

18 Kaczynski, Vlad M., David L. Fluharty (2005). *European Policies in West Africa: Who Benefits from Fisheries Agreements? UNDP Human Development Report 2005, UNDP, NY, USA*

19 Abundant examples of "*Damages to the Coastal and Marine Environment of West Africa,*" are available on the web

all vessels in the GCLME waters, including the monitoring of the legal and illegal fishing vessels that operate in these waters. In Nigeria, in particular, the National Communication Commission (NCC) is providing ICT infrastructures at its Community Resource Centres (CRCs) in order to equip the local and rural (agricultural and artisanal fishing) communities with necessary tools of development.[20]

While the NCC's centres will go a long way in ensuring development at the rural areas of the country, Nigeria and other GCLME countries can enhance the services being provided by such centres by learning, in particular, from the experience of India and its Village Resource Centres (VRCs) as detailed in Chapter VIII.

Weather Forecasting

The provision of meteorological services in Nigeria began in 1892, during the colonial rule by Britain,[21] and later by the former Department of Meteorological Services (DMS) which was part of the Federal Ministry of Aviation. In November 1960, almost immediately after gaining independence from Britain, Nigeria became a member of the World Meteorological Organisation (WMO); it joined that organisation's Global Telecommunications System (GTS) in 1988. The latter receives and transmits satellite-acquired meteorologically-related data around the world in internationally agreed codes to avoid language problems and ensure a uniform standard. By an Act of the Nigerian National Assembly in 2003, the DMS was transformed into the Nigerian Meteorological Agency (NIMET). It uses its 55 manual and 37 automatic weather-observing stations in the country to measure surface weather parameters needed for providing meteorology-related services for aviation safety, agriculture/food security, energy, health, land use, water resources management and other key socio-economic sectors of Nigeria.[22]

Data from the geostationary meteorological satellites of Europe's METEOSAT and the polar-orbiting satellites of the USA (NOAA) as well as those of Russia are used by NIMET to generate local surface weather information. These data are transmitted

---

20 NCC's Community Resource Centres (CRCs) are funded, mostly, by the Universal Service Provision Fund (USPF). The objective of the CRC project is to promote the adoption of ICT at community level for social and economic development. The CRC offerings include, but not limited to conflict and disaster warning and management, improving farmers' access to markets, information centre for community development mobilization, small scale businesses, e-agric, and community health.

21 http://nimet.gov.ng/ (Accessed, June 5, 2014)

22 Aderinto, S. A. (2010). Implementation of Automatic Weather Observing Stations in Nigerian Meteorological Agency, *www.wmo.int/pages/.../P3(20)_Aderinto_Nigeria.pdf – Switzerland.*

to Nigeria through WMO's GTS. The nation's GTS facilities were initially connected to the satellite communication system (SATCOM) of the Nigerian Airspace Management Agency (NAMA) and were used to transmit and receive meteorological data about Nigeria. However, the system broke down immediately after installation and was restored to full service on July 13, 2007.[23]

In 2006, NIMET conceived the Safe Tower Project with the objective of automating air traffic management services at the four major airports in Lagos, Port Harcourt, Abuja and Kano. The Safe Tower facilities at Kano, the last of the four airports to be so equipped, were commissioned on September 15, 2015.[24] The project, began *with donor support*, included the installation of new meteorological observatory equipment at these and other airports in the country.[25] These pieces of equipment, which should enhance weather forecasting and aviation transportation safety in Nigeria, include sensors that measure meteorological variables. The *Terminal Aerodrome Forecast (TAF)* component of the new sets of equipment now enables NIMET forecasters to provide significant weather charts at various flight levels to aircraft pilots. These charts have six-hourly validity. Satellite pictures of jet streams and areas of turbulence as well as atmospheric water vapour content and humidity profile, acquired over the Nigerian air space by METEOSAT and NOAA satellites, are now readily available to the agency to execute the required forecasts.

Individual Nigerians can even now access such processed weather-related information to plan their daily activities including accessing safety information in time of danger, particularly from such natural elements as tropical storms. This information is regularly available in the news media, and on-line at NIMET's web-site. Similarly, aircraft pilots that operate within the Nigerian airspace also depend on such accurate weather forecasts; they need such information to enable them to fly their passengers safely to their respective destinations across the nation and the globe. It is now clear that NIMET has an obligation, to use its newly acquired capability and capacity to provide improved weather surveillance and

---

23 Usim, Uche (2007). *Weather Information Vital for Natural Disaster Mitigation, The SUN News*, Monday August 20, 2007.

24 Nigeria: Permanent Secretary Inspects Kano Safe Tower Project, Daily Independent, Lagos, September 15, 2015.

25 Aderinto, Samuel A. (2010). Improvement to Meteorological Observation in Nigeria, w*ww.wmo. int/pages/.../P5_2_Aderinto_Nigeria.docx - Switzerland (Accessed, February 15, 2014).*

forecast across the country to meet a variety of our daily needs as well as discharge its statutory obligations to Nigeria and its people. However, there is a lingering question: What happens when the donor support for the *Safe Tower Project* is no longer available?

Search and Rescue System

Human safety and the global clamour for a solution, particularly to aviation and maritime accidents and related tragedies, were the major reasons for the October 5, 1988 formal agreement, between Canada, France, the former USSR, and the USA, that established the International Search and Rescue Satellite System (COSPAS-SARSAT). By this agreement, and in accordance with international law, the signatories decided, to use their space assets, in line with the operational architecture shown in Figure 1:2, to implement a global search and rescue programme and provide access to the programme/system, free of charge and on a non-discriminatory basis, to all States and the end-users in distress. Figure 2:4 shows a current map of the COSPAS-SARSAT Mission Control Centres around the world. From its original inception in 1979 to December 2014, the COSPAS-SARSAT System has provided assistance world-wide[26] that led to the rescue of 39,565 persons in 11,070 distress situations that included land, fishing, boating, shipping, snowmobile, snow-avalanche and aviation accidents, as reflected in Figure 2:5. A critical understanding of Figure 2:5 is that any living human being may be in need of COSPAS-SARSAT service, anytime, anywhere, particularly in time of emergency.

Before February 1, 2009, the COSPAS-SARSAT system received distress signals from both analogue and digital beacons. Thereafter, COSPAS-SARSAT stopped processing analogue 121.5 MHz distress signals and switched completely to the processing of only signals from digital 406 MHz Beacons. The latest beacons incorporate the Global Positioning System (GPS) receivers that can transmit highly accurate positions (within about 20 metres) of the objects in distress.

Until January 2004, Nigeria was nowhere near any structured search and rescue programme that could be compared to the COSPAS-SARSAT of today. Only the north-west corner of Sokoto State had the footprint of the nearest COSPAS-SARSAT station and that was the Spanish Mission Control Centre in Maspalomas, Gran Canarias. From Nigeria's record of aviation disasters, it is apparent that for a long time, Nigeria had been in need of such a rescue system.

---

26 http://www.cospas-sarsat.int/en/system-overview/quick-stats (Accessed, July 10, 2015)

For example, between November 20, 1969, when Nigeria experienced its first recorded air crash, and June 3, 2012, the loss of lives due to aircraft–related disasters was staggering.[27] The absence of COSPAS-SARSAT emergency beacons in the country's civil and military aircraft when these accidents occurred contributed to the difficulties experienced in locating the sites of these accidents prior to mounting needed rescue operations. In most cases, an effective rescue effort could have resulted in a number of saved lives. At that time, Nigeria and most other African countries could not organise any meaningful search and rescue programme in times of distress.

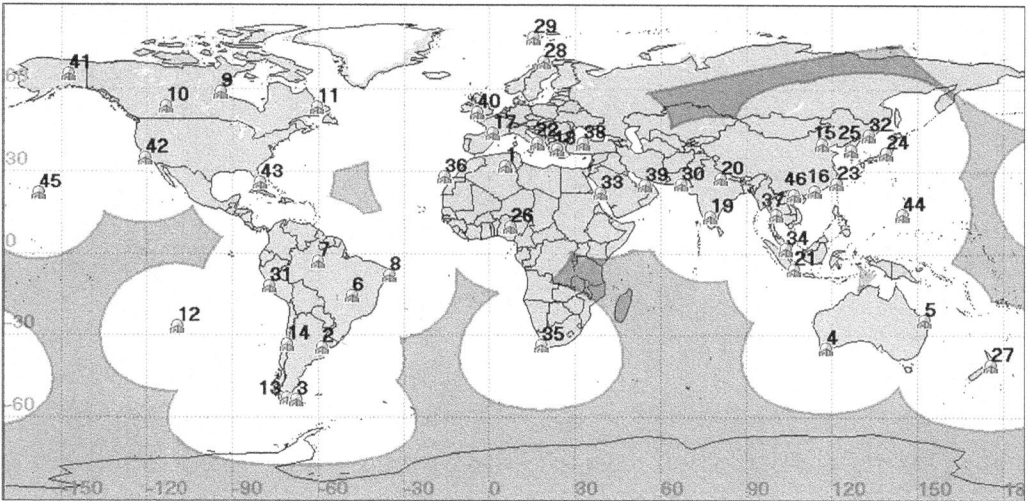

Figure 2:4 COSPAS-SARSAT Mission Control Centres around the World[28] *(Credit: Cospas-Sarsat)*

The mounting number of casualties from recurring aircraft crashes in Africa led the United Nations Office for Outer Space Affairs (UN-OOSA) and COSPAS-SARSAT to explore how to provide the needed knowledge for such a space-related service to interested African countries. A 1998 UN-OOSA workshop was held

---

27 Nigeria recorded its first plane crash on November 20, 1969, when a government-owned DC-10 aircraft on a flight from London crash-landed in Lagos and killed all 87 passengers and crew on board. Between November 1969 and June 3, 2012, over 2,000 persons were killed in plane crashes on Nigerian soil. Sources: (1) (http://allafrica.com/stories/200610301057.html *(Accessed, February 15, 2014)*), (2) *History of Plane Crashes in Nigeria*, Sun News Online (Nigeria), June 4, 2012, and (3) *Over 2,000 Lives Lost in Air Crashes in 40 Years*, Vanguard News, June 3, 2012.

28 http://www.cospas-sarsat.org; http://ben.br1.net/wp-content/gallery/epirb/cospas-sarsat.png

specifically for the benefit of the African countries that were within the footprints of the Maspalomas Station.[29]

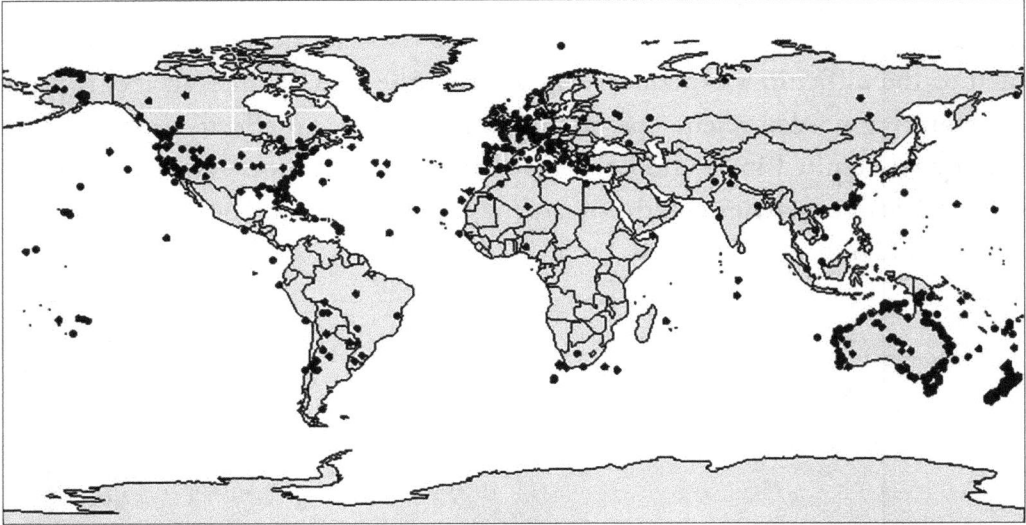

Figure 2:5 Geographical Distribution of all Reported Search and Rescue Events for which Cospas-Sarsat Data was Used (2014) *(Credit: Cospas-Sarsat)*

The goal of the workshop was to encourage these countries to take advantage of such a life-saving programme. A second workshop followed in 1999. The participation of the managerial staff of both the National Emergency Management Agency (NEMA) of Nigeria and the Nigerian Airspace Management Agency (NAMA) at these two workshops, and the related exposures that these staff members had in the process, paved the way for President Olusegun Obasanjo of Nigeria to commission the national COSPAS-SARSAT Mission Control Centre at NEMA's Headquarters in Abuja, on January 13, 2004. This action made Nigeria the fifth country in Africa with COSPAS-SARSAT facilities. The other countries are Algeria, Madagascar, South Africa and Tunisia.

The COSPAS-SARSAT facility in Nigeria is meant to facilitate Nigeria's efforts at locating distressed ships, land vehicles, aircraft and their passengers as well as individuals on foot, such as members of the armed forces, who may be in danger or in distress. The most critical requirement is that those who might be in need

---

29 In collaboration with the European Space Agency and the Spanish National Institute for Aerospace Technology (INTA), UN-OOSA organised *The First UN/INTA/ESA Workshop on Space Technology for Emergency Aid/Search and Rescue Satellite-Aided Tracking Systems,* in September 1998.

of such rescue assistance should have on them the necessary beacon(s) that would transmit their co-ordinates {i.e., geographical location(s)} to the monitoring station via the COSPAS-SARSAT satellites on sight, at the time of need.[30] However, the inability of the nation's COSPAS-SARSAT system to rapidly and successfully aid in search and rescue operations, in times of such tragedies, has called into question the merits of the investment that Nigeria made in this COSPAS-SARSAT station in Abuja.[31] Unfortunately, most of the time, the problem is not one of machine; it is human—either with the airline management and their operators or with the station and its operators. What is needed is how to shed some light on the operation of the system. A mandatory publicly-available semi-annual report of NEMA, to the nation, on the detailed operations of Nigeria's COSPAS-SARSAT System, including requisite information on incidents to which the organisation (NEMA) responded, should enable government authorities to determine its effectiveness in rendering the services that it is mandated to provide the nation and its citizens.

Building a Culture of Peace in Africa

Since the successful launch of the first artificial satellite, *Sputnik-1*, by the Union of Soviet Socialist Republics (USSR) on October 4, 1957, satellites have and continue to function in more ways than was ever envisaged. Although they neither provoke wars nor initiate peace, satellites, being dual-use tools, can serve as accomplices to attain either of these two goals. This is not all—space-acquired data, when accurately analysed and appropriately utilised, can also provide critical input in support of peace-making and peace-keeping efforts in our world. Since the beginning of the space age, Earth observation satellites have played surveillance roles and will continue to do so for years to come. Satellites have been used to monitor the adherence of signatories to multi-lateral and bi-lateral treaties that are associated with tactical and strategic military developments. For instance, they were used to monitor *the 1963 Limited Nuclear Test Ban Treaty* and *the 1972 First Strategic Arms Limitation Talks (SALT-1)*, and will be used to monitor the

---

30 This pre-supposes that there is a corresponding and funded programme to purchase and equip those who need rescue assistance with the appropriate beacons.

31 The Nigerian COSPAS-SARSAT system did not perform as expected on October 22, 2005 when a Bellview Boeing 737 plane with 117 people on board crashed soon after take-off from Lagos, killing everyone on board. Similarly, what role did Nigeria's COSPAS-SARSAT station play when, on September 17, 2006, the Nigerian Air Force Dornier 228-212 aircraft (with 18 people on board including 10 generals and 4 colonels) crashed into the hills of Ushongo village in Benue State, on its way to Obudu in Cross River State?

adherence of Iran to the Joint Comprehensive Plan of Action (JCPOA) international agreement on the nuclear program of Iran which went into force on January 16, 2016.[32] Satellite-acquired information continues to be very instrumental in resolving major differences among nations; it is also contributing to the reduction of tensions that could lead to wars. To be effective as well as serve as a conflict preventative mechanism, satellites should be able to monitor and provide advance warnings of signals that could result in conflicts. The critical elements that make Earth observation satellites attractive for surveillance and intelligence-gathering include their re-visit, real-time and synoptic operational capabilities and the ability of the sensor systems on board these satellites to see beyond the normal range of human vision.[33]

Because of world-wide abhorrence of weapons of mass destruction (WMD) and of large-scale violation of human rights by political leaders in many societies, the global community has acquiesced, in recent times, to the use of surveillance satellites by the United Nations Security Council to monitor, verify and inspect such violations in order to ensure compliance and peace in our world. These practices are possible today partly because there is a tacit understanding among nations that sovereignty does not extend to satellite altitudes and beyond.[34] In addition, satellite signals can be transnational and unfiltered. Satellite data acquisition and transmission also have no respect for national or political boundaries. These and many other attributes have made satellites invaluable tools and essential partners in sustaining the global changing views of sovereignty, today.[35] Thus, satellite-acquired

---

32 The JCPOA was reached in Vienna on July 14, 2015 between Iran, the P5+1 (the five permanent members of the United Nations Security Council—China, France, Russia, United Kingdom, United States, plus Germany) and the European Union.

33 While the first civilian Earth observation satellite, *LandSat-1(USA)*, had a spatial *resolution* of 80 metres, the world is now getting used to 0.5 metre resolution on board such commercial satellites as *WorldView I, II and III*. The term *resolution* is the minimum distance between distinguishable objects in an image.

34 There is no international agreement on where the air space ends and the outer space begins. Under the United Nations *Space Treaty* of 1967, outer space is declared to be free and not subject to national appropriation. The treaty did not define the altitude at which outer space begins and air space ends. All this notwithstanding, and as a result of *emerging scientific consensus*, member States have tacitly accepted an altitude of between 100 and 110 kilometres above sea level to demarcate where the air space ends and the outer space begins.

35 Annan, Kofi (1999). Annual Report of the Secretary-General of the United Nations to the United Nations General Assembly, New York, NY 10017, September 20, 1999.

information continues to be a critical input in the decisions of the United Nations Security Council, particularly in several recent conflicts, such as in Bosnia, Liberia, the Republic of Congo (Former Zaire), Rwanda, Sierra-Leone, Somalia, Sudan and Yugoslavia.

Satellite as a tool for peace was also demonstrated in 1979 when, *Vela 6911*, a USA satellite, alerted the global community to the September 22, 1979, nuclear explosion that took place off the coastline of South Africa. In 1986, *Spot* satellite of France and *Landsat* satellite of the USA also first alerted the whole world and provided significant and incontrovertible information about the April 26, 1986, Chernobyl, USSR, nuclear power plant accident—to-date, it remains the worst power plant accident in human history. Satellites are also the main instruments for monitoring compliance among nations on such legal instruments as *the 1989 Montreal Protocol on Substances that Deplete the Ozone Layer* and *the 1997 Kyoto Protocol to the United Nations Framework Convention on Climate Change,* and for gaining a better human understanding of climate change and its potential consequences for us all. When appropriately deployed, satellites can also serve as instruments of peace in Africa.

Satellites for Peace in Africa

Over the years, Africans and the rest of the global community have lamented over the state of under-development in some parts of the world, particularly in sub-Saharan Africa. These lamentations can be a thing of the past if only all Africans can cultivate, imbibe and commit themselves to a culture of peace. The opportunities are there for Africa to show-case satellites as tools for peace on the continent by engaging the capabilities of space tools to:

- Identify sources of conflicts before they develop into major crises;
- Monitor and verify compliance with the terms of agreement arrived at by the combatants;
- Locate buried land-mines and subsequently de-mine the entire African landscape;
- Monitor and identify mobile units and terrorists that transfer firearms across national borders;
- Track the movements of refugees fleeing from conflict zones so as to locate them as well as provide them with essential support;
- Resolve the conflicts that often develop among contiguous states that share water resources of the same drainage basin;

- Arrest the depletion of Africa's living and non-living resources, both on land and in the oceans around the continent;
- Arrest the degradation of Africa's marine environment and landscape; and
- Resolve pastoral land management problems that are associated with the age-long unregulated pastoral practice of herdsmen particularly in Nigeria and other West African countries. Such an effort is critical to the establishment and sustenance of communal peace in each of the countries of the sub-region.

For those who are not familiar with the problems associated with cattle herding in West Africa, here are some facts. West African herdsmen often let loose their herds to forage on and rampage through private farm lands and properties in villages, urban communities and across national borders. The practice is unregulated, and the herdsmen and those that invest in herding, particularly in cattle herding, often claim that it's a traditional practice from ages past. It is very true that it is a practice that belongs to the past, because, today, it is endangering peace in the sub-region of West Africa as a whole. It is an archaic practice that is fostering much communal strife in the sub-region and often, it results in un-ending conflicts, warfare and bloodshed between the herdsmen and the affected communities—but not the behind-the-scene cattle investors and owners. The practice, rooted in ancient tradition and customs, is neither tenable nor sustainable in 21st century West Africa, a time when communal land ownership has given way to individual land ownership and the establishment of fixed boundaries. While herding is an economic occupation for those who invest and engage in it, however, it is ruining the economic life-line of a large number of others and is sending many of our citizens to their early graves. The governments of the affected countries must arrest the practice and legislate that the herdsmen and the associated investors should confine their herds to their own ranches in their own states of origin; otherwise the traditional practice of cattle herding, will continue to be a major source of communal conflicts, gun-battles and insecurity that may develop into civil wars in the years ahead.

By using the capabilities of Earth observation satellites, appropriate and regulated solutions can be found for the herdsmen and the communities concerned. In the case of Nigeria, for example, both the federal and state governments should work together and use satellite acquired data to generate maps of land-use practices within their jurisdictions. Such maps should be used to regulate pastoral activities within these jurisdictions by apportioning and demarcating adequate pastoral land to individual herd owner(s) in his own state of origin, as well as educate such individuals on how to invest and grow adequate fodder that can nurture and sustain their herds within each newly *fixed* location.[36] Such a land management policy and practice is critical; the land that is available for growing fodder is limited, accordingly, the number of cattle per herdsman should also be limited.

The issues addressed above and more are some of the security-related issues that are contributing to the absence of peace in sub-Saharan Africa today. To be an effective manager of these and other crises within the African continent, the African Union (AU) needs to establish its own independent African Regional Satellite Monitoring Institute (ARSMI). Such an institute should be adequately equipped and staffed as a unit of one of the existing or proposed regional space-related institutions in Africa.

Today, space-derived products and services have become ubiquitous tools for improving our wellbeing and for securing our security and safety. We also know that the trend towards the production and consumption of these products and services will continue to accelerate[37]. What is Nigeria's contribution? As avid consumers and virtually a non-producer of these products and services, we Nigerians need to take a stock of our assets—both natural endowments and human resources (see Chapter X), and determine how we can significantly contribute to the production of these space-derived consumables and services. In order not to mortgage the future of generations of Nigerians to-come, it is imperative that we reverse this embarrassing trend; it is not only scary but it is also an unacceptable situation. Today, the nation can showcase a number smart indigenous ICT professionals who are blazing the trail in their various fields; but they are too few. As is the practice in many other societies, the federal and state governments must provide these Nigerian talents with the enabling environments where they can prove their mettle. Surely, Nigerians and indeed all Africans can meet this challenge! Chapters III and IV that follow trace the initial steps of Nigeria on its space journey.

* * *

---

36 In Mongolia, tracks of land for pastoral purposes are provided to herdsmen in publicly owned grazing land areas at government expense.

37 Abiodun, Adigun Ade (2007). Keynote Address on *Future Trends in the Global Space Arena—Challenges and Opportunities for Africa,* presented at the Second African Leadership Conference on Space Science and Technology, Pretoria, Republic of South Africa, October 2-4, 2007.

SPACE EFFORTS IN NIGERIA

Chapter III

RELATED INTERACTIONS

Space Activities by Nigerian Scientists between 1950 and 1965

Between 1950 and 1965, a number of Nigerian scientists along with their British colleagues, engaged their minds in space-related research in geomagnetism and the ionosphere. Prominent among the Nigerian scientists of that time were those in the Department of Physics at the University of Ibadan (UI), Ibadan.[1, 2]

These efforts coincided with and were inspired by the International Geophysical Year (IGY) which was a pre-occupation of the international scientific community at that time and lasted from July 1, 1957 to December 31, 1958.[3]

Figure 1:1 in Chapter I shows that one of the three geographical lines of interest to the IGY initiative was longitude 10⁰E which transects Nigeria. The 1995 edition of the publication, *Science Today in Nigeria*, confirmed these efforts and provided some details on earlier scientific works of Nigerians, particularly between the 1950s and the

---

1 Originally established in 1948 as an affiliate of London University, University College, Ibadan (UCI) became an independent university in 1962 and was subsequently renamed "The University of Ibadan (UI)."

2 Onwumechili, Cyril A. (1998). The Equatorial Electrojet, Taylor & Francis, United Kingdom, pp. 648.

3 The organizing committee for the IGY encouraged every participating country along the four meridians shown in Figure 1:1 to fill in all the gaps, so that for the first time, in history, simultaneous observations in meteorology, geomagnetism and ionospherics can be made." [Ron Fraser (1961). *Once Around the Sun: The International Geophysical Year*, The MacMillan Company, New York.].

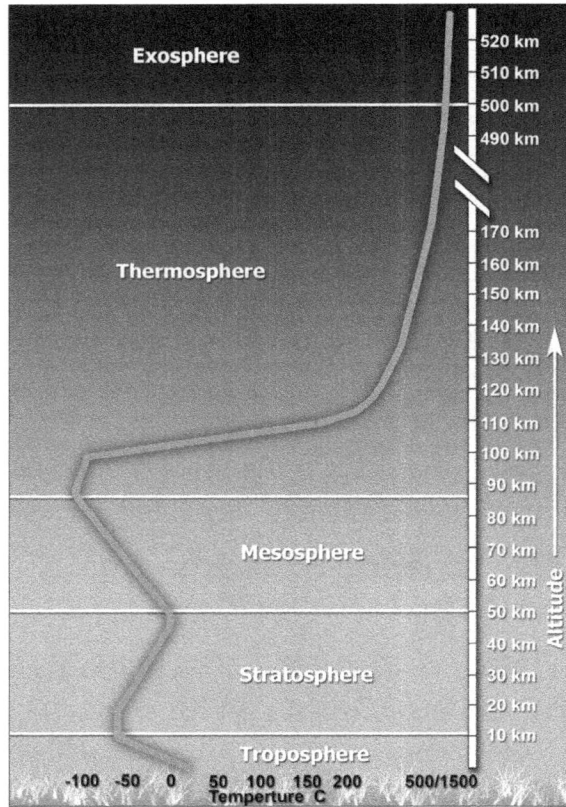

Figure 3:1 Layers of the Earth's Atmosphere
(Credit: Andria Edwards NASA/FAU/CES)

early 1960s.[4] This was the period when these Nigerian pioneering scientists gained an understanding of the structure of lightning discharges in general and of tropical storms in particular. Other studies included geomagnetism, the equatorial electrojet and the equatorial ionosphere, shown in Figure 3:1.[5, 6] In his inaugural lecture, in 1971, Professor Olatunde Aro of the University of Ilorin specifically credited the Department of Physics at the University of Ibadan for establishing an international

---

4 Science Today in Nigeria (1995). A publication of the Nigerian Academy of Sciences, University of Lagos, Lagos, Nigeria.

5 A concentration of electric current in the atmosphere found in the magnetic equator, otherwise known as "equatorial electrojet."

6 Ionosphere—A region of the earth's atmosphere where charged particles (ions) are abundant is called the ionosphere. These ions result from the removal of electrons from atmospheric gases by solar ultraviolet radiation. Extending from about 80 to 300 km (about 50 to 185 miles) in altitude, the ionosphere is an electrically conducting region capable of reflecting radio signals back to Earth.

reputation in the studies of the ionosphere in the period 1950/60.[7] Other scientists have given similar credit to the efforts of the UI physicists in fundamental science, at that time.[8, 9] Among the initial direct applications of these scientific research efforts was the prediction of radio wave absorption in the various layers of the ionosphere.[10] Unfortunately, without a continuing support for such foundational scientific works at the UI, Nigeria's march towards its space destiny in modern times has been a very jagged one. This notwithstanding, there is a need for all Nigerians to understand the world in which we live and the atmosphere (gases/air) that surrounds our planet Earth and serves as our shield by making life here on Earth possible in many ways.[11]

Nigeria-USA Connections

Nigeria's space-related interactions began in 1957 with a number of decisions taken outside the country, namely, in Washington, D.C., USA. That was also the time when Nigerian leaders were in Lancaster House, London, UK, negotiating Nigeria's political independence from Britain. With the successful launch of *Sputnik-1* by the USSR that same year, the United States felt that its super-power status was under

---

7 Aro, Olatunde T. (1982). *The atmosphere of physics and physics of the atmosphere*, The 7th Inaugural Lecture, University of Ilorin, Ilorin, Nigeria, March 18, 1982.

8 Oni, Ebun; (Editor) (1972). Proceedings of the Fourth International Symposium on Equatorial Aeronomy, University of Ibadan, September 1972.

9 Bamgboye, D.K. (1971). *Investigations of North-South Ionospheric Movements at a Low-Latitude Station,* Radio Science, Vol. 6, p. 1051.

10 Oyinloye, J.O. (1980). *Prediction of Radio Wave Absorption in the Ionosphere, D3-1 to D3-13*, in Vol. 4 of Solar Terrestrial Predictions Proceedings, NOAA, ERL, Boulder-Colorado, Ed. K.F. Donnelly.

11 As shown in Figure 3:1, the Earth's atmosphere is made up of five layers:

Troposphere: All that we know about weather occurs in this layer, including lightning; it is the reservoir of all urban pollution and effects of biomass burning; it contains 99% of the Earth's water vapour; and air pressure is at the highest at this layer;

Stratosphere: It contains the ozone layer; the air is not breathable without assistance; Commercial jet planes cruise near the bottom of this layer; and weather balloons operate here as well;

Mesosphere: It is the coldest layer of the atmosphere; most meteors and shooting stars burn up here and it is the layer where supersonic jet planes operate;

Thermosphere: Some satellites, the International Space Station (ISS) operate in this layer; it is the hottest layer of the atmosphere; and auroras occur here; and the

Exosphere: It is the uppermost layer, where the atmosphere thins out and merges with interplanetary space; Air pressure is lowest at this layer; Low Earth Orbiting Satellites operate within this layer; some satellites also operate in this layer, depending on their orbit and mission.

severe test. If the United States were to be reckoned with globally in the newly born space age, so went the reasoning amongst the political class in Washington, D.C. and beyond, it must respond quickly in kind and with a concrete measure of achievement to the challenges posed to it and its allies by *Sputnik-1* and the USSR. The United States did respond with its first successfully launched satellite, *Explorer-1*, which was developed, built and launched, in 1958, by the Jet Propulsion Laboratory (JPL) under contract from the U.S. Army.[12]

University of Ibadan tracked Explorer-1

The United States soon invited Nigeria to play a role in the monitoring of its first satellite. To track the launch and the progress of *Explorer-1* while in space, JPL established and set up portable radio tracking stations in Nigeria, Singapore, and California to receive telemetry and to plot the orbit of *Explorer-1* as it went around the Earth. Since Nigeria was still a colony of Britain at that time, the United States had to seek Britain's permission to establish one of the hypersensitive ground radio-receiving units, known as Microlock, in the Department of Physics at the University College, Ibadan. These were the units used to monitor the progress of *Explorer-1*.[13] "As *Explorer-1* looped around planet Earth, the bullet-shaped satellite not only salvaged the national honour of the United States, but it also yielded up the first major scientific finding of the infant space era: the discovery of the Van Allen radiation belts wrapped around the Earth."[14, 15]

Kano Station tracked USA's Manned Missions—Mercury and Gemini

The United States followed the success of its radio-receiving station in Ibadan in support of its *Explorer-1* mission with a second request, this time in 1960, to

---

12 JPL is a science and engineering research facility of the United States National Aeronautics and Space Administration (NASA); it is managed by and housed within the campus of California Institute of Technology, Pasadena, California.

13 http://astronauticsnow.com/explorer/index.html (Accessed, March 8, 2014)

14 Van Allen Radiation Belts of two doughnut-shaped zones of highly energetic charged particles trapped at high altitudes in the magnetic field of the Earth. The zones were named for James A. Van Allen, the American physicist who discovered them in 1958 using data transmitted by the U.S. *Explorer-1* satellite. The Van Allen belts are most intense over the Equator and are effectively absent above the poles.

15 O'Donnell, Franklin (2007), *Explorer-1* First U.S. Satellite, Jet Propulsion Laboratory (JPL), California Institute of Technology, Pasadena, California, USA.

a politically independent Nigeria. The request was for Nigeria to serve as a host country for one of the United States' tracking stations for "*space vehicle tracking* and *communications.*" On October 19, 1960, barely two weeks after Nigeria obtained its political independence from Britain, Prime Minister Abubakar Tafawa Balewa, on behalf of Nigeria and Mr. John K. Emmerson, Chargé d'Affaires ad interim, on behalf of the United States of America, jointly signed an *18-Article Agreement* in Lagos.[16] With a great fanfare, the Nigerian news media announced the signing of this Nigeria-USA space vehicle tracking agreement to Nigerians and the rest of the world. The new station, known as "*NASA Tracking Station 5,*" is shown in Figure 3:2; it was constructed on the outskirts of the city of Kano.[17]

In signing this agreement, Washington was aware of the USSR's manned space flight programme with a moon landing as its first goal. Gaining a mastery of inter-planetary space exploration requires incremental steps such as aiming first at the Moon, the nearest planet to Earth in the solar system. By beating the Russians to the Moon, the Americans wanted to demonstrate that they could be bold and could aim higher than the Russians—and if successful, it would be a concrete way to showcase America's science and technology (S&T) superiority and prowess.

Figure 3:2 A 1962 Photograph of the NASA tracking station in Kano, Nigeria, for Gemini and Apollo Missions *(Credit—NASA [Historical Reference Collection])*

16 Agreement between Nigeria and the United States of America for the establishment within the Federation of Nigeria of a station for space vehicle tracking and communications. Signed at Lagos, on 19 October 1960, United Nations Treaty Series, No. 5672, 1961.

17 NASA is the United States National Aeronautics and Space Administration.

However, the case of *Sputnik-1* experience repeated itself when, on April 12, 1961, Russia successfully launched the first manned space vehicle, Vostok-1, which completed one orbit, with Yuri Gagarin as the cosmonaut. Twenty-three (23) days later, on May 5, 1961, America responded with the successful launch of *Project Mercury*, which carried astronaut Alan Sheppard as America's first man in space, on a sub-orbital flight that lasted 15 minutes and 28 seconds.[18] The success of these two feats by both the USSR and the USA soon made it apparent to the international community that a space race was on. It was John Glenn who, on February 20, 1962, first completed three orbits for the Americans, on board Mercury spacecraft #13, named *Friendship*.

The Kano station played a major role in John Glenn's first space mission and in a few other USA space travels thereafter; it also served as a forerunner for America's Deep Space Network. Depending on the orbit, the Kano station provided anywhere from a three to a six and a half minute communication window with the Mercury (and later Gemini) spacecraft as it passed over the African continent after leaving the Grand Canary Island coverage area.[19]

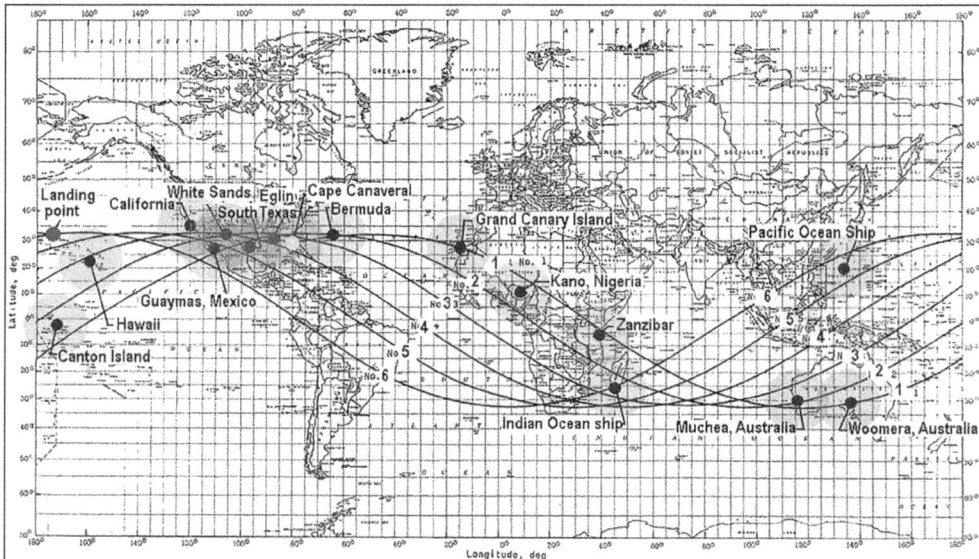

Figure 3:3 A Map of USA's Manned Spaceflight Network (Late 1960s–early 1970s) *(Credit—NASA)*

---

18 https://www.nasa.gov/multimedia/imagegallery/image_feature_1344.html (Accessed, March 9, 2014)

19 Tsiao, Sunny (2008). "READ YOU LOUD AND CLEAR!", National Aeronautics and Space Administration, NASA History Division, Office of External Relations Washington, DC.

The local Hausa villagers living around Kano described the station site as "*the place the Sardauna built to get the message from the stars.*" According to the Americans at the site, the Kano station was a "*hi-tech tourist attraction of its time, where open-door was the norm, drawing the curious from all parts of Nigeria.*"

The Kano tracking station, highlighted along with others in Figure 3:3, was installed between 1960 and March 1961; it provided ultra-high frequency (UHF) and high frequency (HF) relay with the Mercury spacecraft. It also transmitted telemetry parameters to Mission Control at Cape Canaveral, USA, by telephone and telemetry (TTY) through Lagos and via a direct HF communications link with London. That was the only telephone channel opened to Nigeria at that time; as a colony of Britain, Nigeria then had no direct communication with the United States or any other country in the world. The Kano station also provided a teletype-high frequency (TTY-HF) communications link between London and Zanzibar in East Africa.

A flight surgeon and a spacecraft system specialist, both stationed at Kano, monitored the spacecraft telemetry parameters along with both the blood pressure and the ECG[20] of orbiting astronauts, displayed during station contact with the spacecraft. In 1965, the Kano Station also provided another temporary site for STADAN when it briefly housed the equipment that supported the International Satellite Programme for Ionosphere Studies.[21]

That period also coincided with the time when France was testing nuclear bombs in the Sahara desert—an act the African governments and their people vehemently opposed. Most Nigerians strongly believed that the American team that was associated with the Kano tracking station had a link with the French atomic bomb tests in Algeria between 1960 and 1966, and probably contributed to the pre-mature closing of the Kano tracking station.[22] Although the Kano station, labelled a *Cultural Dichotomy* by the Americans, met a critical need at that time, NASA gave the "*evolution of new technologies,*" and the need to revamp its Manned Space Flight Network

---

20 ECG—Electrocardiography—It is the interpretation of the electrical activity of the heart, over time, captured and externally recorded by skin electrodes. It is a noninvasive recording produced by an electrocardiographic device.

21 The Satellite Tracking and Data Acquisition Network (STADAN) which supported many of the USA's unmanned scientific satellite programs, and its Deep Space Network (DSN).

22 France conducted (i) One nuclear test at Reggane, Sahara Desert, Algeria on February 13, 1960 at a time when Algeria was a protectorate of France, and (ii) Thirteen (13) nuclear tests between 1961 and 1966 in Ekker, Sahara Desert, Algeria.

(MSFN), of which the Kano Station was a part, in preparation for the Apollo manned missions, as the official reasons for its closure. The 1964 agreement on the use of the station only covered the period, May 21, 1964 to June 30, 1966 and was subsequently extended to November 18, 1966, to accommodate the landing of Gemini 12, just one week after its splash down. While this drama between Nigeria and the USA was going on, the British tried to convince their African colonies that the American tracking stations in both Kano and Zanzibar were part of an experiment that was harmless in nature and that the stations would contribute to the world's scientific knowledge. But the plea of the British had no effect on the citizens of the affected British colonies or on NASA's decision to eventually close the Kano station.

The attainment of independence in the 1960s by many African countries also offered the United States and the USSR the opportunity to seek new allies and resources for their on-going *cold war*.[23] Senator John F. Kennedy became the President of the United States on January 20, 1961. Thereafter, diplomatic plans to establish a relationship with Africa's most populous and widely perceived most powerful country, Nigeria, took centre stage in Washington, D.C. The USSR equally targeted its own allies within the African continent, beginning with the Congo and Ethiopia.

Nigeria benefits from the African Scholarship Program of American Universities (ASPAU)

In order to build the human capital needed for addressing the overwhelming development challenges that confronted each African country after independence, both the USA and the USSR offered a variety of assistance to their new African partners. In 1961, under the Kennedy Administration, the US introduced the *Peace Corps Program* that deployed many American fresh university graduates to serve in several African countries.[24] Within the same period, the Kennedy Administration

---

23 The period (1947–1991) known as the *Cold War* era followed the Second World War (WWII-1939 through 1945). The period was noted for the continuing state of political conflict, military tension and economic competition between the then Soviet Union and its satellite states and allies on one hand and the Western countries, primarily the United States and its allies on the other. The two sides demonstrated these conflicts in many ways and in various disciplines. The latter included appeals and military aids to *neutral* countries, espionage, propaganda, conventional and nuclear arms race, rivalry at sport events, and technological competitions including the space race.

24 The Peace Corps was established by Executive Order 10924 on March 1, 1961, and authorized by Congress on September 22, 1961, with passage of the Peace Corps Act (Public Law 87-293). The Peace Corps Act declares the purpose of the Peace Corps to be:

(Continued On Next Page)

established the African Scholarship Program of American Universities (ASPAU) in collaboration with a number of governments in Africa, and with the support of USA universities and private sponsors. ASPAU was administered from 1961 to 1975 by the African American Institute (AAI) in Manhattan, New York. It served as a major vehicle that enabled many young African students, including this author, to attain their university education in many first-rate American universities, in various fields of human endeavour, including space science and technology (SST). Upon the completion of their studies, the beneficiaries of the ASPAU programme were to return home to replace the departing Peace Corps participants in their respective countries.

The USSR equally offered scholarships to African undergraduates, including Nigerians, to study at its universities, including the prestigious University of Moscow. The USSR went a step further—it established *The Patrice Lumumba University* in Moscow, initially for the benefit of African students only. This university, which was named in honour of the late Hon. Patrice Lumumba, the first post-independence Prime Minister of Congo, who was assassinated on January 17, 1961, was later renamed the *Peoples' Friendship University* after the cold war. Many of the African students that opted for SST-related education, in both the western and eastern educational institutions, are now contributing, to an appreciable extent, to the evolution of space activities in their respective countries.[25]

Balewa—Kennedy Space Connection

The relationship between Nigeria and the United States also became personal at the highest political level. It culminated in the first-ever satellite-based telephone conversation between two heads of states on August 23, 1963. On that day, the Nigerian Prime Minister, the Honourable Tafawa Balewa, went aboard the United States naval ship, USNS *Kingsport* (T-AG-164) in Lagos, Nigeria; it was the first US satellite communications ship that ever visited Lagos.

But why did the USA select Nigeria and its new Prime Minister rather than Britain and the British Prime Minister for this unique honour especially as Britain

"To promote world peace and friendship through a Peace Corps, which shall make available to interested countries and areas, men and women of the United States qualified for service abroad and willing to serve, under conditions of hardship if necessary, to help the peoples of such countries and areas in meeting their needs for trained manpower."

25 Abiodun, Adigun Ade (1998). *Human and Institutional Capacity Building and Utilization in Science and Technology in Africa: An Appraisal of Africa's Performance to-date and The Way Forward,* African Development Review, Volume 10, Issue 1, 1998, John W. Wiley Publishers.

was a very close ally of the USA? This was probably because, at its independence in 1960, Nigeria was perceived globally as a country with great potential that would make it the Hong Kong of Africa and the Giant of Africa. All this happened just before a Martian rock landed on Nigerian soil.

From Planet Mars to Katsina State, Nigeria

By 1963, its third year as an independent country, Nigeria had passed from being a country that interacted with human beings on space matters to one that interacted with the natural elements, such as asteroids and meteorites, from another planet. Asteroids and meteorites are extra-terrestrial rocks that are mostly composed of carbonaceous, stony and metallic (mainly iron) materials. In 1993, there were about 100 known such natural rocks, greater than 1 km in diameter, that had been detected; that number could reach thousands as new and more advanced telescopes are built and deployed in the coming decades to track these space rocks. For example, as of July 27, 2015, 12,928 Near-Earth objects (NEOs) have been discovered. Some 873 of these NEOs are asteroids with a diameter of approximately 1 kilometre or larger. Also, 1,604 of these NEOs have been classified as Potentially Hazardous Asteroids (PHAs).[26] A PHA is a near-Earth asteroid or comet with an orbit such that it has the potential to make close approaches to the Earth and is of a size large enough to cause significant regional damage in the event of impact on Earth. A collision of any of these asteroids with our home planet could result in major impact craters and attendant destructions here on Earth.[27]

An asteroid impact, 2,000 million years ago, resulted in the oldest known crater on Earth. This crater, which measures 140 to 400 km in diameter, is in Vredefort, Republic of South Africa (RSA). Natural rocks from space have also created impact craters in many other African locations some of which are shown in Table 3:1.[28]

Although the African continent, as a whole, has been spared of any major natural catastrophe from an asteroid or meteorite in recent decades, an alarm bell rang in the afternoon of October 3, 1962 when, without any advance warning, *the stars decided to pay a return visit to the Sardauna* in his own backyard[29]. On that day, a meteorite

---

26 http://neo.jpl.nasa.gov/faq/#howmany (Accessed September 19, 2015)

27 Most of what the world knows today on this topic has come from the United States Space Command.

28 Abiodun, Adigun Ade (2003). *Re-Entry of Space Objects into Nigerian Territory*, The Presidency, Abuja, Nigeria.

29 Sir Ahmadu Bello was both the Sardauna of Sokoto and the Premier of Northern Nigeria from 1960-1966.

Table 3:1 Rentry and Impacts of some Natural Space Objects in Africa

Age (in years) or Estimated Date	Place or Location	Diameter of the Resulting Crater
70 million years	Algeria (Ouarkziz)	4 km in diameter
345 million years ago	Chad (Aorounga)	17 km in diameter
Still to be determined	Egypt (Kebira)	19 km in diameter
One million years ago	Ghana (Bosunmtwi Lake)	10.5 km in diameter
80,000 years ago	Namibia (Roter Kamm)	2.5 km in diameter
2,000 million years ago	South Africa (Vredefort)	The oldest known crater on Earth, 140 to 400 km in diameter

from Mars landed near the Village of Zagami, in the then Katsina Province (now Katsina State) of Nigeria. There are a number of facts we know now about this Martian rock. For a better appreciation of where in outer space the Zagami rock came from, we should note that the mean distance between the Earth and Mars is approximately 78,300,000 km, that is, roughly half the distance between the Earth and the Sun. The Moon, the closest planet to Earth, is only about 380,000 km away. What a journey for one piece of rock! There are, however, some other facts that we do not know. For example, no one knows the original size and weight of the rock the moment it broke off from Mars; the day it broke off from Mars; the duration of its journey from the surface of Mars to the surface of the Earth; and its space track before it landed in Nigeria. It is also worthy of note that between April 1984 and May 2010, the Centre for Basic Space Science, of the University of Nigeria, Nsukka, Nigeria, reported the landing in Nigeria of seven other meteorites (of unknown origins)[30].

The Zagami meteorite landed on October 3, 1962, about 0.75 miles from Zagami Rock, Katsina Province, Nigeria, in a cornfield where the farmer was trying to chase

---

30 Okeke P.N., Okere B.I., Yusuf Naji, Ofodum C.N., Ayantunji B.G., Nasiru Aliyu and Esaenwi S. (*Unpublished*) Studies of Near Earth Objects (Meteorites) which Landed in Nigeria, Centre for Basic Space Science, University of Nigeria, Nsukka,(Undated).

away some crows.[31] "The farmer heard a tremendous explosion and was buffeted by a pressure wave. After a puff of smoke and a thud, the meteorite got buried in a hole about 2 feet deep." Weighing 18,100 grams (40 pounds) on landing, the Zagami meteorite is to-date the largest single individual Mars meteorite ever found on planet Earth.[32] A sketch of the original rock appears in Figure 3:4. Zagami was subsequently deposited at the headquarters of the Nigerian Geological Survey in Kaduna and placed in its museum. In 1963, the 40 pound Zagami rock made its first overseas trip to the Natural History Museum in London for analysis. It was returned to Kaduna, Nigeria, in 1965, after the British museum had sawn off a piece that weighed 234 grams.[33]

About 1988, Robert Haag, a meteorite dealer (see Figure 3:5), obtained a large piece (7,200 gm, i.e. 40%) of Zagami from Kaduna. This appears to be the first major *trade/sale* of the museum in precious stones. And for how much money? Since then, a significant amount of the rock has been cut up and *"distributed globally,"* as can be seen from Figure 3:5[34] below. According to NASA, Zagami has been extensively subdivided into multiple pieces, in recent years.

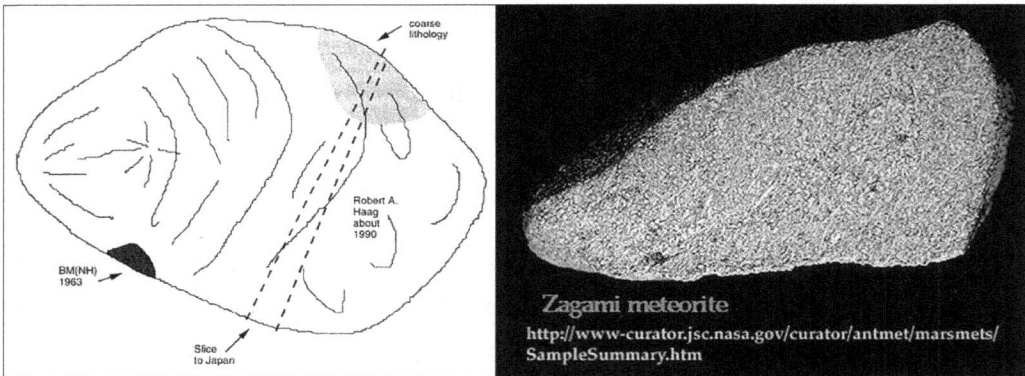

Figure 3:4 A Sketch of Zagami provided by R. Haag who witnessed the sawing off of the sample in 1988. The coarse lithology is referred to in the text as the Dark mottled lithology (DLM) of Zagami meteorite. The sample obtained by the Natural History Museum in London, in 1963, is known as normal Zagami (NZ); it seems to be representative of the main mass. *(Credit: NASA)*

---

31 http://curator.jsc.nasa.gov/antmet/mmc/Zagami.pdf (Accessed, March 15, 2014).

32 Mars Meteorites, Jet Propulsion Laboratory/NASA, Pasadena, California, USA, 1996.

33 Information from The Meteorite Catalogue of Graham *et al* 20 years later in 1985.

34 Mars Meteorite Compendium 2003, http://www2.jpl.nasa.gov/snc/zagami.html (Accessed, March 15, 2014).

It is available for research at the University of New Mexico, USA, and for purchase from dealers in different parts of the world.[35] According to Ron Baalke, in 1996, *"The Zagami meteorite is the most easily obtainable SNC meteorite available to collectors.*[36, 37]

The Zagami Meteorite – Cut Up and Shared Globally

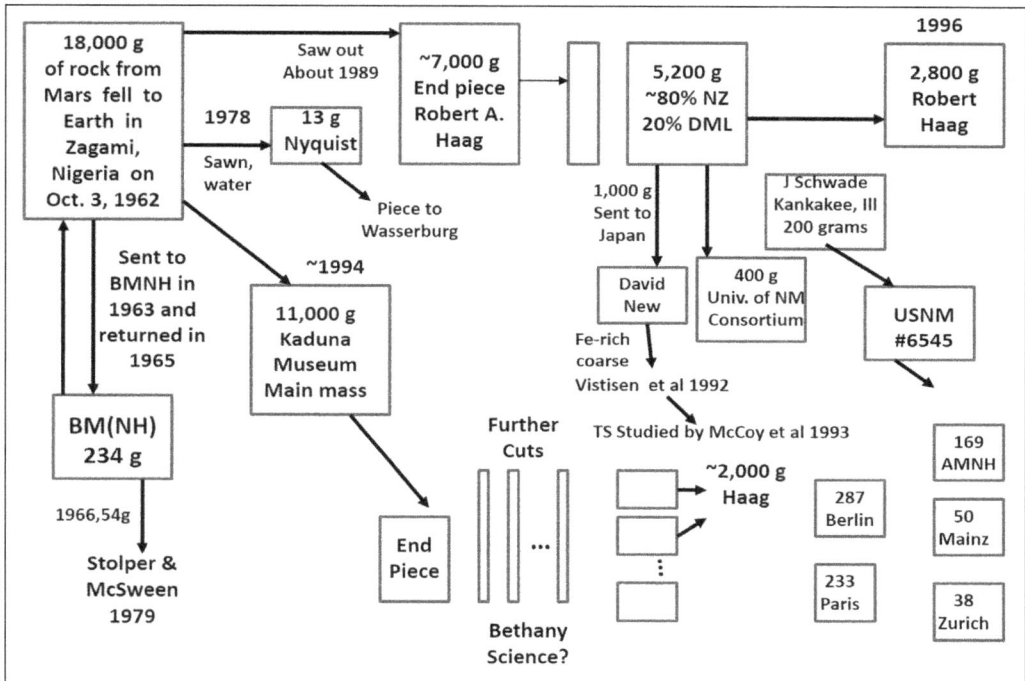

Figure 3:5 A Schematic Illustration of the Cutting up and Distribution of Zagami Meteorite (Credit: *NASA*)

While an amount of Zagami rock is still housed at the Kaduna Museum, Kaduna State, Nigeria, to-date, not many Nigerians have ever heard of any rock from Mars that landed in Nigeria; hence, its fate, since it landed, is also unknown to them. Indeed, there is a host of unknowns about the Zagami meteorite, amongst which are the following:

35 https://curator.jsc.nasa.gov/antmet/mmc/zagami.pdf

36 BAALKE, Ron (1996). Jet Propulsion Laboratory, Pasadena, California, USA - http://www2.jpl. nasa.gov/snc/news2.html

37 *Your Own Little Piece Of Mars: The Odd Truth*, Aug. 16, 2003 - CBS News, New York, USA (Bids for Zagami fragment, which weighed about 6.6 ounces and was about the size of a soda can, began at $450,000 when the online auction started on Sept. 5, 2003 on the Internet site eBay).

- Who first picked up the Zagami meteorite (hereafter referred to as Zagami) from its Katsina landing site on October 3, 1962?
- How did it get to the museum of the Nigerian Geological Survey in Kaduna?
- What happened to Zagami immediately thereafter in Kaduna?
- How and when did Zagami leave Kaduna for the Natural History Museum in London?
- Who, at the Nigerian Geological Survey in Kaduna, around 1988, was responsible for cutting Zagami into pieces, on its return from London, for distribution for scientific research purposes and/or for sale to international dealers as shown in Figure 3:5?
- If Zagami were still in its 18,100 gm (40 lbs.) original state, as it landed in Katsina, what would be its monetary value, today?[38]

According to South Africa's National Heritage Act, No. 25 of 1995, if such a meteorite should land on South Africa's soil, it becomes the property of the state.[39] The existence of a similar legal instrument in Nigeria is uncertain at this time. Meanwhile the Nigerian Geological Survey Agency (NGSA), the successor to the Geological Survey of Nigeria, certainly owes the Nigerian people a detailed account of the fate of Zagami, a piece of national treasure that should have been residing permanently in the nation's museum of natural history as an inheritance of all generations of Nigerians, past, present and future.

* * *

---

38 http://geology.com/meteorites/value-of-meteorites.shtml - At the high end of the pricing scale are unusual meteorite types such as the *diogenite* Tatahouine (landed June 27, 1931 at Foum Tatahouine, Tunisia). A prime specimen will easily fetch $50/gram while rare examples of lunar and Martian meteorites may sell for $1,000/gram or more — almost forty times the current price of gold!

39 http://www.accu.or.jp/ich/en/pdf/c2006Expert_MANETSI_1.pdf — South Africa's National Heritage Act, No. 25 of 1995

Chapter IV

INITIAL EFFORTS

Except for the period from October 1979 to December 1983, when H. E. Alhaji Shehu Shagari served as an elected President of Nigeria, the country and its people endured, from January 1966 to May 1999, all that military dictatorship had to offer, from a variety of military administrations and administrators. It was an era of military coups and counter coups. The author, who left the country on August 2, 1962, for the University of Washington, Seattle, State of Washington, USA, on an ASPAU scholarship returned home for a family visit on December 31, 1967. By then, the military coup of January 15, 1966 had escalated into a full-blown Nigeria-Biafra War. Many friends and colleagues in the United States wondered aloud: Why do you want to visit the country in the middle of an on-going civil war? *"I already made a promise to my dear mother, by letter, that I would be home to see her before 1967 was three-hundred and sixty-five days old. I cannot disappoint her,"* was my true and only response. However, I burst into tears and wept uncontrollably when, at the arrival hall of the Lagos airport, on December 31, 1967, I saw, first-hand, the impact of the civil war on my mother and the rest of my people, all who had come to the airport to welcome me back home; they were all extremely emaciated beyond belief and description.

The military incursions into the political life of the nation also strangled most scientific and technological initiatives of Nigeria's inquiring minds, many of whom actually relocated to more peaceful environments outside the country. After their under-graduate studies, a majority of those who went to the United States on the ASPAU programme, in the early 1960s, also chose to proceed to graduate school or seek employment there.[1] On the contrary, Nigeria's rulers of that period persistently portrayed to the global community that all was well with the country and its people. Thus, on November 14, 1967, while still grappling with a civil war that was, by all accounts, very much crippling from its start, on July 6 of that same year, Nigeria's leaders signaled the country's solidarity with the space community of the world and became party to the United Nations *Space Treaty*. To date, The Space Treaty, otherwise known as "*The Treaty on Principles Governing the Activities of States in the Exploration and Peaceful Uses of Outer Space, including the Moon and Other Celestial Bodies,*" is the most important of all the United Nations legal instruments on outer space. But Nigeria could not field any representation at the First United Nations Conference on the Exploration and Peaceful Uses of Outer Space, held in Vienna, Austria, in 1968, because of its pre-occupation with the civil war that was raging in the land.

At the end of the civil war that lasted through January 13, 1970, the Nigerian government was overwhelmed with all kinds of problems that needed its attention or intervention. Indeed, the magnitude and complexity of the problems foreclosed most national efforts on all fronts. The political climate in Nigeria also reverberated across the oceans into the global community; indeed, it took a while for the country to regain any measure of international credibility. With the gradual return of normalcy in the country at the end of the civil war, space-related initiatives also began to emerge at different academic institutions and government establishments. Areas of interest included fundamental space research, satellite communications, climate research and satellite meteorology, and the observation of the Earth from space. The professional expertise and connections of the drivers of this resurgence dictated most of the orientation of subsequent space-related efforts in the land. A major contributor to the resurgence was the successful launch, by the United States, of the first civil earth observation satellite, ERTS-1 (later renamed *Landsat-1*), on July 12, 1972, and its promise of providing critical data for national development. In 1973, Nigeria concluded that it could benefit from a judicious use of Landsat data. In that same year, it submitted a proposal, prepared by the University of Ife (since re-named Obafemi Awolowo University), to the USA, requesting its space agency, NASA, to use its Landsat satellite to acquire Nigeria's Earth resources data.

---

1 The UN Economic Commission for Africa and the International Organisation for Migration (IOM) estimated that 27,000 Africans left the continent for industrialized countries between 1960 and 1975. During the period 1975 to 1984, the figure rose to 40,000. It is estimated that since 1990 at least 20,000 people leave the continent annually. http://www.un.org/ecosocdev/geninfo/afrec/vol17no2/172brain.htm(Accessed, March 25, 2014)

However, the need for communication services in the country was in greater demand and more urgent.

A Taste of Communicating via Satellites

That was also the era when Nigeria was still yoked to the British communications net-work; in effect, all communications (in-coming and out-going) between Nigeria and the rest of the world had to first be routed through London, the capital of the extant colonial master, Britain. This experience was not unique to Nigeria; it was the same for other former British and French colonies in Africa. Any telephone call from an Anglo-phone colony to a Francophone colony, and vice-versa, had to pass through London and Paris before being routed to its final destination. The use of a telephone in those days was also a luxury that only a few could afford. The communication infrastructure that the nation and its people inherited from the colonial era, and which had received no national attention during the civil war, was also grossly inadequate to cope with the overwhelming post-civil war demands. As could be expected, uppermost among the most immediate needs and demands in the country was the need to re-establish communications with loved ones and family members that were displaced by the civil war, as well as with the international community. Accordingly, one of the first investments on satellite-services the country made was on communication. To address its communication woes as well as burnish its international image, the Nigerian government of the day took a number of space-related steps in 1972 and 1973.

On October 25, 1972, Nigeria and Intelsat signed an agreement to provide external communication links between Nigeria and the rest of the world.[2] This agreement went into effect on September 12, 1973. Thereafter, Nigeria established, at Lanlate (between Ibadan and Abeokuta), its first Satellite Earth Station, shown in Figure 4:1, and thus successfully linked Nigeria with the rest of the world, via Intelsat Satellite VA-F11.

---

2 Intelsat Ltd is a communications satellite services provider. Originally formed as International Telecommunications Satellite Organisation (Intelsat), it is now an intergovernmental consortium that owns and manages a constellation of communications satellites that provide international broadcast services. As of 2007, Intelsat Ltd owned and operated a fleet of 51 communications satellites. Originally established as an Inter-Governmental Organisation (INGO) on August 20, 1964, with 11 participating countries Intelsat provided service to over 600 Earth stations in more than 149 countries, territories and dependencies. By 2001, Intelsat had over 100 members. It was privatized on July 18, 2001 and it changed its name to Intelsat Ltd.

Figure 4:1 Nigeria's firstSatellite Earth Station (at Lanlate) that linked Nigeria with the rest of the world via Intelsat Satellite VA-F11. (*Credit: Ade ABIODUN, 1992*)

Taken together, Nigeria's request for Landsat satellite data and the establishment of the Lanlate station, signalled its interest in space matters and emboldened it to engage in a variety of dialogues with the space faring nations of that period at the appropriate space fora. Such interactions led the country to the doors of the United Nations where, in 1973, Nigeria applied for and was granted membership in one of the most prestigious bodies of the organisation—the United Nations Committee on the Peaceful Uses of Outer Space (COPUOS).[3] In that same year, Nigeria also became party to *The Rescue (of Astronauts) Agreement*. Thus, the nation proudly marched into the 21st century with the rest of the global community, and reinforced its trust in the international legal regime on outer space by becoming party, in 2005,

---

3 GA Resolution 3182 (XXVIII), dated 18 December 1973. Nigeria participated in the deliberations of COPUOS for the first time in February 1976, in Geneva, with a four-man delegation that included I.P. J. Obebe (Ministry of External Affairs), Felix Oragwu (National Council for Science and Technology), Simeon Ihemadu (RECTAS—Federal Surveys, Lagos) and Adigun Ade Abiodun (University of Ife).

to two additional space legal instruments—*The Registration (of Space Objects/Assets) Convention and The Liability Convention.*[4]

But Nigeria's membership in the international-political-space community, in the 1970s, belied its investment in its domestic communication infrastructure. To remedy the situation and increase the country's external communication services, Nigeria built the second communication Earth Station in Lanlate, in the 1975-1980 Plan Period. To accommodate the nation's growing telephone network, which was then projected to increase from 50,000 lines to 750,000 lines at the end of the planned period. It also established, with the aid of additional leased transponders from Intelsat, the Nigerian Domestic Satellite (DOMSAT) system, which provided TV and sound broadcasting services nationally. In the 1980-1990 timeframe, the federal government upgraded the system and thus made it possible to extend telephone and telex services to and between all the then 19 states of the country. By 1999, Nigeria's international communication system, which, at that time, depended on Intelsat satellites, consisted of four gateways: Lanlate, Kujama (near Kaduna), Enugu and Lagos. In December 1995, the telephone density in Nigeria stood at 500,000 lines; in a country which then had an estimated population of about 100 million people, this translated to one telephone line for every 200 people.[5] However, by the year 2000, the theoretical capacity of the number of telephone lines in Nigeria was 800,000 lines while the operational capacity stood at 400,000 lines.[6]

First Attempts at Acquiring Space Assets

The successful deployment of the Intelsat system in the country gave some of the nation's political leaders and government bureaucrats of the day the idea that the time was ripe for Nigeria to own its own independent space-based communication system. Somehow, it seemed that the proponents of such a system forgot about the first successful world satellite communications telephone exchange between Nigeria's first Prime Minister, Hon. Tafawa Balewa and President John F. Kennedy of the USA. Instead of a communication satellite, they opted for a balloon—Aerostat Balloon. This decision was taken during the military rule of General Murtala Mohammed,

---

4 United Nations Treaties and Principles on Outer Space, Document ST/SPACE/11, United Nations, New York, 2002.

5 Abiodun Adigun Ade (1997). *Deregulation of Nigeria's Telecommunication Industry*, Thisday Newspaper, Lagos, (In two parts—Part 1, August 21, 1997, Part 2—August 26, 1997).

6 Information made available at AFRINET 2000, held at Abuja, September 18 to 22, 2000.

who argued that if Nigeria could not be placed on the moon, it could at least be placed in space. "He was so excited by a presentation of some folks from Canada that he signed the then famous Aerostat Balloon Transmission contract with a warning that whosoever opposed him would go the way of those who had given him contrary opinion earlier."[7] Because Gen. Mohammed was committed to use Nigeria's oil wealth to provide a functional telecommunication infrastructure for Nigerians 'in the shortest possible time,' he was misled, for whatever reason(s), by those in charge of the project. Part of the problem also was that the group of advisers around him, somehow 'perceived anything above the ground level to be in space.' Hence, they settled for tethered Aerostat Balloons. The Permanent Secretary, (an engineer by profession), in the Federal Ministry of Communications (FMC), at that time, thought it was a bad idea; but the most senior political official in the ministry, the Commissioner for Communications, who was fully in support of the project, recommended and presented the political decision to Gen. Mohammed.

At that time, most Nigerians knew very little about Aerostat Balloons and about their deployment as make-shift communication tools by the United States during its military engagement in Vietnam. As the war in Vietnam was winding down, the need to find new markets for the balloons became paramount in the agenda of its manufacturers and sales agents. Accordingly, Saudi Arabia, Iran and Nigeria, with over-flowing petro-dollars in their coffers, became targets for Tethered Aerostat Balloons merchandising. The Nigerian government swallowed the bait, but its citizens did not.

To douse the grumbling, nationally, particularly by the country's communication experts, on the efficacy of an Aerostat Balloon as an operational communication tool for Nigeria, the FMC quickly organised a seminar, in 1975, at the University of Ife (UNIFE) on the project. It was such an important event for the Federal Government of Nigeria that the Head of State, General Murtala Mohammed, left his base in Lagos, and attended the seminar. Professor E. Ekundayo Balogun and Professor Gabriel 'Lere Ajayi, both of the Department of Physics at UNIFE, at that time, and other experts, including communication engineers, offered the government sound scientific and engineering pieces of advice during the deliberations. The experts argued that Aerostat Balloons had never been deployed anywhere in the world as an operational communication system; that balloon-based systems were developed and installed, normally, only for short-term emergency use, particularly during military

7 http://www.cyberschuulnews.com/index.htm, Edition 246 (Accessed, September 10, 2012).

engagements and in time of major disasters. Of particular concern to these Nigerian experts was that Aerostat Balloons were very vulnerable to all kinds of adversities and that because of their aerodynamic instability, such a system was not practicable as an operational national communication tool in and for Nigeria.[8] The UNIFE experts and engineers affirmed that the proposed balloon project had a very high probability of failure, and might actually fail within three months of its installation.

But the Federal Commissioner for Communications and his Department of Posts and Telegraphs (P&T) at the Federal Ministry of Communications eventually got the green light from the highest level in the land; the commissioner went ahead and committed the country to the launching of five (5) such balloons for the nation's telephony, broadcasting, telex, radio surveillance, and maritime and mobile communication services. The first of these balloons was installed near Ile-Ife, on the way to Ondo; and *just as the UNIFE experts had predicted, it collapsed within two weeks of its installation.* At a press conference held in commemoration of the World International Telecommunication Day on May 17, 1983, Nigeria's Minister of Communication of the day announced the termination of the Aerostat Balloon contract. He gave no reason(s) for the cancellation of the project and he also foreclosed any questions on his announcement. Nigeria paid a penalty of over US$30 million for eventually cancelling the ill-fated project and lost over US$200million on the project.9

Surprisingly, without any adequate knowledge of the outer space environment, and despite the Aerostat Balloon debacle, the managers of the nation's communication affairs remained undaunted in their subsequent actions; they successfully persuaded the Nigerian government of that era of the need for Nigeria to be there, in outer space, with other nations. Thus, in 1988, and still without a space policy or programme, Nigeria joined the global clamour and asked the International Telecommunication Union (ITU) for its own slot at the geosynchronous orbit (GSO), an imaginary altitude of 35,786 km (22,236 mi) above the Earth's equator.[10] Nigeria made this

---

8 Balogun, Ekundayo E. (1975). *The atmospheric conditions around 10,000 ft. over West Africa, Paper presented at a Seminar organised by the Federal Ministry of Communication on the Aerostat Balloon Project, University of Ife, Ile-Ife, Nigeria, 1975.*

9 Ojo, Tokunbo (2004). Old paradigm and Information & Communication Technologies for Development Agenda in Africa: Modernization as Context, Journal of Information Technology Impact, Vol. 4, No. 3, 139-150, 2004.

10 The geostationary orbit, also interchangeably referred to as a geosynchronous orbit, is about 22,300 miles (35,786 km) above the Earth's surface at the equator. For an observer at a fixed location on Earth, a satellite in a geosynchronous orbit appears to be stationary in the sky; thus the term "geostationary."

request because it wanted to have its own internationally recognised orbital allocation at the GSO, should it decide, in the immediate future, to participate in the space enterprise. On hindsight, that was not a bad idea. If they did not ask then, all the slots would probably have been allocated to other bidders by now. In 2006, ITU eventually assigned Nigeria a slot at the GSO, at 42.5^0E longitude—located at the equator in the Eastern shores of Africa, straight up above the Indian Ocean.

But why so far away from territorial Nigeria, whose westernmost and easternmost geographical boundaries are longitude 2.625^0E and longitude 14.750^0E, respectively? Nigeria probably would have secured an orbital slot that was geographically closer to home except that its leaders' collective understanding and appreciation of the roles of space in human security and development matured very late. The nation woke up to that realization after all the choice slots that would have served Nigeria better had been allotted to earlier comers, on the basis of *first come, first served,* by the same ITU, the Geneva-based international organisation responsible for making such allocations. The first space faring nations operated by this norm of *first come first served'* and continue to occupy the most desirable slots at the GSO, today. How long this condition will last is unknown; those that believe in this principle will still hold on to it as long as it is in their nations' security and economic interests to do so.

The United Nations Space Treaty, which Nigeria adhered to in 1973, made it possible for all nations to own or be entitled to such an orbital position or a real estate in outer space. Articles I and II of the Space Treaty, which the United Nations adopted on December 19, 1966, respectively stated that:

Article I—The exploration and use of outer space, including the Moon and other celestial bodies, shall be carried out for the benefit and in the interests of all countries, irrespective of their degree of economic or scientific development, and shall be the province of all mankind.

Article II—Outer space, including the Moon and other celestial bodies, shall be free for exploration and use by all States without discrimination of any kind, on a basis of equality and in accordance with international law, and there shall be free access to all areas of celestial bodies.

Also in 1988, Nigeria became conscious of the importance of its maritime endowments and assets and the role of space in their management. Before the year was over, it joined the International Maritime Satellite Organisation (Inmarsat). The organisation operates a system of satellites that provide communication services for the world shipping and offshore industries. Through the Inmarsat system, Nigeria began to provide maritime mobile and satellite mobile communication services.

According to Asinugo (1999), Nigerian Telecommunications (NITEL) Plc sought a total monopoly of the new Inmarsat services, *despite very vocal national protests*. The NITEL management used the worn-out arguments of potential revenue loss, even from cities and locations where no NITEL infrastructure existed and where businesses were suffering daily losses due to lack of telecommunication services.[11] NITEL's, patriotic-sounding concern for national security won no converts at the Ministry of Communications; the latter subsequently decided against NITEL's monopoly of the nation's communications networks and authorized the introduction of mobile satellite services in Nigeria beyond the sole control of NITEL.

As the 1980-1990 decade drew to a close, those that promoted the botched Aerostat Balloon project in the 1970s also returned with a more brazen proposal. On hind-sight, they probably reasoned that *time heals all wounds*, and that if Nigeria could ask for and obtain an orbital allocation at the GSO from ITU as well as join Inmarsat, it was almost certain that it would soon procure a communication satellite that would provide its communication services. Either on its own volition, or by invitation from interested party(ies) in Nigeria, Spar Aerospace of Canada soon showed up at Nigeria's shores, this time, with a proposal to the Federal Government to launch a Nigerian communication satellite and place it at the new Nigerian GSO location—with a price tag of US$500 million (space segment only). By then, the nation had a new Hon. Minister of Communications, Engr. Olawole Ige. He quickly set up a committee of Nigerian experts to evaluate the proposal. The committee met for six (6) days in Enugu and *'recommended that the Federal Government should unequivocally reject the proposal in its entirety.'* The Federal Government of the day, albeit, a military one, accepted and adopted this recommendation, without any equivocation.[12]

The Price of Poor Communication Planning

With no immediate action in place that could resolve the national communication strangulation, it dawned on Nigerians that the nation's communication authorities had no long-term plan and development programme to meet their communication needs. The succeeding governments of the day also invested very little in science and technology development, including communications technologies. Gradually, within

---

11 Asinugo, Johnson, (1999), Regulating satellite communications and associated technologies in Nigeria: imperatives for the year 2000—2001, http://www.ncc.gov.ng (Accessed, April, 12, 2013).

12 Personal communication in 2013 with Prof. E. E. Balogun, Dept. of Physics, University of Ife, Ile-Ife, Nigeria. He served as a member of this committee.

three decades, 1970 to 1999, Nigeria became a communication ghost-territory. This was a period when many incompatible telephone parts, components and infrastructures were purchased from diverse foreign sources and installed in the country.[13] In many metropolis of Nigeria, such as Lagos, one could observe, in the 1990s, multiple strands of telephone wires that were wrapped, in several layers, around telephone poles and those that drooped, like thick canopies, over the urban streets. Figure 2:1 in Chapter II provides a graphic example of the nation's communication dilemma. The end-result was a total breakdown, nation-wide, of Nigeria's communication infrastructure and services.

Individuals and businesses had no option but to travel, by pothole-ridden roads, to deliver messages, letters, documents and parcels that could have been effortlessly and successfully transmitted using a telephone, a fax machine or via electronic mail. As a result of the communication-related bad decisions, inadequacies and failures of that era, untold numbers of Nigerians met their untimely death on the nation's death-trap roads. The economic loss was equally colossal and incalculable. One could characterise the monopoly of the Nigerian communications systems and services by NITEL as probably the greatest single national albatross of that period on the nation's development.

The above communication deficiencies and related woes in the land led Professor A. Oyediran, the then Vice-Chancellor of the University of Ibadan to lament, in 1995, in the following words:

> *The major factor [to the nation's underdevelopment] was the recognition that undependable telephones, and unreliable postal system, vast distances and expensive airfares, all isolate African scholars and hinder communications within the African continent and between it and overseas. These difficulties thwart ability to conduct research, share data and results, keep abreast of current scholarship and contribute to solving development problems faced by African countries.*[14]

The nation-wide communication paralysis soon produced a variety of self-help measures, by private companies and diplomatic missions, to fill the communication vacuum in the country. In order to carry out their normal day-to-day activities,

---

13 Abiodun, Adigun Ade (1997). *Deregulation of Nigeria's telecommunication industry, Thisday Newspaper, August 21, 1997 (1), Vol.3, No. 852, page 24 & (2) August 26, 1997, Vol.3, No. 857, page 24.*

14 A Goodwill Message from the Vice-Chancellor, University of Ibadan, Ibadan, Nigeria, to the *Workshop on the Internet: Opportunities, Prospects and Problems for the Nigerian User,* Proceedings of the Internet Workshop, Yaba College of Technology, pp. 15-16, Lagos, March 15-16, 1995.

these entities began to erect their own independent and clandestine communication infrastructure, with direct linkages to the global community, via a variety of communication satellites, albeit, all beyond the control of the Federal Government of Nigeria of that time. How much of Nigeria's national security was compromised, by these do-it-yourself communication infrastructures, was beyond the knowledge of both the government and the Nigerian people. Suffice it to say that the return of the country to civil rule in 1999 brought some order into the communication arena, and with the passage of time, some of these independent communication services dismantled their wares.

Contributions of Nigeria's Tertiary Institutions and the United Nations System

Fundamental Space Research

In spite of the national communication paralysis, and the attendant inability of Nigeria's research scientists to network amongst themselves and with their colleagues within the international community, a few dedicated Nigerians continued their intellectual pursuits in space-related research, albeit, with minimal governmental support. Scientific work continued, particularly at the University of Ibadan (UI), University of Ife (UNIFE), University of Ilorin (UNILORIN), and at the University of Nigeria, Nsukka (UNN).[15] Nigerian scientists also collaborated with their counterparts in Egypt, India, Indonesia, Iraq and Kenya to develop, in 1979, a joint proposal for the development and construction of a *Giant Equatorial Radio Telescope (GERT)* and its *International Institute for Space Science and Electronics (INISSE)*. The latter was to be built in Kenya, near the equator, at a cost of US$15 million, for the benefit of the developing countries.[16] The proposal emanated from a 1979 international workshop that was organised in India, under the sponsorship of UNESCO. At that time, GERT/INISSE was the most comprehensive regional proposal ever developed for fundamental space science and technology research in Africa.[17] Although the

---

15 In 1985, the University of Nigeria (UNN), Nsukka, established The Nnamdi Azikiwe Space Research Centre, under the leadership of Prof. Samuel Okoye, to pioneer studies and research in Astrophysics, Geophysics, Materials Science, Solid State Physics, and Nuclear and High Energy Particle Physics.

16 Swarup, G., T.R. Odhiambo and S.E. Okoye, INISSE and GERT, (unpublished), (1979).

17 Prof. J. B. Aladekomo of the Department of Physics, University of Ife and Prof. Samuel E. Okoye of University of Nigeria, Nsukka, contributed, on behalf of Nigeria, to the development of the GERT/INISSE proposal.

establishment of INISSE and GERT received global positive comments, however no substantive funding materialised for the execution of the proposal. Subsequently, India, the host country for the workshop, showed interest in the outcome of the workshop. Professor Govind Swarup, the convenor of the workshop, got the funding support of his own government and completed the establishment of a GERT-equivalent in India, known today as The Ooty Radio Telescope (ORT).[18] The contributions of ORT to advances in fundamental space research are well known, internationally.

Research works in fundamental space science fields in Nigeria, particularly at the UI, are today being complemented by similar efforts at the University of Ilorin,[19,20] University of Lagos (UNILAG), the Federal University of Technology, Akure (FUTA),[21] and at the Federal University of Technology, Minna (FUTECH). FUTA's work on the 2007 International Heliophysical Year (IHY) served as the contribution of Nigeria to both the IHY and the 50th Anniversary Celebration of the United Nations Committee on the Peaceful Uses of Outer Space (COPUOS).[22, 23]

Climate Research

Today, one of the major pre-occupations of COPUOS is how to advance human collective knowledge of the interrelationship between the Earth's atmosphere and the well-being of its inhabitants. To attain this goal, in 1961, the United Nations General Assembly (GA) called for an international study on *"measures to advance the state of atmospheric sciences and technology in order to improve weather forecasting capabilities and to further the study of the basic physical processes that affect climate."*[24] Translating such a political decision into reality

---

18 Ray Jayawardhana(1994). Science, *Vol. 24, 22 April, 1994, pp 501-502.*

19 Oyekola, O. S. Akin Ojo and J. Akinrimisi (2007). *Vertical drift velocity measurements at F region low latitude ionosphere,* Geophysical Research Abstracts, Vol. 9, 00350, 2007, European Geosciences Union 2007.

20 Adeniyi, J.O., O. Oladipo, D. Bilitza and O.K. Obrou (2004), *Variability of foF2 in the equatorial ionosphere,* Advances in Space Research, Volume 34, Issue 9, 2004, Pages 1901-1906, Elsevier Ltd.

21 Rabiu, A. Babatunde., Adeyemi, B., Ojo, J., (*2005*). *Variability of solar activity and surface air temperature variation in tropical region.* Journal of the African Meteorological Society, 6(3), 83-91. (Kenya).

22 The International Heliophysical Year (IHY), 2007-2008, was planned in commemoration of the 50th anniversary of International Geophysical Year (IGY), 1957-1958.

23 http://www.ihy2007.org/ihy_united_nations.html*(Accessed, November 3, 2013).*

24 UN General Assembly Resolution A/RES/1721 (XVI) of December 20, 1961.

soon became a challenge. The same United Nations came to the rescue and set in motion the necessary mechanisms to achieve that objective. Specifically, it recalled how the International Committee of Scientific Unions (ICSU) organised the very successful 1957-1958 International Geophysical Year (IGY), a collective global scientific research programme that subsequently led to the birth of the space age, as outlined in Chapter I. Thus, it was no surprise that, in 1962, the United Nations General Assembly (GA) formally invited ICSU to collaborate with the World Meteorological Organisation (WMO) to develop a programme of research on atmospheric science.[25]

ICSU and WMO went to work and jointly established a committee that planned what became the Global Atmospheric Research Programme (GARP). The latter was launched, in 1967, as a joint ICSU/WMO programme, supported by data collected by a variety of atmospheric satellites. GARP was originally developed *to observe the global atmosphere on an intensive scale and to have an unprecedented description of it through one annual cycle.*[26] At the first World Climate Conference held in Geneva, in 1979, the participants agreed to transform GARP into the World Climate Research Programme (WCRP), effective in 1980, with the broad objectives of determining how far climate can be predicted and the extent of human influence on climate.[27] Today, WCRP has become an indispensable tool in human understanding of climate change.

Nigeria became a contributing member State to the work of WCRP through a NASA/NOAA-sponsored collaborative research programme that began, in 1992, between the University of Maryland (USA) and the University of Ilorin (Nigeria)[28]. At that time, Nigeria had not formulated a space policy nor established a space programme, but many Nigerian scientists, including those that studied abroad and had acquired some space-related knowledge, felt compelled to make their own contributions. Similar to other space-related initiatives in the country, this scientific partnership on climate research between Nigeria and the USA was the handiwork of Nigerian scientists at the University of Ilorin.

---

25 Abiodun, Adigun Ade (1992). *United Nations Advances Global Environmental Awareness,* IEEE Technology and Society Magazine, pp. 23-29, Spring 1992.

26 WMO Bulletin, vol. 34, no. 4, pp. 320-327. October 1985.

27 *ICSU and Climate Science: 1962—2006 and beyond,* A publication of the International Committee of Scientific Unions.

28 NOAA is the United States' National Oceanic and Atmospheric Administration.

Because of the important role radiation plays in the climate system, WCRP participating entities (including ICSU, WMO, UNEP) agreed to establish a world-wide Baseline Surface Radiation Network (BSRN), shown in Figure 4:2, to continuously measure *radiative fluxes at the Earth's surface.*[29] These BSRN stations, which number over forty (40) world-wide today, provide data for the calibration of the Surface Radiation Budget (SRB) Project and other satellite-based measurements of radiative fluxes. The data collected within the BSRN project are used to detect important changes in the Earth's radiation field (at the Earth's surface) which may be related to changes in climate. Participating satellites include the METEOSAT satellites at the geosynchronous orbit and the NOAA polar orbiting satellites.

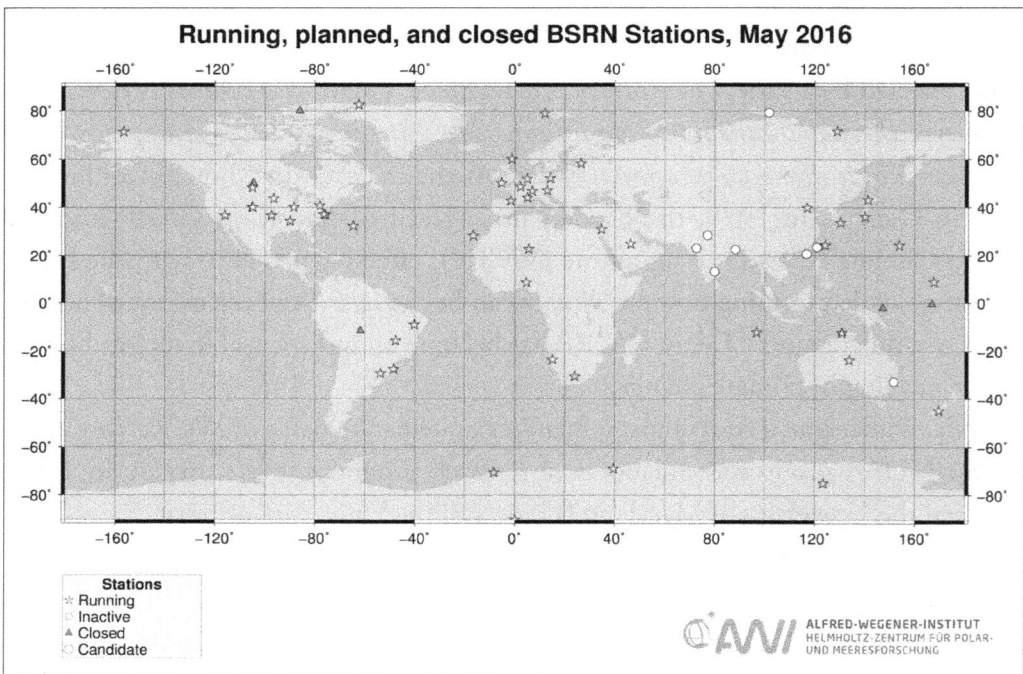

Figure 4:2 Current and planned global BSRN Stations as of May 2016
(Credit: Wolfgang Cohrs, Alfred-Wegener-Institut).

Ilorin, Nigeria, was chosen as one of BSRN's global network stations because it is a climatically important desert region and a transition zone between the Sahara

---

29 At the earth's surface, shortwave radiative fluxes are of primary interest in climate research because they control the total energy exchange between the atmosphere and the land/ocean surface. Information on these fluxes, which is obtained from appropriate instruments on board earth observation satellites, is needed on a global scale to study climate and climate change.

desert and the savannah region of Nigeria (Latitude 08⁰32'N; Longitude 04⁰34'E). The station is located on the campus of the University of Ilorin and managed by its Department of Physics. This region of Nigeria is influenced by the dusty annual harmattan wind which persists for long periods of time and is characterized by steady dust conditions with high aerosol loading; the latter significantly affects the interpretation of satellite observations. Data recorded at the Ilorin BSRN station, include information on aerosols and radiative fluxes. NASA uses these data to support its global climate research programme through the University of Maryland. The data collected by BSRN stations are also essential parameters in long-range weather forecasting. Because of its collaboration with the University of Ilorin, today, the Nigerian Meteorological Agency (NIMET) is a partner to the BSRN programme.[30] On a daily basis, Nigeria and Nigerians are beneficiaries of this WCRP initiative; more importantly, Nigerians need to know that through one of the nation's tertiary institutions, the University of Ilorin, the nation is also contributing to an understanding of a global problem—climate change.

Earth Observation from Space

In 1993, the former University of Ife (UNIFE), now called Obafemi Awolowo University (OAU), brought to light, in Nigeria, the promise of Earth observation from space, not only on the nation's development efforts in general, but particularly on its life support systems.[31] Observation of the Earth from space, otherwise known as remote sensing, offers the global community the opportunities to monitor and manage these life support systems.[32] Remote sensing was introduced into Nigeria in 1973, in a manner similar to how other space-related programmes made their way into the country during that period; that is, by individuals as a result of personal connections to global centres of excellence in the subject area. Professor O. S. Adegoke, formerly of the Department of Geology at UNIFE, was the catalyst for remote sensing in Nigeria. Following the successful launch of the first civil earth observation satellite, Landsat-1, by the United States in July 1972, and upon Prof.

---

30 WCRP Informal Report No. 17/2001, WMO Headquarters, Geneva Switzerland , July 2001.

31 Life-support systems include, but are not limited to air, land, water, agricultural resources and wholesome environment.

32 Observation of the Earth from Space (Remote sensing) is the study, understanding and management of our environment and natural resources using air-borne and/or space-acquired information of the Earth, its land and its water.

Adegoke's recommendation, Prof. H. A. Oluwasanmi, the then Vice-Chancellor of UNIFE constituted, in the first half of 1973, a remote sensing committee.[33] He charged the committee to develop a proposal for Nigeria's participation in the Landsat programme, and in particular for the acquisition of data of Nigeria by the next American satellite, Landsat-2. UNIFE transmitted the proposal, as Nigeria's initiative, through the Federal Government of Nigeria, to NASA in the United States in 1974.

But while UNIFE was still waiting for NASA's response from the appropriate Nigerian government channel, it did not realise that the Nigerian government had received the approval from NASA. The university authorities soon learned that the project had been hijacked by the Federal Ministry of Communications (FMC) of Nigeria. Justification? According to the reasoning of the officials of the ministry, *'FMC is the only entity in Nigeria with any satellite experience, and thus should be the focal point for satellite remote sensing development in Nigeria.'* What experience? As of that time, the only satellite-related assignment that the FMC had ever carried out in Nigeria was its signing, of an agreement with Intelsat, on October 25, 1972, to provide external communication links between Nigeria and the rest of the world; this agreement went into effect almost a year later, on September 12, 1973. How did the FMC officials successfully persuade the nation's leaders that their communication expertise also included knowledge of earth observation from space? Simple! It was a practice-ground for similar antics on subsequent space-related efforts in the land, including the sale, of the Aerostat Balloon concept, addressed earlier, by these same officials, to the then Head of State, Major General Murtala Mohammed in 1975. And that was how UNIFE's role in leading the development of a well-reasoned remote sensing technology programme in Nigeria was overshadowed and summarily truncated.

Thereafter, two proposals on remote sensing development in Nigeria were submitted to the Federal Government.[34,35] While these proposals were still under-

---

33 Professors O.S. Adegoke, E. E. Balogun, Anthony Imevobre, I. K. Jeje, and this author were among the academic members of staff that served in this committee.

34 Abiodun, Adigun Ade (1975). *Remote Sensing Technology in the Development of Nigeria,* Concluding Report of Post-Doctoral Research Fellowship Programme at the Canada Centre for Remote Sensing, Ottawa, Canada.

35 Howard, John. A. and D. C. Nduaguba (1977). *National Centre for Remote Sensing for Nigeria, A technical report (AGD(RS) 3/77) prepared for the Federal Military Government of Nigeria by the Remote Sensing Unit, FAO, Rome.*

going study and review, the Federal Government of Nigeria allocated ₦10 million (US$17.5 million in 1976) to the Federal Ministry of Communication (FMC), as part of the Third National Development plan period (1975-1980), for the establishment of a viable national remote sensing centre with all the necessary infrastructure and manpower.[36] Still unknown was the role of the FMC in the premature release of the funds before the completion of the review of each of the proposals—the same antic as before; indeed, the result of the reviews never saw the light of day.

With money in hand, the ministry immediately sponsored a tour of Brazil, Canada and the USA 'to gain a better understanding of how to develop a remote sensing programme and related infrastructure in Nigeria from the world's leading countries in the technology.' An architect was part of the delegation which included mostly the staff of the ministry; no one in the delegation had any knowledge of the subject that was the focus of the mission.[37] In the absence of any demonstrable performance after two years of funding, the same government that allocated the funds for the programme withdrew the budgeted funds for remote sensing development in Nigeria.

Similarly, in 1976, the Federal Ministry of Agriculture by-passed the then Federal Surveys of Nigeria, without any challenge, and carried out a radar survey of the whole nation because it needed the vegetation and land-use maps of the country. In carrying out this assignment, known as Nigerian Radar Project (NIRAD), the Ministry engaged the services of Motorola Company of USA to acquire the radar (side-looking airborne radar—SLAR) data of Nigeria. That was when radar technology for earth observation was at its teething stage and had not, at that time, become a generally acceptable operational mapping tool.[38] Simply stated, NIRAD was a "guinea pig" project.

Senior officials of Hunting Technical Services (HTS), the British company that was contracted to interpret the Motorola acquired radar images of Nigeria attended, with this author, the 1978 International Conference on Remote Sensing for Decision Makers held at the International Training Centre in Aerospace Surveys (ITC),

---

36 In 1974/75, One Naira (₦1.00) was worth US$1.75.

37 Official Mission Report (1976). *On the Development of a National Remote Sensing Centre in Nigeria*, Report prepared by the Federal Ministry of Communications, FGN, Lagos.

38 Radar is an object-detection system that uses radio waves to identify the range, altitude, direction, or speed of both moving and fixed objects. Radar is used as a mapping tool because radiation from radar penetrates clouds, it permits mapping of the land surface from above the clouds.

Enschede, The Netherlands, and presented the results of their NIRAD project. They publicly and readily admitted, at the conference, and possibly at other fora, the difficulty they had in interpreting and using the radar data acquired by Motorola to develop vegetation maps of Nigeria. The major problem was that the contractor acquired the vegetation data for the NIRAD project during the dry season; hence, there was no contrast in the images acquired. To complete the interpretation stage of the contract, HTS had to use Landsat images of Nigeria, as supplements. In the final analysis, the maps produced were inaccurate and unreliable for any serious use. The NIRAD project contract also called for Nigerians to be trained in the analysis and use of the acquired radar data in national development. The extent to which all these objectives were accomplished remains unknown. Only the records of the Federal Ministry of Agriculture can tell.

A third entity of the government, the then National Council for Science and Technology (NCST), that had responsibility for science and technology development in Nigeria, soon injected itself into Nigeria's remote sensing development agenda. In 1976, the Murtala Mohammed military government dissolved the NCST; it re-emerged in 1977 under Olusegun Obasanjo's military administration as the new National Science and Technology Development Agency (NSTDA) with the same mandate. It also arrived on the scene with its own remote sensing agenda for Nigeria after it had imbibed some lessons from the failure of FMC to launch a national remote sensing programme. In order to foreclose any challenge from other interested quarters, it somehow managed to obtain an official recognition, from the Federal Government, as the national body officially mandated by government to coordinate remote sensing activities and programmes in Nigeria.[39]

However, the new civilian government that came to power in October 1979, under the leadership of President Shehu Shagari, had a different idea and soon replaced NSTDA with a Federal Ministry of Science and Technology (FMST). On December 31, 1983, political conspiracy of many faces resulted in a military coup that soon truncated the Shagari administration, 91 days into its second term. In January 1984, the new Buhari/Idiagbon military administration supplanted FMST with, the newly minted Federal Ministry of Education, Science and Technology (FMEST). Unfortunately, that was not the last government intervention on the re-arrangement of the name of the entity that was supposed to direct the science

---

39 Official Gazette of the Federal Government of Nigeria, No. 4, vol. 64, pp. A45-A52, January 27, 1977.

and technology affairs of the nation. That same year, as a result of another military coup, FMEST disappeared and re-emerged with its old name, FMST, in August 1985 courtesy of the new Babaginda military administration. The subject of remote sensing, as a national programme, resurfaced again at the meeting of the National Council of Ministers in 1988; the council subsequently approved the establishment of a National Centre for Remote Sensing in Jos, Nigeria, with a Satellite Ground Receiving Station as one of its main components. The station was to be established at Kerang, in Mangu Local Government Area of Plateau State; that was one of the sites chosen for the defunct Aerostat Balloon Project. The Jos centre eventually took off, functionally, in October 1995.[40]

To an outside observer, an obvious question could be: How did the goal of developing a viable national remote sensing programme in Nigeria fare in all of these national political upheavals? Certainly, the fortunes and misfortunes of remote sensing in the nation got entangled in the rickety foundation of science and technology in the country. These unending changes in S&T policies, focus and host institutions, at the federal level, were more than enough to derail any type of development agenda in any society. In Nigeria, that certainly happened.

But these internal science and technology indecisions and related confusions did not deter Nigeria from also watching, carefully, with interest, the external on-going wrangling at the Economic Commission for Africa (ECA), in Addis Ababa, between the other African member States and western donors on the establishment and control of an African Remote Sensing Programme (ARSP).[41] Indeed, Nigeria would have been part of the footprints (ground coverage areas) of the two Landsat satellite ground-receiving stations proposed for installation in Ouagadougou and Kinshasa respectively, under the auspices of ARSP, albeit with donor support. However, the project was loaded with a number of problems: it was ill-conceived with vague long-term commitments and it was still-born.

Upon the establishment of ARSP in 1977, member States expanded the mandate of the ECA-supervised regional centres in surveying and mapping, at Ile-Ife, Nigeria and Nairobi, Kenya, to include remote sensing. The centre at Ile-Ife was subsequently renamed *Regional Centre for Training in Aerospace Surveys (RECTAS)*

---

40 http://www.ncrsjos.org/history

41 Abiodun, Adigun Ade (2000*). Development and Utilization of Remote Sensing Technology in Africa,* Journal of American Society of Photogrammetric Engineering & Remote Sensing, pp. 674-686, June 2000.

and that in Nairobi was also renamed the *Regional Centre for Mapping of Resources for Development (RCMRD)*. Thereafter, the inter-governmental experience gained in establishing a continent-wide ARSP was applied in the development of a sub-regional co-operation agreement on the use of space tools to monitor Rivers Niger and Benue. Under the auspices of WMO, and with the support of donors such as the UNDP, European Economic Commission (EEC) and the Organisation of the Petroleum Exporting Countries (OPEC), an Interstate Forecasting Centre (IFC) was established in 1985 within the framework of the HYDRONIGER Project, for the benefit of the nine (9) Niger Basin Authority (NBA) member countries.[42] The immediate objective of the project was to establish a real time hydrological forecasting system covering River Niger and its major tributaries. Today, 95 METEOSAT stations (19 of the stations are in Nigeria), equipped with Data Collection Platforms (DCP) are stationed along the River Niger and its major tributaries in the framework of the Niger-HYCOS Project.[43, 44] The latter is being implemented by the Niger Basin Authority (NBA); Nigeria is represented in NBA by its Federal Ministry of Water Resources.[45]

Contributions of the United Nations System and the European Space Agency

On the whole, the United Nations system has been instrumental in Africa's initial space awareness. From 1978 through 1999, UN-OOSA, ITU, FAO, and WMO as well as ESA contributed to space-related mid-level capacity and capability development in Nigeria and in many other African countries. During that period, the United Nations system and the European Space Agency (ESA), individually and collectively,

---

42 The nine (9) member states of NBA are Benin, Burkina-Faso, Cameroon, Chad, Côte d'Ivoire, Guinea, Mali, Niger and Nigeria.

43 The Meteosat series of satellites, placed close to 0^0 longitude, at the geostationary orbit, are meteorological satellites that are operated by EUMETSAT. EUMETSAT (*European Organisation for the Exploitation of Meteorological Satellites*) is an intergovernmental organisation established in 1983 through an international convention agreed to by a current total of 26 European funding Member States: Austria, Belgium, Croatia, the Czech Republic, Denmark, Finland, France, Germany, Greece, Hungary, Ireland, Romania, Italy, Latvia, Luxembourg, the Netherlands, Norway, Poland, Portugal, Slovakia, Slovenia, Spain, Sweden, Switzerland, Turkey, and the United Kingdom.

44 HYCOS stands for Hydrological Cycle Observing System

45 Final Report of WMO Regional Workshop on improved meteorological and hydrological forecasting for floods in west and central African countries, Niamey, Niger, 4—6 April 2006

organised training courses, workshops, and seminars, and provided technical advisory services in such areas as remote sensing, satellite meteorology, communications, natural disasters management, basic space science, micro-satellite development and search and rescue operations of those in distress.

Training Courses, Workshops and Seminars

The first of these activities was the "*United Nations Regional Training Seminar on Remote Sensing Applications*; it was co-sponsored by the United Nations Environment Programme (UNEP) and Sweden, and held in Nairobi, Kenya, from 4 to 16 September 1978. The *United Nations Remote Sensing Training Course* was held, next, at the University of Ibadan, Ibadan, Nigeria, in November 5-23, 1979. Subsequent United Nations space-related activities held in Nigeria, in particular, included those on *Sustainable Development* at Obafemi Awolowo University, Ile-Ife in 1985, on the United Nations Inter-regional Meeting of Experts on Space Science and Technology and its Applications within the Framework of Educational Systems held at the House of Representatives, Lagos, Nigeria, in 1987, and on *Basic Space Science* at the Ogun State Gateway Hotel, Otta, Nigeria in 1993.

These and other activities of UN-SAP succeeded in whetting the appetite of aspiring scientists and engineers for a higher level of education, particularly at the local level, a desire that subsequently won the approval of the delegates at UNISPACE-82.[46] Other United Nations contributions in Africa included the COPINE Project that metamorphosed, in Nigeria, into NITDA, the establishment of the African Regional Centre for Space Science and Technology Education (ARCSSTE)[47] at Ile-Ife, and the knowledge imbibed from the Maspalomas COSPAS-SARSAT workshops and exposures that led to the 2004 establishment of the COSPAS-SARSAT Station, under NEMA, in Abuja; this last programme is covered in Chapter II.

---

46 UNISPACE-82 was the Second United Nations Conference on the Exploration and Peaceful Uses of Outer Space held in Vienna, Austria, August 9-21, 1982. (UN Document A/CONF.101/10).

47 ARCSSTE-E and CRASTE-LF, (the latter is the French-speaking centre in Rabat, Morocco), and three others in India, Brazil-Mexico, and Jordan were established by the United Nations for the expressed purpose of developing, through in-depth education, an indigenous capability for research and applications in the core disciplines of space science and technology, namely: Remote Sensing and Geographical Information Systems (RS & GIS), Satellite Communications (SATCOM), Satellite Meteorology and Global Climate (SATMET), Space and Atmospheric Sciences, as well as data management.

The COPINE Project

In 1993, it appeared that the communication deficiencies alluded to earlier in this chapter, particularly among Africa's research scientists, in twelve African countries including Nigeria, would be rectified by a United Nations initiative known as COPINE.[48] The initiative originated from the United Nations Office for Outer Space Affairs (UN-OOSA), Vienna, Austria. Copine was the off-shoot of the United Nations Space Applications Programme (UN-SAP) at UN-OOSA to translate the conclusions and recommendations of the United Nations Regional Conference on Space Technology for Sustainable Development in Africa, held in Dakar, Senegal, October 25-29, 1993, into a practical and an operational programme for the developing countries. One of the key recommendations of the conference, which was attended mostly by African professionals, requested the United Nations:

> *To put in place a functional satellite-based communication programme that would improve the existing state of information exchange among African professionals in Africa.*[49]

In early 1994, UN-SAP and ESA established a partnership that conceived and subsequently developed COPINE initially, for the benefit of participating African member States. COPINE was developed as a vehicle that would strengthen collaboration between and among selected universities and institutions within Africa, as well as with their counterparts in Europe and the international community. It was also developed to provide a vehicle for the transfer of know-how and technology in a number of priority application fields, such as health-care, food security, communications & information exchange, and resource management. Switzerland offered its main national communication hub, at Leuk, as the international communication gateway for COPINE.

After over three years of project planning, preparations and fund raising, COPINE soon became a regional project that did not enjoy needed commitment from the

---

48 COPINE aimed at establishing a cooperative satellite information network linking scientists, educators, professionals and decision makers in Africa. The first phase of operational COPINE foresaw an information exchange network with interactive capability, linking urban and rural institutions and centres in 13 African countries (Botswana, Ghana, Kenya, Malawi, Morocco, Mozambique, Namibia, Nigeria, South Africa, Tanzania, Tunisia, Uganda, and Zimbabwe) and selected hospitals, universities/institutions and documentation/information supply centres located in Europe. If successful, it was envisaged that it should be replicated in the other regions of the world.

49 A/AC. 105/562—Report of the United Nations Regional Conference on Space Technology for Sustainable Development in Africa held in Dakar, Senegal, October 25-29, 1993.

benefiting region—Africa. The participating African countries pinned their funding hopes for COPINE only on anticipated support from donors and other external funding sources, without an appreciation of "donor fatigue" that had been weighing down these same donors over the years. More serious was the fact that political leaders in several of the benefiting countries did not appreciate the relevance and importance of COPINE to Africa's social and economic development. A number of the key foreign aid donors also attached too many constraints to their offers. COPINE could not withstand all the burdens and the related pull-and–push tactics. In the end, the project inevitably collapsed.

Offshoots of COPINE soon became a reality. Just as the 1979 GERT-INISSE initiative that was sponsored by UNESCO metamorphosed into *The Ooty Radio Telescope project* in India, the COPINE initiative of UN-OOSA became the vehicle used in launching, on April 18, 2001, *The Nigerian Information Technology Development Agency (NITDA),*[50] *under its pioneer Director-General, Prof. Gabriel Lere AJAYI; he was the COPINE focal point in Nigeria. Similar transformations of COPINE exist today in some of the other African and European countries that participated in the development of COPINE.*

United Nations-Affiliated Regional Centres for Space Science and Technology Education

The most important recommendation of UNISPACE-82 on UN-SAP was the decision of delegates to unanimously expand the mandate of the programme "*to develop indigenous capability in space science and technology and its applications at the local level.*" Subsequent consultations, meetings, and related conferences by UN-SAP, with member States resolved that:

> "*In order for developing countries to effectively contribute to the solution of global, regional and national environmental and resource management problems, there was an urgent need for a higher level of knowledge and expertise in the relevant disciplines by educators as well as by research and application scientists in those countries. Such capabilities, they further noted, could only be acquired through long-term intensive education.*"[51]

---

50 The National Information Technology Development Agency (NITDA) was established to create a framework for the planning, research, development, standardization, application, coordination, monitoring, evaluation and regulation of Information Technology practices, activities and systems in Nigeria.

51 A/AC.105/703—Regional Centres for Space Science and Technology Education (Affiliated to the United Nations), United Nations, June 16, 1998.

In support of the above-mentioned objectives, the General Assembly, in its resolution 45/72 of 11 December 1990, endorsed the recommendation of COPUOS that "the United Nations should lead, with the active support of its specialized agencies and other international organisations, an international effort to establish regional centres for space science and technology education in existing national/regional educational institutions in the developing countries."

UN-SAP conceived each centre as an institution that should offer the best possible education, research and applications programmes, as well as opportunities and experience to the participants in all its programmes. Thus, the principal goal of each centre is the development of the skills and knowledge of university educators and research and applications scientists, through rigorous theory, research, applications, field exercises and pilot projects in those aspects of space science and technology that can contribute to sustainable development in each country. The initial programmes of each centre should focus on: remote sensing; meteorological satellite applications; satellite communications; and space and atmospheric sciences; these have been expanded in 2013, to include global navigation satellite systems (GNSS). Its data management unit should be linked to existing and future relevant global databases. Each centre should also foster continuing education programmes for its graduates and awareness programmes for policy and decision makers and for the general public.

UN-SAP also recognised that in order for the centres to become model institutions that are respected both within their regions and around the world, they would need to meet internationally recognised standards. To promote the achievement of these aims, it developed model curricula in each of the scientific fields identified above, with input by prominent educators around the world, for use at each of the centres.

Beginning in 1995, as each region embraced the UNISPACE-82 mandate, United Nations-Affiliated Regional Centres for Space Science and Technology Education for in-depth education of specialists soon found homes in Asia (India, 1995), Africa (Morocco, and Nigeria, 1998), Latin America (Brazil and Mexico, 1997) and the Middle-East (Jordan, 2012). The centre in Nigeria is located on the Campus of Obafemi Awolowo University, Ile-Ife and it caters to all English-speaking African countries. Morocco hosts the centre in Rabat for the French-speaking African countries.

Overall Impact of the Efforts of the United Nations System and ESA

The aforementioned efforts of the United Nations system have produced a number of qualified educators and professionals who are now gainfully employed at the universities, government establishments and within the private sector in the developing world, and are contributing to the development and growth of their respective countries, particularly in the fields of communications, meteorology and in the application of earth observation data for sustainable development. These United Nations-sponsored activities also offered participants the opportunities to initiate the establishment of national and regional space-related organisations, such as the Nigerian Remote Sensing Society (NISORS), the African Association of Remote Sensing of the Environment (AARSE) and the African Leadership Conference on Space Science and Technology for Sustainable Development (ALC). As shown in Chapter V, the contacts and interactions that developed between official Nigerian representatives and their international counterparts, in the margins of these United Nations system–sponsored activities, paved the way for the subsequent space-related actions that Nigeria took at the dawn and in the first decade of the 21st century.

* * *

Chapter V

SPACE-BOUND NIGERIA

At the dawn of the 21st century, Nigeria officially launched its space journey. Prominent among the first steps the government took were the establishment of a space organisation, the National Space Research and Development Agency (NASRDA), and its immediate pursuit of the *Nigeria Satellite Project*. Before that time, as shown in Chapter IV, a number of establishments of government and academic institutions had initiated some space-related activities in the country. The Nigerian Civil War of July 6, 1967 to January 13, 1970, understandably, foreclosed Nigeria's participation in the First United Nations Conference on the Exploration and Peaceful Uses of Outer Space, held in Vienna, Austria, August 14—27, 1968. By the time the global community re-convened for the Second United Nations Space Conference, known as UNISPACE-82, in 1982, Nigeria publicly made its first national pronouncement on its expected participation in space activities.

Establishing a Space Organization

Nigeria's justification for going into space first appeared in its national paper it submitted to the UNISPACE-82 Conference Secretariat; it read as follows:

> *The objectives of the Nigerian government in the field of outer space are to take full advantage of space science and technology for the purpose of development, inter alia, in agriculture, mining, physical and geological surveys, forestry, fisheries, communications and broadcasting, meteorology and research... Government places premium on education, training, manpower development and the provision of infrastructure on a sound basis for a take-off in this field.*[1]

---

1 Nigeria's National Paper to UNISPACE-82, UN-OOSA UNISPACE-82 Records: National Papers, 1982.

After the conference, the government took no immediate action, neither in policy formulation, in programme development, nor in the mobilisation of the Nigerian scientific community to act on the above pronouncement. However, on November 12, 1992, over a decade later, Nigeria joined the global community to celebrate the 1992 International Space Year (ISY); the latter marked the 35th Anniversary of the first successful launch of an artificial satellite, *Sputnik-1*, into space.[2] The National Agency for Science and Engineering Infrastructure (NASENI), an agency (parastatal) of the Federal Ministry of Science and Technology, organised a one-day ISY commemorative seminar, in Lagos, to revive the nation's interest in space science and technology.[3] The scientific community welcomed the seminar and it was well attended. Professor Peter Olu Adeniyi of the University of Lagos and Dr. Vernon Signhroy of the Canada Centre for Remote Sensing, Ottawa, delivered the two invited keynote lectures, followed by a very extensive and intensive discussion.[4] This author served as the moderator of the seminar.

The enthusiasm generated at the seminar led NASENI to take the next logical step; it initiated and spearheaded the drafting of a national space policy. Unfortunately, the process hit an insurmountable road-block and the drafting effort stalled for good, resulting in an inconclusive document. Space science and technology, as a national undertaking in Nigeria, was re-awakened in 1998. That was when all member States of the United Nations, including Nigeria, received the official invitation from the world body to participate in the Third United Nations Conference on the Exploration and Peaceful Uses of Outer Space (UNISPACE III), July 19-30, 1999.

As it did in 1982, Nigeria participated in UNISPACE III preparatory events that UN-OOSA organised in Africa, Europe and Asia. At these events, the Nigerian

---

2 Abiodun, Adigun Ade (1992). *Earth viewing from space: Potential benefits for sustainable development in Nigeria*, Report of the National Seminar on Space Science and Technology to mark the International Space Year (ISY), held at the Senate Chambers, National Assembly Complex, Tafawa Balewa Square, Lagos, Nigeria.

3 Mission of NASENI—To establish and nurture an appropriate and dynamic Science and Engineering infrastructure base for achieving home initiated and home sustained industrialization through the development of relevant processes, capital goods and equipment necessary for job creation, national economic wellbeing and progress. NASENI was established in 1992 as a parastatal of the Federal Ministry of Science and Technology.

4 Discussants at the Nigeria ISY event included Prof. D. M. J. Fubara (University of Port Harcourt, Port Harcourt), Prof. D. O. Adefolalu (Federal University of Technology, Minna), Prof. G. O. Ajayi (Obafemi Awolowo University, Ile-Ife), Mrs. Oluyinka Adekoya (Deputy Surveyor-General of the Federation, Lagos) and Dr. R. C. Bob-Duru (University of Nigeria, Nsukka).

delegation, mostly from NASENI and the Federal Ministry of Science and Technology (FMST), established a variety of contacts with international entities. Amongst these contacts were companies that marketed micro-satellites as indispensable tools of development and touted micro-satellites as the cheapest way to space for the developing countries. Not featured in these marketing messages were the essential prerequisites and the science and technology readiness required for being an active space participant. Nevertheless, Nigeria and a few other developing countries swallowed the bait.

Being among the first African countries to be associated with any type of space activity was uppermost in the minds of our space-bitten bureaucrats of that era. The government officials that initiated the contacts with the micro-satellite companies cited above were anxious to consolidate a form of understanding, no matter how rudimentary, with these same foreign companies before UNISPACE III convened. Thus, the status of Nigeria at UNISPACE III soon became a pre-occupation in Abuja. Those in the know agreed that if Nigeria were to attend UNISPACE III, it should do so with all the honours it deserved. What will assure this? NASENI managers successfully persuaded the authorities in Abuja on the exigency of establishing a national space institution prior to the conference.

The interactions with satellite manufacturers cited earlier led to a groundswell of support in Abuja for a *space-bound Nigeria*. On May 5, 1999, two and a half months before UNISPACE III, the FEC established the National Space Research and Development Agency (NASRDA) without any complementary space policy and related programme that should guide the work of the newly established space organisation. Did the FEC ask critical questions about the nation's scientific and technological readiness to shoulder and sustain the nation's space aspiration? In approving the establishment of NASRDA, the FEC stressed the need for Nigeria to participate in the space enterprise and it also clearly specified that:

Nigeria would develop, design, build and launch its own (indigenous) satellite.

Buy Technology from Abroad or Develop and grow it at Home?

In an effort to demonstrate its seriousness on and commitment to its new space agenda, in February 2000, the government also appointed a new Senior Special Assistant (to the President) on Space Science and Technology (SSAP-SST).[5] The

---

5 This author, Adigun Ade Abiodun, was appointed Senior Special Assistant (to the President of Nigeria) on Space Science and Technology) from March 2000 to June 2003.

President charged him with the responsibility to address, energetically, the following question:

How can space science and technology (SST) advance the development process in and contribute to the growth of Nigeria?

Achieving the president's space objective presupposed that the government would provide an enabling environment which had been missing for long under various military regimes. There were great expectations that the perceived commitment of the nation to civilian rule and space science and technology, on the eve of the 21st century, would overturn the *business-as-usual* approach that characterised the military era and also permeated the entire nation. The new way of doing things, many had expected, should also usher in the era of consultations, through genuine public discourse, on issues of national importance, including space.

Because space is a multifaceted tool, the Secretary to the Government of the Federation (SGF), Chief U. J. Ekaette, issued and sent a circular to all relevant Federal Government establishments, parastatals and research and academic institutions;[6] it apprised them of the appointment and responsibilities of the SSAP-SST. These responsibilities included the following:

- *Evaluate and appraise the status of all on-going and planned space science and technology-related activities (i.e. programmes and capabilities—both human and institutional) in Nigeria;*
- *Determine the benefits that would accrue or have accrued to Nigeria, to date, in employing these capabilities and programmes in the interest of the country's development efforts;*
- *Assess the present and future trends of Space Science and Technology within the international community including its roles in improving human conditions here on Earth, and determine the potential for collaboration with and support for Nigeria within the community; and*
- *Formulate policy options for streamlining and harmonising Space Science and Technology activities in Nigeria including programmes that could ensure a successful and beneficial participation of this nation in this global enterprise as is the practice in other countries.*

As the SSAP-SST, I soon coined a title for all of the above, namely: "The Presidential Assignment on the effective development and utilisation of space science and technology in Nigeria," called "*The Presidential Assignment*," for short and began

---

6 Federal Government Circular, Ref. No. SGF 52/S5/50, dated April 17, 2000.

to work immediately on the assignment. With the knowledge and consent of the presidency, I initiated a four-stage process to accomplish the goals set before me; these included:

- Undertake a series of information exchange visits with cognate ministries, parastatals and academic and research institutions in the nation;
- Embark on working visits to a number of space institutions abroad in order to:
  ◊ Exchange views and knowledge on future global trends in space exploration and utilisation for the benefit of humankind; and
  ◊ Discuss collaboration opportunities in specific SST areas that are critical to the attainment of Nigeria's space programme objectives.
- Undertake a fact-finding mission to the UK and in particular to Surrey Satellite Technology Ltd (SSTL), University of Surrey, United Kingdom, in order to gain a first-hand knowledge of its genesis, capabilities, on-going and future programmes; its role in the on-going space and technology developments efforts in Nigeria, the nature of its commitment(s) to Nigeria's space future and a better understanding of the on-going negotiation on NigeriaSat-1. The Nigerian team should also visit the facilities of the Space Innovations Ltd (SIL), Newbury, Berks, United Kingdom. SIL, which had associated itself with Rutherford Appleton Laboratory to carry out space-related research and development activities, at that time, offered a competitive micro-satellite delivery package similar to that of SSTL.
- Organise one or two national meetings of Nigerian Experts Consultative Group –NECG (i.e. resource persons) and the user community[7] on the on-going efforts on SST both in Nigeria and globally. The meeting(s) would also take into consideration input from the information exchange visits in Nigeria, working visits to overseas space establishments including the reports

---

7 24 Nigerians that were knowledgeable in the different fields of space science and technology were invited by the SSAP-SST to serve as resource persons and deliberators at the meetings.
(**Resource Persons:** Prof. Joseph Adesola Adedokun, Prof. Daniel Oladele Adefolalu, Prof. S.A. Adekola, Prof. Peter Olu Adeniyi, Prof. Gabriel Lere Ajayi, Abdul-Hakeem Ajijola, Engr. Johnson Asinugo, Engr. D. G. Awoniyi, Prof. Ekundayo E. Balogun, Dr. Efiong Emmanuel Ekuwem, Dr. Omotayo A. Fakinlade, Engr. Oluyemi Falomo, Prof. Fatai B. A. Giwa, Dr. Sebastian Patrick Hayatu, Lt. Cdr. Ademola Arowosafe Mustapha, Archt. Adetokunbo Obayemi, Dr. Anthony A. Obarefo, Gp Capt. Osita Obierika, Dr. Imoh Bassey Obioh, Wing Cdr. N.E. Offor, Prof. Pius N. Okeke, Prof. Funso Olorunfemi, Prof. V.O.S. Olunloyo, and Col. Suleiman Isa Wali.
Staff of the Office of SSAP-SST that serviced the NECG Meetings: Dr. Akin Fapohunda, Ms. Chinyere Nzeduru, Suleiman Abubakar Takuma, Bola Salami and Dr. Adigun Ade Abiodun.

of the mission to the UK. Thereafter, the meeting(s) would propose vital space research and application programmes that can successfully address the short, medium and long-term needs of Nigeria.

The first NECG meeting was organised and held on June 29, 2000, as designed, with input from all the national sources indicated above; it was formally opened by the SGF. In his address in Annex I, which focused on the goals, objectives and mechanism for developing Nigeria's space capabilities and utilising same in our nation's development process, the SGF charged the experts to take cognizance of the following:

- Programmes in and experiences of other countries such as Brazil, India, South Africa and the Republic of Korea, and the lessons therefrom;
- Development of intermediate and long-term research and application programmes that can successfully address the needs of Nigeria;
- Rationalization of institutions, programmes and activities in order to bring harmony and orderly execution; and
- Identification and optimal utilisation of the country's large pool of competent and experienced science and technology expertise, both locally and abroad, and the development of new talents, that could be engaged in carrying out the nation's development agenda.

The meeting had before it the Report of the Fact-finding Mission to the UK and a number of documents that the FMST provided on the Nigerian satellite project. The Director-General of NASRDA also addressed the meeting on the work of his organisation.

Excerpts from the Report of the Mission to the UK

The report of the Mission to the SSTL facilities at Surrey, UK, noted that the Nigeria Satellite Project was far advanced as a 3-pronged commitment by Nigeria to:

- *Purchase an SSTL-built micro-satellite for Disaster Monitoring to be launched into Earth's low orbit by a Russian launcher;*
- *Purchase an SSTL-built mini-satellite to be launched into the Geostationary Orbit (GSO) by a Russian launcher for communication purposes in Nigeria; and to*
- *Participate in SSTL's MoonRise Satellite Mission. Within this Project, Nigeria will fly its experiments and instruments to the Moon on SSTL's satellite.*[8]

---

8 Purported benefits for Nigeria, as outlined by both FMST and SSTL, in Nigeria's participation in *SSTL's MoonRise Satellite Mission* are: Demonstrate Nigerian Space Capabilities; First African Nation to the Moon; Generate public enthusiasm in Nigeria; National pride; Invest in Nigeria's future—education and excitement for the young; and Stimulate interest in science and technology.

At the pre-contract stage, SSTL had claimed that the Nigerian Earth Observation satellite, NigeriaSat-1, would be unique and that Nigerian scientists, through SSTL's Know-How Technology, would participate in its development, design and construction. However, under scrutiny by the fact-finding mission to the UK, SSTL finally admitted to the mission that:

- *The Nigerian satellite was going to be part of a constellation of five proposed Earth observation satellites, including the one intended for Nigeria;*
- *These satellites, including Nigeria's, would be built with off-the-shelf components; and*
- *The Nigerian satellite, NigeriaSat-1, would be built ONLY by SSTL personnel.*

But were these conditions disclosed to the FEC in Abuja before Nigeria entered into the satellite project contract? Today, we know that NigeriaSat-1 is not unique and that there was no input from the Nigerian scientists that were stationed at SSTL's facilities when it was being built. Based on its findings at Surrey, the mission also concluded that:

"The satellite (NigeriaSat-1) in question is a pre-packaged offer from SSTL. It does not benefit from specifications that could have been determined by the Nigerian user community."

Observations of the Nigerian Experts Consultative Group

In the Interim Report on the Presidential Assignment, dated July 3, 2000, the NECG noted, *inter-alia*, the following, as its observations on the FMST, NASRDA and the Nigeria Satellite Project:

- FMST has not addressed the myriad of essential preparations and infra-structural requirements that must be in place, locally, for the successful take-off of any technology, particularly a new and advanced technology such as a satellite programme. FMST has also not correctly determined the immediate priorities of this nation in the field of space science and technology. The acquisition of satellites, albeit for unproven and untested missions, should not be the first activity of FMST or any other organ of the Federal Government of Nigeria, in its quest to give Nigeria a space capability;
- Today, Nigeria finds itself in the situations presented above, partly because the lead agency for space in Nigeria, the National Space Research and Development Agency (NASRDA), has no cognate scientific and technological competence and experience needed to successfully formulate, plan, develop, implement, maintain and supervise a project such as the Nigerian Satellite

Project. NASRDA also has not engaged the user community in Nigeria in all the aspects of this project. Similarly, NASRDA totally side-lined and ignored the Nigerian community of experts in the field of space science and technology, and did not tap the rich and varied talents of this community for the project. It is thus apparent that Nigeria has been and continues to negotiate with SSTL from a scientifically and technologically weak position on this project; and

- The Nigerian Experts Consultative Group (NECG) hereby strongly recommends that Mr. President should URGENTLY suspend, forthwith, any further action(s) or activities by FMST and its partners on the FMST's sponsored Nigerian Satellite Project. Mr. President can revisit the development of a viable space science and technology programme in Nigeria after the Office of SSAP/ SST has completed the on-going total review of the nation's space science and technology activities and has presented to Mr. President, for due consideration and approval, a harmonized and streamlined space science and technology programme for the development and growth of Nigeria.

Nigerian Space Programme: *A Blue Print for Scientific and Technological Development*

As promised on my appointment as the SSAP-SST, I submitted to the President, on September 26, 2000, my main report, titled: Nigerian Space Programme: A Blue Print for Scientific and Technological Development—Blue Print, for short. The key proposal in the document stated as follows:

Nigeria should build indigenous competences in developing, designing, and building hardware and software in space technology including rockets, satellites and antennas for scientific research and practical applications.

To accomplish the above, the report highlighted a number of issues that must be addressed by government, namely: that Nigeria and its stunted development are manifested in all spheres of human endeavor in the country, and that the *science and technology* sector is perhaps the worst off, vis-à-vis, what ought to have been, in comparison to the achievements of other countries such as Brazil, India, Republic of Korea (South Korea) and South Africa. The report categorized Nigeria' problems as follows:

(a) Unfocused policy objectives and poor awareness;

(b) Poorly established weak institutions;

(c) Lack of continuity in Project Execution;

(d) (Under and Non-) Utilisation of Human Resources;[9] and

(e) Funding constraints.

On funding of the nation's space programme, the same report cautioned the Nigerian administration in the following words:

"Today, the space enterprise has become one of the critical and fundamental foundations of industrialisation, and will be much more so in the foreseeable future. By its very nature, space enterprise engenders global collaboration and competition. Nigeria should welcome collaboration and co-operation at the international level in order to develop and enhance its own space capability. However, it should not expect any other country to subsidise its (Nigeria's) space ambition, since by its own committed efforts, Nigeria stands to compete, in the immediate future, with these same collaborators in the international market, particularly within the African market."

Buy Technology Abroad or Develop and grow Technology at Home

From the above, it was clear that there was no national consensus on how Nigeria should embark on its space journey. For the FMST and NASRDA, that journey should begin with the overseas purchase and launching of satellites aimed at acquiring the necessary Earth observation data needed for addressing the immediate problems of the nation. On the contrary, the SSAP-SST and the Nigerian Experts Consultative Group (NECG) stressed that:

- *In the nation's quest to acquire a space capability, the immediate acquisition of satellites should not be the first activity of FMST, or any other organ of the Federal Government of Nigeria; and that*
- *The first priority should be the rebuilding of Nigeria's science and technology capability in order to provide a solid foundation for its space programme.*[10]

That Nigeria was bent on buying a satellite, irrespective of the soundness of the mounting arguments against such a purchase, was a very bewildering experience. Have we forgotten the rush to acquire the tethered Aerostat Balloon and the national

---

9 On human resources utilisation, the report noted that "It has become the norm in our country, Nigeria, to lament the dearth of qualified personnel. However, the overwhelming number of competent Nigerians that have passed through the very best institutions in the world and are gainfully employed in many major science and technology establishments globally, suggest otherwise. The problem is under-utilisation, mis-utilisation and non-utilisation of our own very competent human resources in our nation's development agenda."

10 *Observations* from the Interim Report on the Presidential Assignment

embarrassment Nigeria went through thereafter? This time, a Nigerian satellite bought from Britain! A flurry of questions bubbled up within the 24 Nigerian experts invited to deliberate on Nigeria's space future in June 2000. Why are we putting the cart before the horse? Don't we need to first develop a national space policy along with the associated programmes? Where is our knowledge of space technology that will enable us to contribute to the definition of the satellite's specifications? Where is Nigeria's capacity and capability to analyse and judiciously utilise, in decision-making, the data from the satellite it was going to buy from SSTL? But Nigeria had not analysed nor made use of, in any way, the large array of data of Nigeria acquired by USA's Landsat –1 and –2 satellites at the request of the Government of Nigeria, in 1973, they also noted. Most of these concerns and questions were captured in the *Interim Report on the Presidential Assignment.* Before taking a critical decision, such as buying a satellite, the NECG noted, a genuine public debate would have been helpful. But for the Nigerian administration of that era and others before it, these were just too many expectations and questions; expecting a respectful hearing on these and similar concerns was a long shot. Even when the news media accommodated the opinions of members of the public on the pages of the local newspapers on other critical national issues, is anyone out there, in government, listening? For the Nigerian experts, it was *déjà vu*, again.

The SSAP-SST and the NECG Investigated

In an attempt to bend or break the SSAP-SST and the Nigerian Experts Consulta-tive Group (NECG), because of their stand against the Nigerian satellite project, the FMST/NASRDA and their collaborators on the satellite project mounted two simul-taneous attacks. The first consisted of authorised and unauthorised releases/leakages of *twisted and massaged* information in a variety of press releases, advertisements and insti-gated "Brown Bag Journalism" against the integrity and credibility of the person and the office of the SSAP-SST. And according to the Post Express Newspaper of August 31, 2000, "there was also concerted pressure on the Government of Nigeria from the British Government."[11] To hone in their *bend or break* effort, the FMST/NASRDA, in co-operation with other vested interests, also secured the consent of the President to set up a Special Committee, under the auspices of the Minister-of-State (FMST), to

---

11 (i) "Obasanjo cancels $700 million Space Project," Guardian Newspaper, Front page, July 16, 2000; (ii) Publicity Advertisement on SSTL, Punch Newspaper, Page 16, August 16, 2000; (iii) Nigeria's Space Project under Threat, Post Express Newspaper, August 31, 2000; (iv) Banigo gives reasons for resignation (*as Minister FMST), Vanguard Newspaper, page 7, November 21, 2000; and (v) Alleged resignation of Dr. Abiodun, Daily Trust, March 13, 2001.*

review (i) The SSAP-SST document, titled: The Nigerian Space Programme: *A Blue Print for Scientific and Technological Development*, and (ii) The Interim Report on the Presidential Assignment; this particular report contained an analysis of the FMST (NASRDA)/SSTL—sponsored Nigerian Satellite Project.

On October 12, 2000, the Special Committee submitted its report to the President through the Minister of State (FMST). Excerpts from the Report of the Special Committee on the SSAP-SST document titled: Nigerian Space Programme: *A Blue Print for Scientific and Technological Development* are reflected in Annex II to this chapter. Surprisingly, there was no smoke-screen in the Special Committee Report that could be held against the SSAP-SST and the NECG. On the contrary, judging *only* from the contents of Annex II and the Report of the Special Committee, it was almost certain that Nigeria was ready to do the right thing on its space aspiration. But unknown exigencies dictated otherwise.

The Buy and Purchase Agreement

Nigerians woke up in the morning of November 7, 2000 to learn, from the front pages of most of their national newspapers, of the decision of their government, on the Nigerian Satellite Project. That was when and how the Federal Government of Nigeria announced the "*Nigerian/United Kingdom Collaboration in Space: Signing of Contract Agreement, Monday, November 7, 2000*".

Analysis of the Agreement

As the SSAP-SST, I received a copy of the signed Agreement only through the kindness of the National Security Adviser who asked me to review and analyse it. Within forty-eight hours, I completed the assignment and submitted a copy of my *Analysis of the Agreement (Analysis)* to the NSA.[12] The *Analysis* showed that the Agreement was one-sided, that it overwhelmingly favoured SSTL, and that several of its provisions/clauses were detrimental as well as inimical to the interest of Nigeria. The Agreement was also a package that hindered technology transfer from SSTL to Nigeria. Several of its clauses portrayed a master (SSTL)—servant (FMST/Nigeria) relationship. The Agreement clearly showed that SSTL had exploited, to the fullest, the naivety, inexperience, incompetence and lack of knowledge of the subject matter on the part of the Nigerian Government officials who negotiated the contract signed

---

12 *Analysis of the Purchase and Sale Agreement between the Federal Ministry of Science and Technology (FMST) and Surrey Satellite Technology Limited (SSTL),* Entered into on November 7, 2000 in Abuja, NIGERIA

on November 7, 2000 between FMST/NASRDA and SSTL. The Executive Summary of the *Analysis* concluded as follows:

> *"For the avoidance of any doubt, the Federal Government of Nigeria is welcome to subject this Agreement and the Analysis presented herein to a further independent review by an international committee of experts familiar with the execution of space projects."*

With the *Analysis* delivered to the NSA, I thought I could leave Abuja in peace by the third-week of December 2000 for a well-deserved long vacation. But don't be so sure was literarily written all over the message I received on January 3, 2001 from the Presidency, summoning me back to Abuja for a high-level January 8, 2001 meeting with the President, senior cabinet members and NASRDA Director-General. In short, the deliberations at the meeting un-earthed, once more, the canker-worms in the Nigeria (FMST)/SSTL Agreement to the extent that the meeting concluded with the President setting up a 6-man "Presidential Committee on Nigeria's satellite project;" but he also chose the Minister of FMST as the Chairman of the newly established Presidential Committee.

The Presidential Committee met five times; it had a mandate to propose revisions to the Agreement as appropriate, taking into consideration the SSAP-SST's *Analysis*. It concluded its work and signed-off its report to the President on Friday, February 9, 2001. At its final meeting, the Chairman of the Committee, the Minister of FMST, invited all the members of the Committee to append their signatures to the document as its unanimous report. Before signing the report, I indicated that it was not a unanimous report, that it was not in Nigeria's best interest and that my signature should only be seen as an affirmation that I saw and read the report. I subsequently signed the report with the following proviso: *"I have signed this document simply to signify that I saw and read the document. I do not agree with some of the critical elements contained therein."* Thereafter, I submitted my *Minority Report* on the Committee's work on the Agreement to the President.

The biggest problem was the revised Agreement which was not in Nigeria's best interest. For example, the final recommendations did not make the revised Agreement any better than the signed Agreement. Although the Committee noted that the signed Agreement *"suffered from lapses and inadequacies,"* its own suggested remedies were equally inadequate and not in the nation's interest. The Committee's report also failed to reflect the various fundamental differences that surfaced among the members of the Committee, in the course of the Committee's deliberations, on a number of key clauses in the Agreement. This was a deliberate tactic aimed at selling the Committee's report to the President as a unanimous report, but it was not. Even when individual Committee members asked that the different and specific views expressed at the meetings of the Committee be part of the report, the final report did not reflect any such differences of opinion among its members. In the *Minority Report*, I highlighted some of the differences that were left out of or glossed over by the Committee's final report; these included the following:

- The trampling upon the sovereignty of Nigeria; and
- Warnings against expectations that any technology will accrue to Nigeria from the Agreement.

◊ *Trampling upon the sovereignty of Nigeria*

The *Analysis* had pointed out that the clause,

"This Agreement shall be governed construed and performance thereof shall be determined in accordance with the laws of England and the Parties hereby agree to submit to the jurisdiction of the English courts,"

trampled upon the sovereignty of Nigeria as an independent country, more so, since Nigeria paid its money to SSTL for this satellite. Furthermore, similar agreements that were signed by SSTL with a number of other countries contained no such clause. Rather, the laws and the courts of these other countries were the ones stipulated to prevail and adjudicate any disagreement in their respective agreements with SSTL. Unfortunately, in the case of Nigeria, the final document issued by the Presidential Committee on the Nigeria-SSTL Agreement accepted the above odious SSTL proposal and added the following as a rider:

"The Agreement shall be governed by Nigerian laws for items to be supplied in Nigeria and SSTL (Nig.) Ltd to be held liable for such items."

The problem with the above modification by the Committee is this: Which component(s) of the satellite was to be manufactured in or supplied by Nigeria? The answer then was none; today it is still none. Since no components of the satellite were to be manufactured in Nigeria, if anything goes wrong with any aspect of the project, the Nigerian government would have to argue it in the English courts and the adjudication determined in accordance with the laws of England. This addition, by the Committee, trivialised both the interest and the sovereignty of Nigeria.

◊ *Warnings against expectations that any technology will accrue to Nigeria from the Agreement*

The *Analysis* also alerted the Nigerian administration that there were "over-reaching as well as subtle and clear signals and warnings from SSTL, (in the

Agreement), against expectations that any technology will accrue to Nigeria as a result of this Agreement." Surprisingly, The Presidential Committee consequently noted the same thing in the revised Agreement and it finally acknowledged *"lapses, inadequacies and errors in the FMST/SSTL Agreement, errors that if not corrected, could be inimical to the development of space capability in Nigeria."* These errors included restricted licenses granted by SSTL to Nigeria and Nigerians, and SSTL's claim to all the copyrights and entitlements to the designs, inventions, discoveries and patents of the Nigerian/ SSTL trainees, while at SSTL and after they would have graduated from SSTL training programme.

In addition, as signed on November 7, 2000, the Agreement stipulated that:

"FMST shall not remove or alter any copyright or other proprietary on any of the Know-how."

In essence, this particular clause foreclosed Nigeria's ability to modify the design and software codes it was to receive from SSTL, codes which are very critical to a successful technology transfer and subsequent technology development in Nigeria. But on comparing this Nigerian/SSTL Agreement with similar agreements that SSTL entered into with other countries, it became apparent that the Nigerians that negotiated this Agreement had subjected Nigeria to SSTL's practice of *double-standard*. The equivalent clause in each of the agreements that SSTL entered into with these other countries read as follows:

"Country X shall be entitled to build upon and/or modify the SSTL technology, Know-how and Designs transferred to Country X under the Agreement and to use the same for non-commercial satellite projects, missions, and/or applications."

The Presidential Committee reviewed and compared the above two clauses, and it studied the objection to this Nigeria/SSTL clause, as contained in the *Analysis of the Agreement*. The Presidential Committee subsequently accepted its own error; thereafter, it offered the following opinion:

"The Committee considers this issue very significant indeed as it presents the kernel of Technology Transfer,"

and subsequently went on to propose the following amendment to the clause, as follows:

"SSTL hereby grants to FMST for a period of (10) years from the commencement date a non-exclusive and non-commercial license to use, in part or in full, the know-how, design and source code, subject to the following conditions (in the Agreement):"

But an informed mind would recognise that there are three major differences between (i) what the Committee called "the kernel (i.e., the core, heart or essence) of Technology Transfer," (ii) what FMST/NASRDA subsequently negotiated with SSTL for Nigeria, and (iii) a comparable clause in the Agreement that SSTL entered into with each of the other countries cited above as Country X. Specifically:

- Country X can build upon SSTL technology, Know-how and designs, Nigeria cannot;
- Country X can modify the SSTL technology, Know-how and designs, Nigeria cannot; and
- Nigeria can use SSTL technology, Know-how and design just as Country X can also, but the difference is that there are no limitations in or restrictions on the duration and how Country X uses these elements. However, in the case of Nigeria, the use of these same elements is subject to a number of limitations.

In short, within the framework of the Agreement, the amendment the Committee proffered did not give Nigeria the opportunity to openly develop any technology from the country's association with SSTL's technology know-how and designs.

In addition, there were other major concerns; these included the misleading of government, emphasising the spirit and not the letter of an official government agreement, and the price that Nigeria paid because the Nigerian Satellite Project was enveloped in secrecy.

(a) The Federal Government was misled

In the Executive Summary of its report, the Presidential Committee exerted pressure on the Nigerian government to consummate the Agreement as signed on November 7, 2000. Specifically, the Committee jolted the government into action as well as impressed upon it that:

"Government must treat this project with some urgency so as to position itself properly, as there are signs that an international movement is underway to restrict access to satellite imagery by developing countries."

Unfortunately, the government fell for this scam.

But, let us get the facts straight on satellite imagery/data accessibility. Article I of the United Nations Space Treaty, which Nigeria acceded to in 1973, grants all countries the right to explore and use the outer space for their own benefit; no country has ever prevented any other country, industrialised or developing ones, from doing so. In that same year, 1973, Nigeria requested the United States to use its Landsat -1 and -2 satellites to acquire data of the Nigerian landscape and its resources, but, to-date, how these data sets were used and for what purposes remain mysteries.

In 1986, the United Nations General Assembly, with Nigeria in attendance, unanimously endorsed the Remote Sensing Principles.[13] Its Principles XII and XIII state as follows:

- *Principle XII*—As soon as the primary data and the processed data concerning the territory under its jurisdiction are produced, the sensed State shall have access to them on a non-discriminatory basis and on reasonable cost terms. The sensed State shall also have access to the available analysed information concerning the territory under its jurisdiction in the possession of any State participating in remote sensing activities on the same basis and terms, taking particularly into account the needs and interests of the developing countries.

- *Principle XIII*—To promote and intensify international cooperation, especially with regard to the needs of developing countries, a State carrying out remote sensing of the Earth from space shall, upon request, enter into consultations with a State whose territory is sensed in order to make available opportunities for participation and enhance the mutual benefits to be derived therefrom.

Above all, remote sensing data do not fall into the same category as nuclear bombs or weapons of mass destruction. These and other related indisputable facts discredit any such "*...international movement...[that could be]...underway to restrict access to satellite imagery by developing countries.*"

(b) Emphasise the Spirit and not the Letter of this Agreement

After rejecting as many modifications/revisions as possible in the original text of the signed Agreement, the Chairman concluded our review exercise with the following final remarks:

> "*We (i.e. the members of the Presidential Committee) should not lose sight of the objective of the Agreement. Whatever is in the Agreement now, we may not change. Whatever is discovered will ultimately be claimed. Surrey (SSTL) can lay a claim to it, and Nigeria can lay a claim after members of the FMST team leave Surrey. We should emphasise the spirit and not the letter of this Agreement. We should accept the current stipulation as spelled out in the Agreement.*"

---

13 UN Document ST/SPACE/11—United Nations Treaties and Principles on Outer Space (2002), New York, NY, pp. 44-47

With such a closing remark by the Chairman of the Committee, the nature of the government decision to follow on the Agreement itself was a foregone conclusion. The Committee's review report also changed almost nothing from the text of the signed original Agreement. Such a position by the Committee could have been predicted right from the beginning of its work, particularly, since the same minister who signed the Agreement, on behalf of the government, also served as the Chairman of the Committee that reviewed that same Agreement. That double role of the chairman was more than enough to silence most of the dissenting views of most members of the committee.

(c) The Nigerian Satellite Project was enveloped in Secrecy

On the whole, the revisions proposed, as "corrections" to be inserted in the already signed Agreement, were lopsided in favour of SSTL and heavily weighted against Nigeria. Nigeria became party to this Agreement without a very careful study of all its clauses. In addition to the NECG's observations cited on page 93, much secrecy that enveloped the Nigerian Satellite Project foreclosed such a careful study. The crux of the matter was that this Agreement, a national document, became such a secret and private property of its managers that experts that should and could have been helpful in thoroughly analysing its terms were side-lined and kept in the dark.

The National Space Policy and Programmes

The nation soon moved beyond the Agreement whose focus was the Nigerian Satellite Project, to developing a national space programme which should be guided by a national space policy. This is necessary, because a policy defines what is being aimed at, the nature of the budget that will see it through, and the political commitment that is critical to the success of the programme that translates the policy into action. The document, 'Nigerian Space Programme—*A Blueprint for Scientific and Technological Development*,' offered the nation a sound path to accomplishing these objectives. In concert with the October 12, 2000 recommendations of the Special Review Committee on the above document, excerpts of which are shown in Annex II to this chapter, President Obasanjo subsequently directed the FMST to develop and prepare a National Space Policy and The Nigerian Space Programme for consideration and approval by the Federal Executive Council (FEC).

The ministry (FMST) single-handedly developed the *National Space Policy and Programmes* using 'Nigerian Space Programme—*A Blueprint for Scientific and Technological Development*' as a major working document to achieve that goal. On

July 4, 2001, the Federal Executive Council approved the National Space Policy and Programme document. Although the 'buy and purchase' Agreement with SSTL for NigeriaSat-1 had been consummated by that time, nevertheless, the National Space Policy and Programme document still boldly declared the following as the space vision of government:

To make Nigeria build indigenous competence in developing, designing and building appropriate hardware and software in space technology as an essential tool for its socio-economic development and enhancement of the quality of life of its people.[14]

How has Nigeria lived up to the above declaration? Chapter VI that follows highlights the state of S&T that was supposed to support the nation's space aspirations. Chapter VII chronicles Nigeria's satellite acquisitions to-date and in the immediate future; it also examines the 25-Year Road Map on Nigeria's Space Mission, a successor to the 2001 National Space Policy and The Nigerian Space Programme.

* * *

---

14 Nigerian Space Policy and Programme, Federal Ministry of Science and Technology, Abuja, July 2001.

Annex I

EXCERPTS from the OPENING ADDRESS & REMARKS
AT THE FIRST EXPERTS-CONSULTATIVE MEETING
ON
SPACE SCIENCE AND TECHNOLOGY IN NIGERIA
29 – 30 JUNE, 2000
ABUJA, NIGERIA

by

Chief Ufot EKAEETE, Secretary to the Government of the Federation; and
Dr. Adigun Ade ABIODUN, Senior Special Assistant (to the President) on Space
Science and Technology (SSAP-SST)

Address by the Secretary to the Government of the Federation

As you are all aware, the present administration is determined and desirous of harnessing every potential endowment bestowed by the Almighty on this nation. An aspect of this quest relates to the use of manpower and their expertise in various fields of human endeavour.

One practical demonstration of this commitment on the part of Government is the appointment, earlier this year, of Dr. Adigun Ade ABIODUN as the Senior Special Assistant to the President on Space Science and Technology (SSAP-SST). As I indicated in my Circular Ref. No. SGF. 52/S5/50 of 17th April this year, the President has appointed Dr. ABIODUN to:

- Evaluate and appraise the status of all on-going and planned space science and technology-related activities (i.e. programmes and capabilities–both human and institutional) in Nigeria;
- Determine the benefits that would accrue or have accrued to Nigeria, to date, in employing these capabilities and programmes in the interest of the country's development efforts;
- Assess the present and future trends of Space Science and Technology within the international community including its roles in improving human conditions here on Earth, and determine the potential for collaboration with and support for Nigeria within the community; and
- Formulate policy options for streamlining and harmonizing Space Science and Technology activities in Nigeria including programmes that could ensure

a successful and beneficial participation of this nation in this global enterprise as is the practice in other countries.

On behalf of the President, I seize this opportunity to thank him for taking up the offer to come home and assist in the development of this potentially giant nation. In accepting this appointment, the Senior Special Assistant to the President on Space Science and Technology let it be known to the President that he would identify the Nigerian talents in the different fields of Space Science and Technology with a view to engaging them in the accomplishment of the responsibilities I just outlined.

The presence of this body of space experts in this room here today is a testimony to the many blessings our creator has showered on our nation. This Administration strongly believes that the quality of your recommendations along with practical strategies for their implementation will be commensurate with the calibre of scientists and technologists that are present here today. In this connection, I wish to remind you all of the words of President OBASANJO in his message to the Conference of Directors of Civil Service in Africa, this past Tuesday, here in Abuja, wherein he stated that …**Conferences would be worth-while if they were directed at solving the ordinary problems of the living…**

I am aware of the capabilities of space science and technology that can be harnessed to solve many problems in our country – communications, search and rescue operations, water resources management, agriculture and food security, education, health-care, environmental management, weather forecasting and climate studies, transportation, surveying and civil works, mitigation of disasters, census enumeration, national security and defence, just to name a few.

As you deliberate on the goals, objectives and mechanism for developing Nigeria's space capabilities and utilizing same in our nation's development process, you should take cognizance of the following:

a) Programmes in and experiences of other countries such as Brasil, India, South Africa and the Republic of Korea, and the lessons therefrom;

b) Development of intermediate and long-term research and application programmes that can successfully address the needs of Nigeria;

c) Rationalization of institutions, programmes and activities in order to bring harmony and orderly execution; and

d) Identification and optimal utilization of the country's large pool of competent and experienced science and technology expertise, both locally and abroad, and the development of new talents, that could be engaged in carrying out the nation's development agenda.

In so doing, I should reiterate that this Administration is looking forward to conclusive results soonest and sufficiently early enough to guide the budget for the year 2001.

Remarks by the SSAP-SST

As you all may be aware, Nigeria first participated in space activities in the 1960s; that was when the United States built a monitoring station in Kano which it used, among many others, worldwide, to monitor the progress of its space exploration programme. The latter subsequently resulted in the landing on the Moon, by Neil Armstrong, in 1969. For political reasons, Nigeria demanded that the USA should dismantle the Kano station, and the USA complied. Today, a similar station, built by the USA in Ouagadougou, has become the communication life-line of Burkina Faso, and the one built in Cotopaxi, Ecuador, has also been modified to serve as an earth observation station for receiving, from a variety of satellites, the data being acquired over 21 Caribbean and South American countries, by these satellites. The only station on the African soil today that is performing a similar function as that of Cotopaxi, for the benefit of southern African countries, was built by South Africa outside Johannesburg. By the end of this year, Egypt should begin its construction of a similar station for the benefit of north and northeast Africa and the Middle East countries. In our Third National Development Plan (1975-80), the Federal Government of Nigeria allocated N10 million15 for the establishment of a National Remote Sensing Centre with all the necessary manpower and infrastructure, including the building of a receiving station similar to the one in South Africa today. After three years of very bad performance by those in charge of the project, the Government withdrew the allocation, and today, we as a nation are suffering from the inaction of that era. We should also note that the ability of scientists and engineers at Stellenbosch University to develop, design, model, test and build a micro-satellite that was successfully launched on February 23, 1999, has placed South Africa among the space-capable countries of the world.

But can we as a nation benefit from our effective participation in space activities? Nigeria certainly has the human and material resources necessary to participate actively and successfully in this enterprise. Twenty Nigerian competent and active space scientists and engineers are here in this Conference Hall with us all, working with me on the on-going Presidential Assignment on space science and technology.

---

15 In 1975, Nigeria N1.00 = US$1.75

For example, all of us are daily users of space information such as weather forecast that is presented on our TV sets. Today, satellites make it possible for us to communicate, daily, by telephone, fax and Internet with one another here in Nigeria and with many more globally. We have experienced major aircraft disasters (e.g. by ADC and the Military aircraft accidents) within our borders. The absence of necessary beacons that could have conveyed the precise locations of these and other ill-fated aircraft to satellites that are dedicated to International Search and Rescue Operations, for subsequent rescue operations, led to the loss of many precious Nigerian lives.

Other aspects of our national lives that could benefit from space activities include Agriculture, land-use, forestry and food security; Air, land and maritime navigation; Assessment and management of natural resources (air, land, minerals and water including fisheries and marine resources); Information exchange, transportation and tourism; Energy development and management; Education and health-care, including rural and urban development as well as census enumeration; Environmental assessment, management and control; National Defence Security operations and the Judiciary.

Today, Mr. President has recognized the role that space science and technology (SST) can play in advancing the development process in and growth of Nigeria. Accordingly, early this year, he appointed a Senior Special Assistant to the President on Space Science and Technology (SSAP/SST) to appraise our national situation vis-à-vis the international efforts in this ever-expanding field, to formulate programmes and propose institutional arrangements that will ensure that we, as a nation, derive maximum benefit from our effective participation in this enterprise.

Nigerians will believe that space science and technology is relevant in our national development efforts, if we can demonstrate, in a practical way that this field of human endeavour can enhance the quality of life in this nation. With this meeting, we are taking one of the necessary steps that should result in the development of a *Blue Print on future space activities for the benefit of Nigeria.* Please note that Mr. President and our National Assembly are anxiously waiting for this a document.

Developing an effective mid-and long-term space programme for Nigeria requires input from and interactions with Nigeria's experts, institutions and overseas partners. The beneficiaries of Nigeria's space activities as well as the legislators that will approve necessary funding for such activities must be fully informed. We must not fail our President and the Nigerian People. They are all counting on us. And on their behalf, I wish to thank you all for accepting to contribute, selflessly, to the attainment of these goals.

* * *

Annex II

EXCERPTS

from

The Report of the Special (*Review*) Committee on Nigerian Space programme: *A Blue Print for Scientific and Technological Development*,

1. On the Nigerian Space Programme document, the Special (Review) Committee noted, in its report, as follows:

    (i)   The Committee is of the view that the SSA/SST's report on Nigerian Space Programme is a good working document and should be adopted as a Blue Print for achieving national objectives in the area of space science and technology.....(Conclusion and recommendations Page 27)

    (ii)   The SSAP/SST's report *(NIGERIAN SPACE PROGRAMME........)* should be adopted as Blue Print for the acquisition and development of space technology by the nation ....(Conclusion and recommendations, page 28)

    (iii) The Nigerian Space Programme document produced by Dr. A. A. ABIODUN and Nigerian experts was carefully studied by the Committee and found to be of high standard. (Page ii, para. 5 of Executive Summary)

    (iv)   The Committee is convinced that if properly implemented and adequately funded, the proposed Blue Print would yield the desired results. (Page ii, para. 5, Executive Summary)

    (v)   The Committee is of the opinion that the report was well prepared and embodies what the country should embark upon in various areas of space activities, without further delay. (Page 21, para. 9(1))

Chapter VI

THE STATE OF SCIENCE AND TECHNOLOGY FOUNDATION

Today, the future of space is being propelled by technological advances as well as unparalleled human imagination and ingenuity that are inspiring a variety of activities. The goal is not only to develop new products and provide improved services for our daily needs here on Earth, but also to expand and advance human knowledge of the universe and reach to the stars. To attain these goals, such countries as Brazil, Saudi Arabia, South Africa and South Korea are staking their future and are consistently investing their fortunes in ambitious long-term research and development activities. The backbone of these efforts is a high level of competence in science and technology (S&T).

An equally ambitious Nigeria must also first invest in a credible level of science and technology programmes as the foundation needed to power and sustain its economic development activities, part of which is its space aspiration. Nigeria would also need to work closely with other African countries on space and other issues that affect the wellbeing and future of Africa as well as partner with the global community to address issues of common and global interest. Since the tool of such collaboration and cooperation is world-class S&T, it is thus necessary to ascertain the state and adequacy of the nation's S&T. Both the *Nigerian Experts Consultative Group* (NECG) that worked with the SSAP-SST in 2000, and the Special Review Committee that reviewed its work on behalf of government, raised a similar concern about the state of S&T in Nigeria at that time. But it was a critical requirement the Federal Executive Council (FEC) did not address, when it approved the establishment of NASRDA in

1999. What, then, is Nigeria's S&T standing? And how does Nigeria compare with its neighbours on the African continent?

Globally, there are a number of standard indices used in adjudging the state of science and technology (S&T) as well as the state of development in a given country. These indices include the quality of mathematics and science education, percentage of the nation's GDP that is committed to research and development (R&D), who pays for the R&D and who actually carries out the R&D activities. Other factors include the state of the libraries in the country—both the public ones and those at the research institutions—the number of scientific and technical publications in certified journals, the state of old and new technologies in the country, the number of patents and trademarks awarded to individuals and entities in the country, the access of the citizenry to computers and telephones, particularly broad-band, and the impact of information and communication technologies (ICT) on access to basic services. A number of figures in this chapter show how Nigeria compares, particularly with six African countries, namely, Algeria, Egypt, Ghana, Kenya, Morocco and South Africa, and others, on some of these indices. But, first, let us examine and understand the role of S&T in a nation's development.

Science and Technology defines our World

As a proven driver of global growth and development, S&T is universally acknowledged as a tool that can be applied to improve human condition. Nigerians, particularly members of the nation's academies, have advocated for such an under-standing by the nation's decision makers in order to move Nigeria and its people forward. In spite of such entreaties, over many decades, Nigeria and Nigerians have not experienced any appreciable impact of S&T on the nation's development. It is thus appropriate to revisit what science and technology is all about.

By definition, science is the development and accumulation of knowledge about a particular process or thing, through study, observation and research including experimentation and experience. Technology is the practical tool for translating the accumulated scientific knowledge into economic productivity—a capacity that depends on the individual and the institutional character of a nation. Globally and on a daily basis, from the time we humans wake up to the time we return to our beds at night, a significant part of our activities and the devices that serve and support our existence here on Earth are either related to or are end-products of science and technology processes. Simply stated, those countries that can innovate and generate the knowledge needed to translate raw materials into useful products that meet the needs of its own people as well as for export are classified as industrialised countries. The countries that have little or no industrial base to translate their own raw materials into useful products, but instead mostly buy, from abroad, the goods and services that their people need and consume at home, are referred to as developing countries. There is a third group, the emerging countries—these are countries that are beginning to transform themselves from developing to industrialised status. In short, the class a given country belongs to is dictated by its contributions, through its creativity and innovations, or lack thereof, to development efforts here on Earth.

John Stansell reinforced that class distinction when he assessed the economic power and prosperity of his country and proudly reminded us, in his 1979 commentary on *Britain and innovation*, that:

> *Everyone knows that we are an industrial country, that our wealth is based on adding value to raw materials, and that our trained engineers are our lifeblood.*[1]

Laying the Foundation for the Industrialisation of Nigeria

The above reminder, by Stansell, captures the realities of Nigeria's relationship with the rest of the world before and after independence, as well as today. That was when Nigeria's raw materials, such as cocoa and crude oil, were (*and are still*) feeding the factories in Britain, and in many other countries that in turn, continue to export the finished products back to Nigeria, at exorbitant prices. Over 50 years ago, Nigeria inherited, from its colonial master, a legacy that had no science and technology foundation that could sustain the nation; it was one that suppressed all local technological initiatives and creativities. Instead of promoting the development and nurturing of indigenous industries, the British favoured the importation into Nigeria of manufactured products solely from British companies, such as Paterson Zochonis, John Holt and Gottschalks and other companies with British stamp of approval, such as K. Chellaram and J. Chanrai, both of India. The testimony of Harold Smith, a former British colonial officer in Nigeria, from 1955 to 1960, offers an insider's confirmation of this type of relationship between Britain and its former colonies.[2]

---

1 John Stansell (1979). *Britain and innovation,* New Scientist, February 15, 1979, p. 458.

2 Mike Akpan, (2014). *The Political Culture Britain Gave Nigeria, Real News Magazine*–(http://real-newsmagazine.net, Accessed, January 10, 2015)—Extracts from Harold Smith's Autobiography—Blue Collar Lawman.

But the Nigerian leaders at independence, in 1960, had enough foresight and were not unaware of the overseas transformation of Nigeria's raw materials into finished goods for subsequent consumption in Nigeria. In Nigeria's interest, they were determined to reverse the trend and keep Nigeria's raw materials at home for processing into manufactured products by Nigeria's factories. To that effect, industrial estates that would pave the way for the industrial transformation of Nigeria sprouted up all over the country, beginning in 1958, with the establishment, by the Federal Government, of the Sabo Industrial Estate in the then suburb of Lagos. The realities on the ground also shaped Nigeria's first post-independence national development plan that spanned the period, 1962-68. The plan demonstrated that the post-independence leaders knew and understood S&T as a critical tool of economic development and national growth.

At that time, a major national pre-occupation was the drive, by Nigerians, to take the destiny of their new nation and its peoples into their own hands. Human resources development in S&T at home and abroad took centre stage and education became the cornerstone of that effort. In the words of Chief U. J. Ekaette, the Secretary to the Government of the Federation from 1999-2007:

> *Indeed, the government of the day [at independence] along with the majority of Nigerians placed education on the highest priority list of the nation because it represents our lifeline for national survival, the best guarantee for genuine human and material development, and the surest avenue for the achievement of the values of social justice.*[3]

Education, industrialization and gainful employment were among the key economic targets of the government. Prominent amongst the pre-independence S&T institutions that were established in the 1950s was the tripartite residential college system, known as the Nigerian College of Arts, Science and Technology (NCAST), with branches in Ibadan, Zaria and Enugu. These three branches of NCAST were later upgraded to university-level institutions and respectively became the University of Ife (since re-named Obafemi Awolowo University), Ahmadu Bello University, Zaria and the University of Nigeria-Enugu Campus. To strengthen the S&T capabilities of Nigeria, the Federal Government also established the Federal Emergency Science School at Onikan, Lagos, in order to increase the intake, by

---

3 Ekaette, U.J., (2001). *Policy and value systems in higher education, An Acceptance Speech on the occasion of his conferment with the degree of Doctor of Science (Honors-Causa) and on behalf of Honorary Graduands of the University of Uyo, Uyo, Akwa Ibom State, June 9, 2001.*

NCAST, of science and technology (S&T) students as well as enhance the quality of S&T education in the country.

From the early 1960s through the early 1970s, governments in Europe and North America threw their hats into Nigeria's human resources development ring. They offered scholarships and admissions to a couple of thousands of Nigerian students for education and training in science and technology-related disciplines, in their colleges, institutes and universities.[4, 5] That Nigerian students were readily admitted into these institutions was a testament to the quality of the academic preparations attained in Nigeria's primary and secondary schools in the 1960s through the mid-1970s. Consequently, Nigerian students admitted into the universities in such countries as Australia, Canada, Germany, United Kingdom, United States and the USSR, at that time, were accorded special recognition of one kind or another, such as advanced placement in many overseas institutions. The Federal and the Regional Governments in Nigeria also offered scholarships, on a merit basis, to many students for studies at home and abroad. It was indeed an era that witnessed much investment in Nigeria's future professionals, including scientists and engineers.

As shown in Chapters IX and X, because of these earlier collective educational investments, today, there is scarcely any field of human endeavour in which Nigerian scholars, men and women alike, have not made their mark. Nigerians, with no trace of science (as we know it today) in their upbringing, rose up to the challenge of acquiring all the necessary knowledge and skills in various fields of human endeavour such as physical, chemical and biological sciences, engineering, architecture and health sciences, as well as in humanities and social sciences.

Planning for Gainful Employment

The leaders of that era had a vision for Nigeria and were conscious of the need to gainfully utilise the talents of the new Nigerian graduates in the development process.

---

4 Prominent among these scholarship programmes were (i) UK—*Commonwealth Scholarship and Fellowship (CSF) Plan*, funded by the UK government, (ii) *USA*—(a) *The African Scholarship Programme of American Universities (ASPAU)*, (b) *The African Graduate Project (AFGRAD) and (c) The African Training for Leadership and Advanced Skills (ATLAS*—funded by the United States Government and a number of private American foundations and administered by the African American Institute (AAI), New York, from 1961-1975.

5 Abiodun, Adigun Ade (1998). *Human and institutional capacity building and utilization in science and technology in Africa: An appraisal of our performance to-date and The way forward,* African Development Review Journal. Vol. 10, No. 1, June 1998, pp. 10-51, African Development Bank, Abidjan.

They subsequently responded with bold steps and planned meaningful employment opportunities for the graduates, including those in the S&T fields. At the request of these same leaders, Britain invited the International Bank for Reconstruction and Development (IBRD), the precursor of the World Bank, to prepare a master plan for the economic development of Nigeria. Two S&T-related institutions featured in the subsequent proposal of the IBRD, namely, the Nigerian Industrial Development Bank (NIDB) and the Institute of Applied Technical Research now the Federal Institute of Industrial Research (FIIRO). The functions of both the NIDB and FIIRO included the following:

The NIDB was [established and] funded to provide medium and long term finance at lower/viable interest rates for industrial projects that utilize local factor endowments notably local raw materials. The FIIR was [established and] mandated to carry out investigations and researches into Nigerian raw material resources for use in local industries. More specifically, those materials that have the potential of being utilized as industrial inputs should be identified and characterized for suitability in processing, and to source the technology required, endogenously or externally. Technologies imported should be adapted for use under local conditions and tested for performance.[6]

Nigeria's main raw materials of that era included groundnut, cocoa, cassava, cotton, gum Arabic, rubber, soya bean (Glycine Max) and palm oil—they still are, today. The country was also a significant producer of coal, bauxite, gold, iron ore, leather and textiles. That was also the time when Nigeria was among the world's leading exporters of palm oil, groundnut and cocoa beans, and a time when the Nigerian economic landscape was dominated by the groundnut pyramids in northern Nigeria, the cocoa warehouses in western Nigeria, and the palm produce stores in the Eastern region of the country. "In the 1960s, Nigeria produced over 60% of global palm oil exports, 30% of global ground nut exports, 20-30% of global ground nut oil exports, and 15 % of global cocoa-bean exports."[7] Also at that time, agriculture accounted for more than 60% of Nigeria's GDP and 70% of its exports. Nigeria fed itself and was a large net exporter of food without the benefit of today's modern agricultural machinery and related technologies.[8]

---

6 Akinrele, I. Adedayo (2004). *Technology, resource endowment and national development,* The 9th Annual Guest Lecture of Stephen Oluwole Awokoya Foundation for Science Education, Lagos, Nigeria, March 15, 2004.

7 World Bank—Overview of Nigeria, African Business Seminar, October 3, 2014.

8 Federal Ministry of Education (2005). *Nigeria Education Sector Diagnosis—A Framework for Re-engineering the Education Sector,* Education Sector Analysis Unit, Federal Ministry of Education , Abuja, Nigeria, May 2005; Project Coordinator: Dr. G.A.E. Makoju (Mrs.).

Nigerians of that era were also mostly rural people from humble backgrounds; they were not born with *silver spoons in their mouths*. Sound education was inculcated in the Nigerians of that era as a guarantee of a *meal-ticket and a promising future*. Nigeria subsequently made considerable gains in education, nation-wide, and in laying the foundation for the industrialization of a great nation. A large pool of educated and skilled Nigerians soon became available; they had the confidence that their federal and regional governments would provide individual and collective opportunities for them to contribute to the growth of the great nation Nigeria was perceived and destined to be. However, to-date, this has not materialized. What went wrong?

Derailment of the Nation's Goals

For a variety of reasons, '*things fell apart*.' The first blows were the military incursions into Nigeria's national arena and political make-up for a period of over three decades, beginning on January 15, 1966. Except for a brief interlude, 1979-1983, military rule in Nigeria from January 1966 to May 1999 eroded most of the earlier gains in the nation's development plan. Accountability went out of the window and Nigeria went through its own *Wild, Wild West*. The post-independence period before the January 1966 military coup also remains a very beautiful but painful reminder of what Nigeria could have been. In his inaugural address, on May 29, 1999, as civilian President, Chief Olusegun Obasanjo bade farewell to military rule and the woes that accompanied it in the following words: "*Our national nightmare is over.*" But is it really over?

The Pre-occupation with Crude-oil

Let us recall that exactly ten years before the first military coup took place on January 15, 1966, in Nigeria, an exploration begun by the oil company, Shell Darcy (later renamed Shell-BP), resulted in the discovery of crude oil at Oloibiri, in the Niger Delta, on January 15, 1956. By 1985, this first oil field was producing 5,100 barrels of oil per day (bpd).[9] Today, Nigeria's crude oil production is in excess of 2 million bpd. The continuing gush of the crude oil has made Nigeria, Africa's biggest oil producer; it has been exporting crude oil for over 55 years for the benefit of the world's major economies, particularly the industrialised countries. But according to the people of Oloibiri, "*they have not seen much of the money made in the 55 years of*

---

9 Odularu, Gbadebo Olusegun (2007). *Crude oil and the Nigerian economic performance,* Oil and Gas Business, 2007.

oil production."[10] The same can be said for other acclaimed national investments on different aspects of the nation's economy, particularly on basic infrastructure, albeit, with no demonstrable returns. Some have proclaimed oil discovery in Nigeria as a curse, others have seen it as a blessing. One fact remains incontrovertible—Nigeria has not been the same ever since.

Leap-frog, Do not re-invent the Wheel

The income from crude oil blinded Nigeria into believing that money could buy anything it wanted from the world markets. That belief was evident in the remarks of the then Nigerian Minister of Science and Technology, on November 23, 1979, at the closure of the United Nations/FAO Remote Sensing Training Course that was held at the University of Ibadan. The minister in question had just visited Germany, a country he described as '*the land of experts*,' where he saw and would soon import, for use in Nigeria, '*a machine that could do everything for us here in Nigeria*'[11] This German machine, to be purchased with the funds earned from the sale of Nigeria's crude oil in the international market, was the minister's justification for turning down the application of three University of Ibadan young chemistry research scientists and lecturers for a research grant; he described the young scientists as *upstarts*. But more accurately, these were promising young individuals with great potential to contribute to the economic development of their father-land and who were denied the opportunity by an S&T decision maker that was unwilling to invest in home-grown expertise. I reminded the minister, who earned a PhD in civil engineering in the United States, that "engineers and scientists often use a number of experiments and projects as guinea pigs to gain requisite knowledge and experience before becoming experts. To be an expert, you must try and try again. How are we going to grow and nurture experts in Nigeria if we do not invest in their scientific development, aspirations, and associated risks and failures?" I asked, and also added: "If they eventually succeed, we can take credit for the fact that Nigeria has its own experts who can be called upon if a similar problem arises in the future. This would be preferable to running overseas for foreign advisers who have very little or no knowledge of what ails us." But all my entreaties,

---

10 *The day oil was discovered in Nigeria*, BBC News, Tuesday, 17 March 2009 (Accessed, June 30, 2014).

11 Personal discussion by the author with the Nigeria's Minister of Science and Technology, at the Premier Hotel, Ibadan, on November 23, 1979, at the closing reception given by the Federal Government of Nigeria in honour of the African participants at the United Nations/FAO Regional Training Course on Remote Sensing Applications for National Development, University of Ibadan, Ibadan, Nigeria, November 5-23, 1979.

including my request that *making Nigeria also* a land of experts should be very high on the priority list of his ministry, did not change his mind.

This interaction with this minister has stayed with me ever since. It was (and still is) inconceivable, for me, that a Nigerian decision maker, at the level of a minister, with the responsibility for the nation's S&T policy and programmes, was an advocate of technology transfer from abroad, in the form of imported goods and services, at the expense of technology development at home. The problem was that every one, at home and abroad, wanted a piece of the Nigerian pie—its oil money. That was also when the newly independent countries got indoctrinated into both how not to *reinvent-the-wheel* and how to leap-frog fundamental steps in development.

Since then, we have invented very little, and have done much leap-frogging. After imbibing such a belief, Nigeria subsequently embraced technology transfer and relentlessly used its oil funds in the pursuit of *turn-key* projects of every kind to meet its development needs and goals. But we did not learn the critical lessons that enable technology transfer to survive and succeed in its recipient environment. First and foremost, such a recipient environment must nurture a science and technology culture. The latter, which is very critical for the survival and continuing exploitation of what is being transferred, is not available in Nigeria. Transfer of technology works among the industrialized countries because they have attained that level of technology development that enables them to retool the parts in most of the machines/equipment they import from anywhere else.[12] But we embraced technology transfer that made no provision for the development and nurturing of a science and technology culture and related institutional capacity for S&T development; today, we are in the dark alley of under-development where there has been no role for the S&T initiatives of educated and trained Nigerians. The negative consequences of these actions have been incalculable.

Unlike the immediate post-independence era, when there was much opportunity for educated and skilled Nigerians to contribute to their nation's development agenda, engagement of foreign consultants became the order of the day, not only in Nigeria, but also in most of sub-Sahara Africa. Nigeria was also not producing any reasonable part of its daily needs, and the nation steadily became wholly dependent on the goods and services of other societies. From the late 1970s, many highly skilled African graduates and professionals, particularly those with science and technology background,

---

12 Abiodun, Adigun Ade (1998). *Human and institutional capacity building and utilization in science and technology in Africa:...*African Development Review Journal. Vol. 10, No. 1, June 1998, pp. 10-51.

subsequently came to the dire conclusion that *their talents remain unrecognised and therefore unutilized, unchallenged and hence underutilized, or misdirected and therefore misapplied.*[13] They also concluded that if they were *to manifest any real growth and reach their [respective] higher potentials, their creativity would need nourishment from their environment.*[14] Unfortunately, for most of these individuals, Nigeria did not provide such an environment. And in order to avoid being a parasite within the society, a large number of those with the S&T knowledge, who had the opportunity to do so, checked out of the country for other lands. This was the genesis of The Diaspora; today, the checking-out has not abated.

Inability to add Value to Our Raw Materials

Meanwhile, the crude oil and what would make it flow abundantly and uninterruptedly also became the national pre-occupation. And after so many years of operating four *turn-key* oil refineries, we are still unable to properly maintain them without seeking external assistance at prohibitive costs. Accordingly, in the land that is awash with crude oil, fuel shortage and unrelenting dependence on imported refined petroleum products have become the norm rather than the exception. Our singular focus on oil also led to the abandonment of those activities, particularly agriculture, where the nation once excelled.

The nation's palm oil and rubber plantations, cocoa and coffee farms and cotton and groundnut fields were deserted and abandoned for the quick money to be made from oil-related tasks; everyone, who could, strove to be oil merchant or a political juggernaut. In those agricultural areas where Nigeria was once among the leading producers, such as cocoa, it was no longer a force to be reckoned with. By 1982, agriculture contributed only 17% of Nigeria's GDP as compared to 60% of GDP in the early 1960s. And instead of being a net food exporter, we are, today, a huge importer of foods from many corners of the globe, with an annual import bill that rose to ₦1.7 trillion (US$11 billion) by 2012.[15, 16]

---

13 Abiodun, Adigun Ade (1981), *Technology development in Africa*, AFRICA, No. 113, pp 56-57, January 1981.

14 Bhattasali, B.N. (1972). *Transfer of Technology among the Developing Countries,* Asian Productivity Organisation, Tokyo.

15 Osagie, Crusoe (2011). Nigeria Spends N991billion on Rice, Wheat Importation—Minister of Agriculture, ThisDay Newspaper, August 15, 2011.

16 Address of Dr. Akinwunmi Adesina, Nigeria's Minister of Agriculture, at AGRIKEXPO 2013, Lagos, Nigeria.

Similarly, as shown in Figure 6:1, among the nine world leading cocoa producing countries in 2012, Nigeria was in the last position, in adding value and industrially transforming its raw cocoa beans it produced that year, into cocoa shells, cocoa paste, cocoa butter and powder and chocolate.[17]

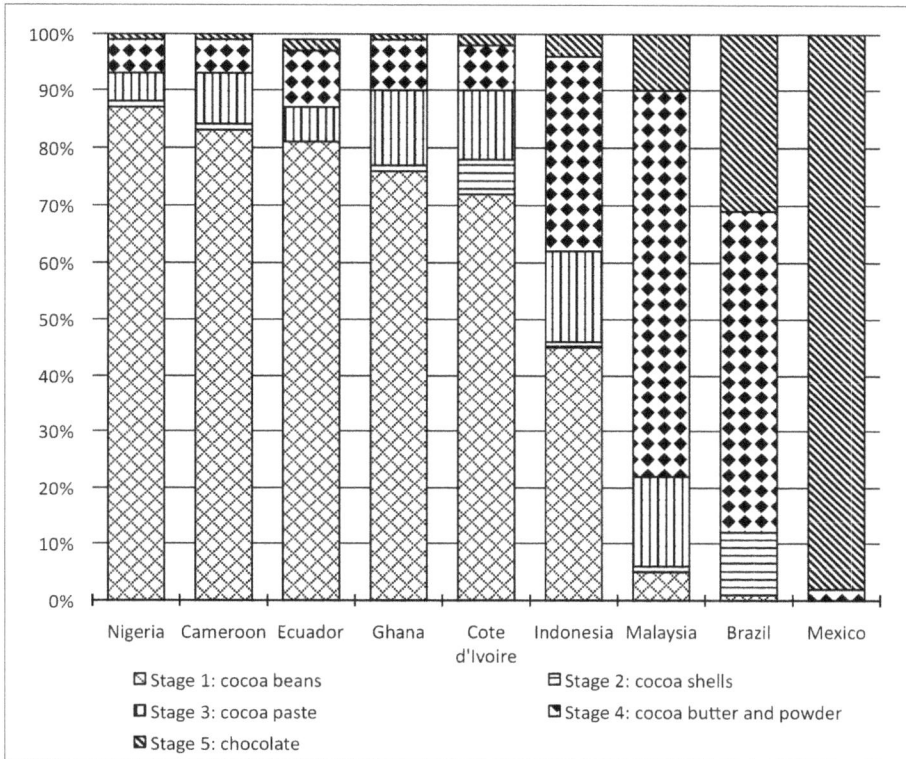

Figure 6:1 Value-added content of selected developing countries' total cocoa exports, 2011 (%). *(Source: ECA/AU Africa Economic Report (2013))* Accessed, November 2, 2014.

With money coming in from the sale of around 2 million barrels of crude oil a day, the expectation was that the nation's tertiary institutions and research establishments would be among the beneficiaries—but that is yet to happen.

Underfunding of Tertiary Education

Unlike most of the other countries, available evidence shows that for a long time, Nigeria did not commit itself to knowledge generation, skill development and

---

17 Africa Union (AU)—Economic Commission for Africa (ECA) Africa Economic Report, 2013 (Accessed, Nov. 2, 2014).

a judicious utilization of the nation's human resources. Indeed, among the 20 largest emerging economies in 2014-2015, Nigeria stood at the 17th position, while its public tertiary institutions nose-dived from a world ranking of 122 in 2010-2011 to 132 in 2014-2015, out of 144 countries.[18]

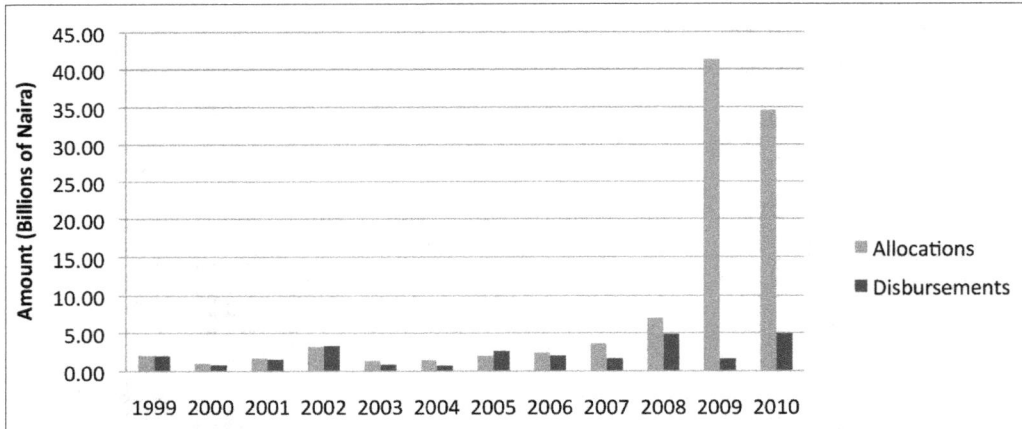

Figure 6:2 TETF Allocations and Disbursements to State and Federal Universities, 1999-2010[19] *(Source: http://www.tetfund.gov.ng)*

Many have finger-pointed a myriad of problems including chronic underfunding of tertiary education, as exemplified by Figure 6:2, as the key causes of the decay and lack of productivity in the nation's educational systems and research

---

18 The Global Competitiveness Report 2014–2015.

19 The main source of income available to the Tertiary Education Trust Fund (TETF) is the two percent education tax paid from the assessable profit of companies registered in Nigeria. The funds are disbursed for the general improvement of education in federal and state tertiary educations specifically for the provision or maintenance of:
- Essential physical infrastructure for teaching and learning
- Institutional material and equipment
- Research and publications
- Academic staff training and development and
- Any other need which, in the opinion of the Board of Trustees, is critical and essential for the improvement and maintenance of standards in the higher educational institutions (Source—http://www.tetfund.gov.ng—Accessed, August 15, 2015).

establishments.[20, 21, 22] Today, all hopes are pinned on the new administration that took over power in May 2015 to effect a change.

Underfunding of Research and Development (R&D)

While all of the above have focused exclusively on Nigeria's public tertiary institutions, there is also the need to examine the state of research and development (R&D) at the research-focused establishments of government, such as the agricultural, engineering, bio-technology and space research institutes and agencies, among others.

Funding of R&D in each and every nation is critical because our collective future, as global inhabitants that occupy the same planet, depends on our creativity innovations. That creativity is nurtured through R&D is no longer in doubt. Over the years, the problem, in Nigeria, has always been how to translate the results of R&D into products and services that can have economic impact and desired improvements in the quality of life of Nigerians.[23] That process requires policies, national budgets and operational programmes that can be intentionally implemented, over sufficiently long periods of time. The overriding concern and the unceasing complaint of Nigeria's research establishments and their staff echoed those of the tertiary institutions—that the Nigerian government has not provided necessary funding support that will nurture creativity in Nigeria and in Nigerians.

The 2010 review of fifteen (out of 21) Nigeria's Institutes of Agriculture reinforced all of the above concerns; the review was jointly conducted by the US-based International Food Policy Research Institute and the Agricultural Research Council of Nigeria (ARCN). The reviewers found that because of underfunding, the institutes suffer from decaying infrastructure and lack of qualified staff; they also noted that the institutes were failing to innovate, collaborate with farmers or monitor and

---

20 Federal Ministry of Education (2005). *Nigeria Education Sector Diagnosis—A Framework for Re-engineering the Education Sector*, Education Sector Analysis Unit, Federal Ministry of Education , Abuja, Nigeria, May 2005 (Project Coordinator: Dr. G.A.E. Makoju (Mrs.)

21 *Report of the Committee on Needs Assessment of Nigerian Public Universities*, Council Chamber State House, Abuja, November 1, 2012.

22 Humphreys, Sara with Lee Crawford (2014). *Review of the literature on Basic Education in Nigeria,* EDOREN—Education Data, Research and Evaluation in Nigeria, Abuja, Nigeria.

23 Odhiambo, Thomas R. (1994), *How can the scientific community support deep-rooted development in Africa:* Whydah, African Academy of Sciences Newsletter, Vol. 3, No. 8, June 1994.

evaluate their own work.[24] A 2016 report *"Agri[cultural] research spend[ing] fails to lift farmers' productivity (in Nigeria),"* also called attention to the underfunding issue. The report noted that only 10% of the budget was going for research while 90% was going to support the institutes' 13, 000 staff members.[25] A critical issue cited in the report is that *"most of the money allocated to these institutions in the budget doesn't get to the institutions. One-third of the total allocations finally gets to these institutions. The money gets into private hands, instead of the research institutions."* The case for the National Agency for Science and Engineering Infrastructure (NASENI) appeared in a Leadership Newspaper editorial as follows: *"It is ironic that past administrations have spent resources on job creation rhetoric while the very engine designed to spur that process is deliberately relegated to the back burner."*[26] The editorial concluded by urging President Buhari to "give this agency the serious attention it deserves."

As shown below, Nigeria's historical record of funding provided for R&D justifies these concerns. Parts of that record include the following:

- In 1980, the Organisation of African Unity (OAU) recommended a guideline of 1.0% of GDP for S&T activities; the guideline was agreed upon by all member States, including Nigeria, as a major component of the 1980 OAU Lagos Plan of Action. However, from 1980· to 1992, funding for science and technology stood at 0.15% of Nigeria's GDP. In that same period, Tanzania funded its S&T activities to the tune of 1.5-3.5%, while India and Malaysia respectively registered 2.0% and 6 .0% of their GDP.[27]

- Between 1990 and 2004, research funding by the government stood at 0.02% of GDP in Nigeria as compared to 0.92% of GDP in South Africa in 2005.[28, 29] Tunisia and Morocco spent over 1% and 0.8% of GDP on R&D respectively.

24 Ragasa, Catherine, Suresh Babu, Aliyu Sabi Abdullahi and Baba Yusuf Abubakar (2010). Strengthening Innovation Capacity of Nigerian Agricultural Research Organisations, Discussion Paper 01050, The International Food Policy Research Institute (IFPRI), December 2010.

25 Okojie, Josephine (2016). Agric. research spending fails to lift farmers' productivity (Nigeria). BusinessDay, Lagos, Nigeria March 16, 2016.

26 *NASENI And The Quest For Industrialisation*, Leadership Editors, http://leadership.ng/opinions/editorial/458409/, — *Sep 4, 2015 (Accessed, December 15, 2015).*

27 UNESCO Statistical Year Book, 1998 and 1992, Paris, France (Percentage of GDP devoted to Research and Development).

28 Field Work Data (2005). National University Commission (NUC) of Nigeria, Abuja, Nigeria.

29 Science and Engineering Indicators (2010).United States National Science Foundation, Washington, DC.

In several other countries, R&D expenditure approached the 1% of GDP mark—these included Brazil at 0.9%, India 0.7%, Iran 0.7%, Malaysia 0.7%, Chile 0.7% (UIS 2007).[30] The U.S. R&D/GDP ratio was 2.7% in 2007 and has fluctuated between 2.6% and 2.8% over the past 10 years.[31]

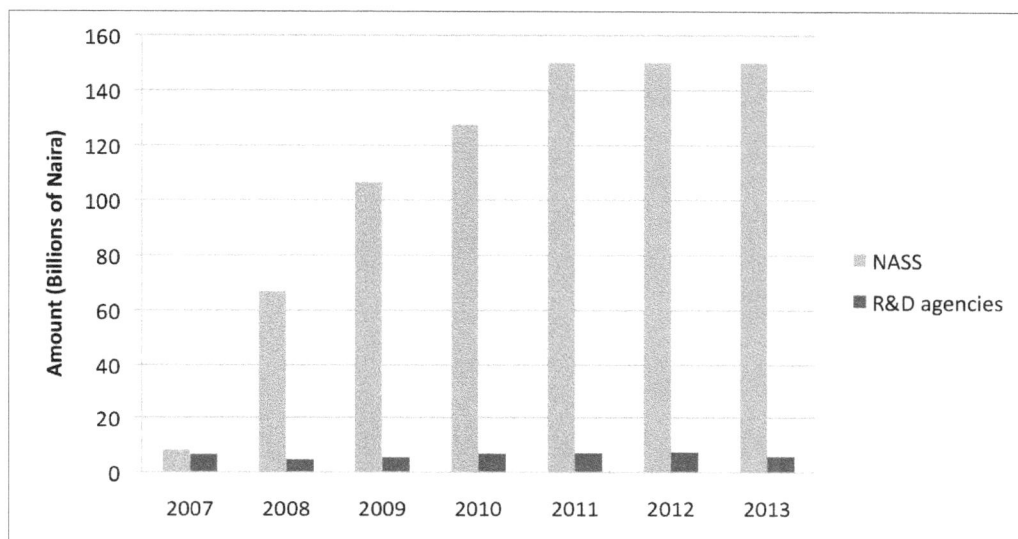

Figure 6:3 2007 to 2013 Budget Allocations to the National Assembly (NASS) and the key R&D establishments of government *(Source: Extracted from Nigeria's Federal Government Annual Budget records)*

- Figure 6:3 also illustrates the funding dilemma of S&T and related R&D in Nigeria; it shows the combined funds allocated, between 2007 and 2013, to a number of science and technology related establishments of government in Nigeria as well as to the Nigerian National Assembly (NASS) during the same period.[32] Characteristic of the allocations to the R&D establishments of government are their inconsistent nature, as the funding alternated between inconsequential increases and downgrades from one year to the next. But

30 Ellis, Simon (2008).*The Current State of International Science Statistics for Africa*, The African Statistical Journal, Volume 6, May 2008.

31 Science and Engineering Indicators (2010).United States National Science Foundation, Washington, DC. USA.

32 The science establishments consist of the National Space Research and Development Agency (NASRDA), National Agency for Science and Engineering Infrastructure (NASENI), the Biotechnology Institute (BIOTECH) and the National Mathematical Centre (NMATHC).

while the NASS approved a pittance of a budget for the R&D establishments of the nation, it nevertheless ensured that its own budget allocations enjoyed astronomical increases in each succeeding year. For example, a close look at the NASS allocations showed an incredible rise from ₦8.284 billion in 2007 to ₦127.800 billion in 2010; this amounted to an increase of 1,543% during the four year period, or an average of 386% increase per year. That was not all; the escalating financial allocations to the NASS later skyrocketed to a ₦150.00 billion level, annually, for each of the four subsequent years, 2011 through 2014; an unrelenting national campaign against the financial compensations of political office holders brought it down to N115.00 billion in 2015. With these allocations, Nigerian legislators are among the highest paid political officers in the world.[33]

Other Sources of R&D Support

Most of what is called research in Nigeria is done at the universities and at the national R&D establishments. The research input of the private sector (industries and the non-profit establishments) is paltry. In comparison with those in other countries, foreign industrial establishments that operate in Nigeria do not fund any measurable research locally, in Nigeria; instead, they conduct their research activities at their respective parent headquarters abroad.[34] After many years of complaints by Nigerians, against such practices, Shell and Total eventually instituted some collaborative programmes with a limited number of Nigerian universities, in support of their oil operations in Nigeria.[35, 36] External sources of funding for local research support also are irregular, and the universities themselves continue to struggle in their development of other independent and sustainable means for generating revenue.

Subsequent Problems in Nigerian Universities

In a society that aspires to be knowledge-driven, one would commend the vision of the authors of the 1977 National Education Policy (revised in 1998) that stipulated that:

---

33 Premium Times, Nigeria August 18, 2015.

34 Fadokun, James B. (2010). University research capacity in Nigeria and the challenges of national development in a knowledge-based economy, Journal of Social Scientist,, 5(1), 1-23.

35 http://www.shell.com.ng/environment-society, (Accessed October 30, 2014).

36 http://www.ng.total.com/06_total_nigeria, (Accessed October 30, 2014).

- A greater proportion of expenditure on university education shall be devoted to science and technology; and that

  Not less than 60% of [admission] places shall be allocated to science and science-related courses in the conventional universities and not less than 80% in the universities of technology; (Section 55, a—b).[37, 38]

Unfortunately, except for a short interlude, thereafter, the patterns of enrolment and graduation in the nation's universities have indicated otherwise; indeed, they were in sharp contrast to the above encouraging policy guidelines that were in favour of building a science and technology culture in Nigeria. Indeed, the reverse seemed to have been in effect since the introduction of the education policy, whereby more than 60% of students are being admitted for programmes in the humanities and less than 40% are pursuing science and engineering education. In essence, *there have often been gaps between rhetoric and reality, promise and provision, investment and productivity in science education in the country since the National Education Policy went into effect in1977.*[39] The aforementioned gaps of old are canyons, today.

At the programme level, there are no adequate funds to purchase teaching and learning materials in such core science fields as mathematics, physics, chemistry, biology and integrated science, and to equip the science laboratories, in the nation's secondary schools and universities, particularly the public institutions. While practical work is universally accepted as being central to the teaching and learning of science and in the development of pupils' understanding of scientific processes and concepts, the nation has reached a stage where many of its science teachers are not skilled enough to teach science subjects nor adequately equipped to organise and lead the related laboratory exercises.[40] The end result is that many students are receiving only theoretical science and technical education with *"alternative to practical"* as a substitute

37 Saint, William, Teresa A. Hartnett, Erich Strassner (2004). *Higher Education in Nigeria: A Status Report*,World Education News and Reviews, September 1, 2004.

38 Arikewuyo, Olalekan (2004). *Democracy and University Education in Nigeria: Some Constitutional Considerations,* Journal of the Programme on Institutional Management in Higher Education, OECD, Vol. 16. No. 3, pp. 121-133.

39 Balogun, T. A. (1982), *Science, Society and Science Teaching Effectiveness in Nigeria,* Journal of Science Teacher Association of Nigeria,2(1), pp. 14-20.

40 Adeyemo, Sunday (2010). Teaching/Learning of Physics in Nigerian Secondary Schools: The curriculum transformation issues, problems and prospects, International Journal of Educational Research and technology, Vol. 1 [1], June 2010, pp. 99-111.

for real practical laboratory work.[41] Today, less than 20% of university entry students are opting for the sciences and performance in the sciences have plummeted. These are just the *tip of the iceberg*.

In addition to the perennial problem of inadequate funding, other major areas of national concern include the following: Infrastructural decay, loss of university autonomy, overburdened academic staff, declining morale of staff and students, declining academic productivity, problem of institutional governance, the dead-lock between the Federal Government of Nigeria (FGN) and the Academic Staff Union of Universities (ASUU), with the students, the academic staff, the universities and the Nigerian society as the victims of the dead-lock. Here are some of the key issues.[42, 43, 44, 45, 46,]

- **Inadequate infrastructures and infrastructure decay:** The 2012 Report of the *Committee on Needs Assessment of Nigerian Public Universities* confirmed, very graphically, the dilapidation of the nation's public universities and the attendant poor teaching and learning environment therein. The latter includes, *among others*, poor research and library facilities, and non-functional instructional and infrastructural facilities.

- **Loss of university autonomy:**   The long period (35 years) of military rule in Nigeria not only adversely affected the psyche of the citizenry, it also militarized the university system in the country. Many academics were dismissed, retired and

---

41 In the absence of actual laboratory equipment, materials and specimens, students receive their science education (in biology, physics and chemistry) and are subsequently tested, through illustrated drawings etc. without actually conducting live laboratory experiments designed to reinforce such an education.

42 Olorode L (2001). Democratic imperatives and higher education in Nigeria: The quest for social justice. Proceedings of the 12th General Assembly of the Social Science Academy of Nigeria, 29-36.

43 Arikewuyo, Olalekan (2004). *Democracy and University Education in Nigeria: Some Constitutional Considerations,* Journal of the Programme on Institutional Management in Higher Education, OECD, Vol. 16. No. 3, pp. 121-133.

44 Report of the Committee on Needs Assessment of Nigerian Public Universities, Council Chamber, State House, Abuja, November 1, 2012.

45 Ezenyilimba, Emmanuel (2015). *The Challenges of Tertiary Management Education and Strategic Learning Panacea in Nigeria,* International Journal of Academic Research in Business and Social Sciences June 2015, Vol. 5, No. 6.

46 Controversy Over ₦1.3tr Fund For Nigerian, The Daily Independent Newspaper, Lagos Nigeria, May 4& 7, 2015.

even jailed unjustly by the past military junta. Similarly, many student leaders were arrested, detained or dismissed from universities without being subjected to clearly documented disciplinary procedures. In the process, university governance became unpredictable and university finances were often in shambles."

- **Overburdened academic staff:** There has been an explosive student population with no matching increase in the academic staff population. This has been compounded by the establishment of new universities although the nation is still struggling to meet the needs of the existing ones. Consequently, the workload of the academic staff on duty has risen sharply with associated deteriorating staff/student ratios.

- **Decline in academic productivity:** The loss of autonomy soon translated into a decline in morale of both the staff (academic and administrative) and students with consequent negative impact on teaching and research efforts, and associated productivity. With the passage of time, incentives and research productivity, teaching excellence and associated innovation dwindled, and in some cases, disappeared.

- **Declining morale:** The waning attractiveness of academic careers in the absence of meaningful research activities and the difficulty of maintaining one's intellectual capability while in isolation from rapid global advances in disciplinary knowledge. Salaries and other benefits are not relevant to prevailing economic conditions, and, on several occasions, the little that is offered is not paid when due.

- **Problem of institutional governance:** One of the leading problems at the universities is the quality of leadership and governance which affects the prioritization of resource allocations and limited funding. Accountability, quality assurance and monitoring of performance also do not receive the attention they deserve. Management innovation is spoken of but not pursued nor practiced. Similarly, little attention is given to institutional operations in terms of graduate and research output, while admissions are based on quotas rather than merit. Since Nigeria cannot continue to operate in isolation, the nation is faced with the challenge of how to bring its own higher education system into concert with higher education practices within the global community.[47]

---

47 Ezenyilimba, Emmanuel (2015). *The Challenges of Tertiary Management Education and Strategic Learning Panacea in Nigeria,* International Journal of Academic Research in Business and Social Sciences June 2015, Vol. 5, No. 6.

- **Relationship between the FGN and the ASUU:** For many years, ASUU has been asking the FGN for changes in the educational policy and funding that could result in the enhancement of the nation's education. Amongst ASUU concerns are urgent need to upgrade the facilities at the academic institutions, proper funding for research, and a redress of the poor salary and conditions of service. But it took 20 years (13 years after the country returned to civilian rule) and 19 strikes before ASUU finally got a national hearing on its grievances. But by then, much damage has been done, as attested to by the hearing's submission, titled: *Report of the Committee on Needs Assessment of Nigerian Public Universities.* The report, which vindicated ASUU, concluded that: *"Physical facilities for teaching and learning in Nigerian Universities are Inadequate, Dilapidated, Over-stretched/over-crowded, and Improvised."* Unfortunately, up to the time that President Buhari's administration took office on May 29, 2015, ASUU claimed that the government had not released any part of the ₦1.3 trillion it promised as a condition for ending the 2013 strike; this was an agreement reached between the FGN and ASUU following the release of the report cited above.

Consequences
For the Nation:
- Continuing decline in the quality of education in the country;
- Inability of the nation to produce the industrious and efficient workforce needed to meet the objective of its development agenda, including Vision 2020 and the Millennium Development Goal;
- Inability of the nation's private sector to recruit qualified employees among local applicants. The alternative is for them to shell out the funds needed to retrain the graduates to a level that will make them employable and enable them to perform effectively and productively on the job;
- Possible relocation of industrial establishments into neighbouring countries with superior educational system , and more job-ready graduates;
- Loss of revenue to Nigeria and its academic institutions, as Nigerian students spend, in excess of ₦400 billion, annually, to pursue their academic goals and dreams in foreign universities.[48]

---

48 Adesulu, Dayo (2015). *Why Nigerian students patronise foreign schools,* Nigerian Vanguard Newspaper, October 22, 2015—In 2009 alone, Nigerian students spent a total of ₦246 billion on education in the

(Continued On Next Page)

For Nigerian Universities and their Staff:

- The quality of education imparted and received within a non-conducive environment, as prevails in Nigeria's universities is universally adjudged as poor;
- There has been an increased rate of labour attrition at the universities with an estimated 30% of approved academic positions being presently vacant in a number of them;
- Research output has plunged, the desire to teach and to learn plummeted, and the quality of education declined;
- Low ranking/rating of Nigerian universities regionally and globally. *The top Nigerian university [University of Ibadan] was ranked* [1296 by Ranking Web of Universities], *and that is very uncomfortable. We hope that we get to 1000 soon. But at the moment, it is only South African and Egyptian universities that are in the top 1000 of the world, no Nigerian university and we cannot be celebrating that;*"[49,50]
- Nigerian universities, which used to attract many international students, have lost ground to other tertiary institutions abroad, particularly to South Africa. They are also losing their credibility, at home, as academic institutions-of-choice among aspiring Nigerian students; and

Nigeria's universities may not be able to attract collaboration with top ranking academic and research institutions in the world.

For the Students:

- *The fear and danger of being irrelevant:* A major challenge for an aspiring student is how to remain focused and productive in *an industrial strike* environment. When the strikes are called-off, many students often find that they are not able to cope with the subsequent flood of academic assignments.

United Kingdom. In 2014, according to the Nigerian Vanguard Newspaper, about 75,000 Nigerians were said to be studying in Ghana, paying about US$1 billion annually as tuition fees and upkeep, as against the annual budget of US$751 million for all Nigeria's federal universities. The amount was about 70 per cent of the total allocation in 2008 to all federal universities.

49 *Our education standard has embarrassingly fallen—TET-FUND Scribe [*Prof. Suleiman Elias Bogoro*]*, The Nation, September 12, 2015 http://thenationonlineng.net (Accessed, November 12, 2015). The second and third Nigerian universities in this global ranking were Covenant University, Ota (2027) and Obafemi Awolowo University, Ile-Ife (2119).

50 2015 Ranking of the world universities—http://www.webometrics.info/en/Africa/Nigeria].

It is also not uncommon that many initially promising students are not able to make the grade at graduation. At that time, such students are confronted with un-employability and the possibility of their own irrelevance both to themselves and the Nigerian society.

• *Be prepared for post-graduate re-training:* Because of today's poor quality of education in the nation, graduating from a university and being able to land a job, immediately thereafter, in Nigeria, is becoming a thing of the past. Nigeria's university graduates of today most likely would need to undergo a one-year retraining programme before they can be offered employment, particularly in the nation's private industry. That is the operating requirement from the Dangote Industries,[51] which already operates the Dangote Academy to re-train the nation's graduates for 12 months before they are employed. Shell similarly offers a 12-month re-training programme, known as Shell Intensive Training Programme (SITP), for fresh graduates before they can be considered for employment. Chances are that many other companies will follow their examples.

Measures of Nigeria's Productivity

The persistent financial neglect and associated mismanagement that were cited above have eroded the foundation of the nation's education, research and development. The attendant consequences elaborated upon earlier and more have negatively impacted the productivity of Nigerians and the overall growth and development of the nation. Nigeria's position, in comparison with six other African countries, Algeria, Egypt, Ghana, Kenya, Morocco and South Africa, on a number of globally accepted measures used in adjudging the state of a nation's S&T capability and related productivity, are illustrated in Figures 6:4 through 6:8 of the Annex to this Chapter. The indices are (i) Figure 6:4—The number of scientific and technical published research works, (ii) Figure 6:5—Number of patents held, (iii) Figure 6:6—Number of internet users (per 100 people), (iv) Figure 6:7—Number of fixed broadband subscribers, and (v) Figure 6:8—Amount of ICT exports.

A significant part of the failings underscored by these figures can also be attributed to *"lack of implementation of the country's Science, Technology and Innovation Policy,"*[52]

---

51 As of 2015, Alhaji Aliko *Dangote* was regarded as the most successful businessman on the African continent. His main business areas are in commodities: cement, sugar, flour and petroleum.

52 *Nigeria: Why Nigeria is not making progress in Global Innovation Index,* ThisDay, Lagos, June 29, 2014.

to the nation's minimal efforts in knowledge generation and innovation, in its inability to add much value to its raw materials, and its weak infrastructures. We must also recognise that within the international community, these indices influence the choice of educational, research, business, industrial and political partnerships and associations. Unless these measures significantly improve, these and other indices may constitute a bottleneck for Nigeria in case it seeks membership in a number of specific international organisations, such as BRICS,[53] MIKTA,[54] and OECD[55]—entities whose focus includes mutual cooperation on science and technology, research and development, economic programmes and global security.

Impact on Nigeria's Industries

A large measure of a nation's productivity is the state of its industries and its ability to transform its raw materials into finished goods and marketable products. Akinrele reminds us that:

> *The commercialisation of research findings is dependent upon the existence of such services (elements) as product and processes development, pilot plant, plant design and installation, process adjustment, advice on manufacturing, operations, quality control and product and process improvement.*[56]

Akinrele also noted that the Federal government established both the Raw Materials Research and Development Council (RMRDC) and the National Agency for Science and Engineering Infrastructure (NASENI) to commercialise the results of industrial research. He concluded that their mandates notwithstanding, both RMRDC and NASENI do not have the composite disciplines enumerated above in their corporate profiles. Accordingly, the effectiveness of the two bodies is limited in translating research results into goods and services. In the absence of any meaningful

53 BRICS—An association of five major emerging national economies, consisting of Brazil, Russia, India, China and South Africa, It had its first Heads of State Summit in 2009.

54 MIKTA—An association of five medium level powers, consisting of Mexico, Indonesia, South Korea, Turkey and Australia; they first met in 2013.

55 OECD—The Organisation for Economic Co-operation and Development (OECD), with a membership of 34 countries today. It was originally established as a European organisation in 1960. It subsequently expanded and its membership now spans the globe, from North and South America to Europe and Asia-Pacific.

56 Akinrele, I. Adedayo (2004). *Technology, resource endowment and national development,* The 9th Annual Guest Lecture of Stephen Oluwole Awokoya Foundation for Science Education, Lagos, Nigeria, March 15, 2004.

industrial production at home, Nigeria's purchases from abroad are arriving daily and are crowding the nation's seaports and airports.

Nigeria's industrial landscape is also plagued by many other problems, particularly the cost of doing business in Nigeria; on top of the list are unstable electricity and the high cost of automotive gas and diesel fuel, a necessity to power manufacturers' stand-by generators and boilers. These negative energy factors constitute major road-block to the nation's economic development. In mid-October 2013, the Director-General of the Centre for Management Development noted that, *"Currently, there are about 60 million generators in Nigeria at the ratio of one per household of 2.5 people with an annual spending of ₦ 1.6 trillion,"* while the national power grid serves as stand-by.[57] That was in 2012. In 2015, Nigeria's energy situation was up-dated as follows: *Unreliable power supply leaves Nigerians with no other option than oil-based expensive generators—60 million Nigerians rely on generators on which they spend an average of ₦3.5 trillion a year ($17.5 billion US dollars).*[58] The electricity crisis has been given wide publicity in Nigeria and abroad, and has been credited for the exodus of Nigeria's industries into neighbouring countries, particularly Ghana. Some concrete figures might shed light on the energy debacle. According to the World Economic Forum, the electricity production (Kwh/person) in the following seven African countries is as follows: South Africa (5,033), Egypt (1,972), Algeria (1,356), Morocco (776), Ghana (451), Kenya (187), and Nigeria (165).[59] But there are other compelling reasons.

In addition to energy-related problems, the comparative indicators shown in Figure 6:9 (venture capital opportunities, quality of mathematics and science education, and quality of government delivery of on-line services, amongst others) provide the justifications for the exodus of the companies. According to the Manufacturers Association of Nigeria (MAN), over 820 manufacturing companies closed down in Nigeria between 2000 and 2008 because of electricity problems and the absence of industrial raw materials.[60] To-date, major companies such as Dunlop Nigeria Plc and Michelin have taken similar steps; others are in the contemplating mode. Cadbury

---

57 Daily Independent, Abuja, Nigeria, June 6, 2013.

58 Wiebusch, Elisabeth (2015). Nigeria's Achilles Heel http://www.energytrendsinsider.com) (Also posted as *Nigeria's Future Depends On Electricity* at http://oilprice.com by Robert Rapier on February 25, 2015).

59 World Economic Forum, Executive Opinion Survey, 2013 and 2014 editions.

60 Borodo, Alhaji Bashir (2010). Address as President of the Manufacturers Association of Nigeria the Association's 37th Annual General Meeting, Lagos, Nigeria.

Nigeria Plc, Unilever and the International Institute of Tropical Agriculture (IITA) opted for sizable staff retrenchment because of the high cost of production, decaying infrastructure as well as the ravaging global economic recession."

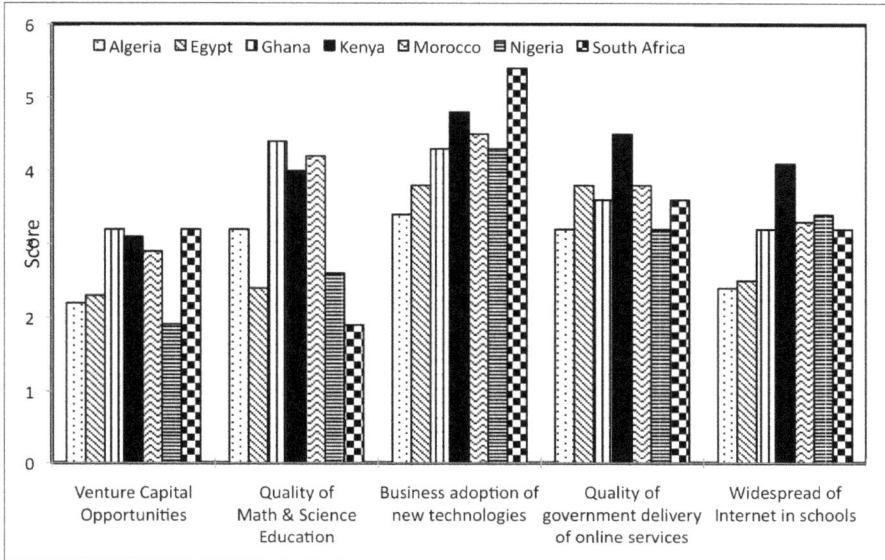

Figure 6:9 Some of the main factors that govern the location of industries *(Source: World Economic Forum, Executive Opinion Survey, 2013 and 2014 editions)*

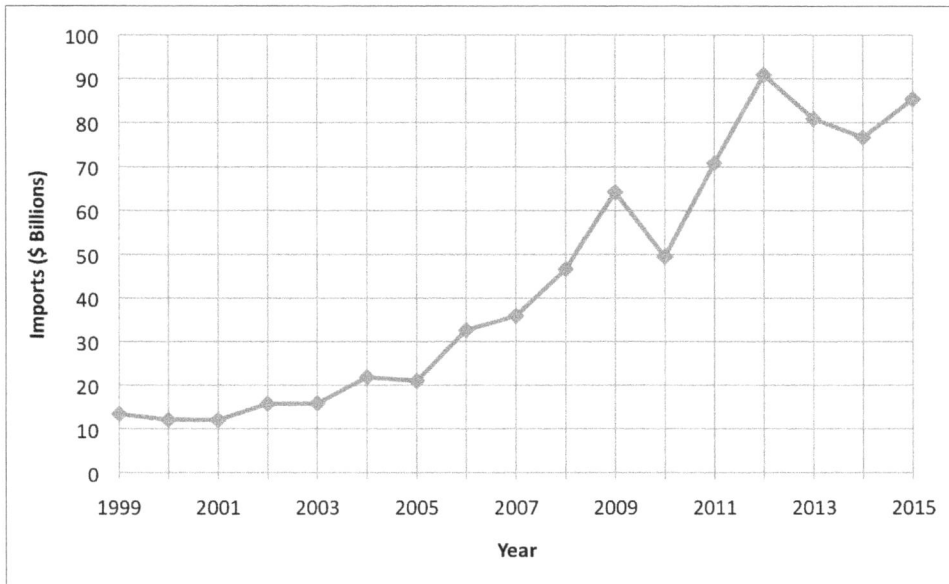

Figure 6:10 Nigeria's import of goods and services from 1999—2015 *(Source: Data.worldbank.org)*

Consequently, Nigeria continues to decline from an industrialising country to a trading economy—with the subsequent influx of manufactured goods, as shown in Figure 6:10, particularly from China, Europe and Asia and other economies, into the country, to compensate for our local non-productivity. These imports include all manner of tools, equipment as well as goods and services which have been sold to the nation's leaders as technology transfer that would foster technology development at home. But that is not happening because Nigeria has not developed the technological capacity to keep its turn-key projects and imported equipment running, without going back to the original overseas source(s) for more parts (hardware and software) and labour (otherwise known as experts).

A major victim of this approach to development is the failed Ajaokuta Steel Company Ltd (ASCL) which was originally established in 1971, with a design capacity similar to that of the Pohang Steel Company Limited (POSCO) in South Korea, which was established in 1968. Today, ASCL is yet to be completed; thereafter, it will go into an effective operation.[61] Meanwhile, Pohang Steelworks was completed in 1983 after four expansion projects producing 9.1 million tons of crude steel per year.[62] Like iron and steel, an automobile industry is a major technological driver and a catalyst to a nation's economic development. Many of the newly industrializing countries, such as Brazil, India, Indonesia, Malaysia, Mexico, South Korea and [South Africa] are following that route.[63] But in Nigeria, "four of the [nation's] six automobile plants have collapsed and the two surviving ones are operating at less than 10 per cent capacity; the three paper projects have closed down; the five steel projects are in comatose; the petrochemical and fertilizer plants are operating epileptically and the small scale industrial base has collapsed."[64]

Responsibility for the Nation's S&T Debacles

The responsibility for the state of S&T in Nigeria can be found in the analysis of the pronouncements of senior officials of the Federal government that Nigeria was

---

61 *Buhari passionate about completion of Ajaokuta Steel*, Vanguard, Lagos, December 15, 2015.

62 Today, POSCO operates 23 overseas subsidiaries and affiliates in 13 countries; it also operates 25 domestic subsidiaries and affiliates in areas such as construction, architecture. http://www.posco.com.pl/pages/39.

63 Agbo, Cornelius O. A. (2011). *A critical evaluation of motor vehicle manufacturing in Nigeria*, Nigerian Journal of Technology, vol. 30, no. 1, March 2011.

64 Okafor, Emeka E. (2013). Corruption and implications for industrial development in Nigeria, African Journal of Business Management, August 7, 2013, http://www.academicjournals.org/AJBM.

rethinking its failed science policy, that Nigerian science needed better links with industry, that Nigeria's agricultural research institutes were failing, and that Nigeria's academies do not contribute enough to development. [65, 66]

Science and Technology Policies

First, let us take a look at the nation's S&T policy documents. Any policy will succeed when its aims are clear and succinct, when it is budgeted for and when the nation that develops the policy also commits itself to its implementation. The nation's past S&T policies were criticised on many counts, the primary objection being that they lacked legal status, thus they were not enforceable. According to the critics of the third Nigeria's S&T policy[67], it was put together without wide consultation. It also did not support the goals of other policies, such as Nigeria's two major plans—the National Development Plan and Nigeria's Vision 2020—which aim to turn the country into a major economy within this decade. The government that called for its development not only failed to implement the policy as designed, it also did not allocate the budget stipulated in the policy document for its implementation.

The revised S&T policy which was approved by government in February 2012 also contains a number of troubling issues. High on that list is its call for the establishment, at each state level, of 'State Science, Technology and Innovation Council (SSTIC) and the Ministries of Science and Technology.' This proposal reminds me of a lecture given, in 1974, at the University of Ife (renamed Obafemi Awolowo University), by Prof. Anthony Adegbola, of the university's faculty of agriculture, on '*Food Science*.' He reminded his audience of the array of agriculture-related sciences at the university without any corresponding food to show for it. Similarly, Nigeria does not need any more bureaucracies and institutions that shuffle files around. The nation needs institutions and establishments that can produce actual goods and services that will meet the needs of the Nigerian people. This should be feasible without establishing S&T ministries and related administrative personnel and overhead burdens, at each state level. In Brazil, the State of São Paulo Research Foundation (FAPESP)

---

65 Abutu, Alex (2010). *Nigeria rethinks 'failed' science policy,* SciDev.net, September 17, 2010 (Accessed, August. 5, 2013).

66 Abutu, Alex (2011). *Nigeria's science minister calls for demand-driven R&D,* SciDev.net, March 8, 2011 (Accessed, August. 5, 2013).

67 The first and second National Science and technology Policies were developed in 1986 and 1997 respectively.

offers a worthy example that is elaborated in Chapter IX of this book. Furthermore, the list of members of the proposed National Council on Science, Technology and Innovation (NCSTI), as described in the revised 2012 S&T policy document, does not include any cognate S&T tertiary institution, the fountain of the nation's S&T knowledge. Similarly, both the Nigerian Universities Commission and the Committee of Vice-Chancellors that actively participated in the development of the policy are also missing in the list. If anyone qualifies to serve on the council, the honour belongs to the representatives of the key S&T cognate institutions in Nigeria.

The Academies

The national academies, and in particular, the Nigerian Academy of Sciences, the Nigerian Academy of Engineering, the Social Sciences Academy of Nigeria, the Nigerian Academy of Education and the Nigerian Academy of Letters were not spared the rebuke of the minister of science and technology. The admonitions he proffered against them, in 2011, included the following:

- *Academies have existed for over a decade but their impact [has]not been felt so far in the efforts to accelerate the country's development;*

- *In most countries, national academies serve as the engine room for development but the situation in Nigeria is different. Here the academies are reluctant to take up challenges [in spite of] all the knowledge and experience that abound within them; and*

- *Nigeria's academies should collaborate more closely with the government and play a more critical role in advising on sustainable development.*[68]

These are very serious charges, but first, let us get to know the academies. They are made up of very hard working and capable individuals that are very rich in knowledge and skills that sometimes go un-recognised and therefore often untapped and un-utilised in a number of countries. Membership in each academy is normally based on academic, scholastic and professional accomplishments as judged by peers. In countries such as the United States, the United Kingdom, India, and South Africa as well as in Nigeria, the national academies are such establishments. In each of these countries, these academies are non-governmental organisations of distinguished scholars who, when given the opportunity, usually play a pivotal role in shaping the S&T–related and other policies and programmes of their respective nations.

---

68 Abutu, Alex (2010). *Nigeria's academies 'don't contribute enough to development*, SciDev.net, March 26, 2010 (Accessed, August 7, 2013).

For example, the United States National Academies, in their mission statement, declare aspects of the role they play as follows:

"Provide expert advice on some of the most pressing challenges facing the nation and the world. Our work helps shape sound policies, inform public opinion, and advance the pursuit of science, engineering, and medicine."[69]

In furtherance of these goals, in 2010, the National Academies Press, the publishing arm of the USA's National Academies, published a book titled *America's Future in Space—Aligning the Civil Space Program with National Needs;* it was authored by the National Research Council of the National Academies[70].

Members of the United States Congress always employ the services of their congressional staff to consult with such academies, research their books and related publications and use their findings to prepare briefing materials on issues of national importance, such as S&T and space. In turn, the congressmen and women study these briefing materials and familiarize themselves with all the related issues in order to be able to make informed contributions to congressional floor debates. Such informed minds shape the subsequent approval or rejection of policies, programmes and budgets of various national entities including the National Aeronautics and Space Administration (NASA) and the National Oceanic and Atmospheric Administration (NOAA) and their affiliated institutions and centres.

To gain additional understanding of the role of academies in other parts of the world, first, let us note that the Academy of Sciences of South Africa (ASSA) is, by law, obligated to address South Africa's parliament once a year. At such a time, the ASSA submits to the parliament an annual report on the Academy's vision of science and technology needed to power the nation's economy and ensure national safety and security of the Republic of South Africa. In Uganda, also by law, the funds for running the Academy of Sciences of Uganda is part of the country's annual budget for science and technology development in that country. In both Uganda and South

---

69 http://www.nationalacademies.org/(Accessed, August 8, 2013).

70 The United States National Academies is made up of three organisations: National Academy of Sciences, National Academy of Engineering and the National Academy of Medicine. The National Research Council was organised by the National Academy of Sciences in 1916 to associate the broad community of science and technology with the Academy's purposes of furthering knowledge and advising the federal government. Functioning in accordance with general policies determined by the Academies, the Council has become the principal operating agency of the National Academy of Sciences, the National Academy of Engineering and the National Academy of Medicine in providing services to the government, the public, and the scientific and engineering communities.

Africa, the governments provide needed support and facilities that enable these national academies to function as mandated by law.[71] The 1958 Scientific Policy Resolution of India, which is still current, proclaims and declares, *inter alia*:

> *It is only through the scientific approach and method and the use of scientific knowledge that reasonable material and cultural amenities and services can be provided for every member of the community....... The Government of India has decided to pursue and accomplish these aims by offering good conditions of service to scientists and according them an honoured position, and by associating indigenous scientists with the formulation of (national) policies.*"[72]

Nigeria has its own equivalent academies; and as it is the practice in many other societies, both the Nigerian Academy of Sciences and the Nigerian Academy of Engineering were invited to contribute to the development of the 2012 S&T policy document. The 'foreword' in the 2012 S&T policy document gave credit to the contribution of the Nigerian Academies of Sciences and Engineering in the preparation of the policy document; it also listed them as members of National Council on Science, Technology and Innovation (NSCTI). However, an analysis of the functions of the NCSTI outlined in the policy document showed that the Nigerian academies can do much more. They should be given opportunities to serve the nation in similar categories as other academies in other countries, such as:

- Provide continuing education to the nation's decision-makers and members of the Nigerian parliament on the role of science and technology in development;
- Have a statutory obligation to inform the nation, annually, through the parliament, on the state of science and technology in Nigeria;
- Participate in related budget formulation; and
- Contribute, at other levels (federal and state), as appropriate, to key policy advisory activities of government that are within and beyond the mandate of the NCSTI.

The Challenge

Given all of the above, what is Nigeria's fall-back position as a nation? Every Nigerian should be challenged by the view that characterizes Nigeria as '*that country*

---

71 Personal communication on April 15, 2010 with Dr. Akin Adubifa, Former Executive Secretary of The Nigerian Academy of Sciences, Lagos, Nigeria.

72 Scientific Policy Resolution 1958, Doc. No. 131/cf/57 of 4th March 1958/13th Phalguna, 1879, Department of Science and Technology, Ministry of Science and Technology, Government of India, New Delhi.

which has everything and produces nothing;' a pronouncement that aptly reflects the consequences of the nation's persistent failings over the years. Blame it on underfunding, lack of accountability, bad governance and many other factors; the fact is that in the past three or more decades, the S&T neglects addressed in this chapter, including the many layers of other local Nigerian factors, have stifled the nation's development and its effort to invest in its own future. These neglects have also undermined the ranking of Nigeria's universities and research institutions, both in Africa and in the world. The educational systems and the institutions established to generate and develop knowledge would need to be urgently revamped before the nation can productively embark on its developmental journey. That system is in tatters, with many casualties to show for it. That is *our nightmare, today.*

Nigeria cannot continue to stand still and let the world go by. Indeed standing still means being left farther and farther behind; inaction is also not a fall-back position. The future and the well-being of millions of Nigerians depend on our efforts today in embracing science and technology as the indispensable and incontrovertible tool of development. How to revive it in Nigeria, as shown in Chapter IX, is our next challenge. Meanwhile, the nation's experience on its space journey, without the benefit of a sound S&T foundation, is the focus of the chapter that follows, Chapter VII.

* * *

Appendix to Chapter VI—Standard indices used in assessing the state of S&T and development in a given country[73]

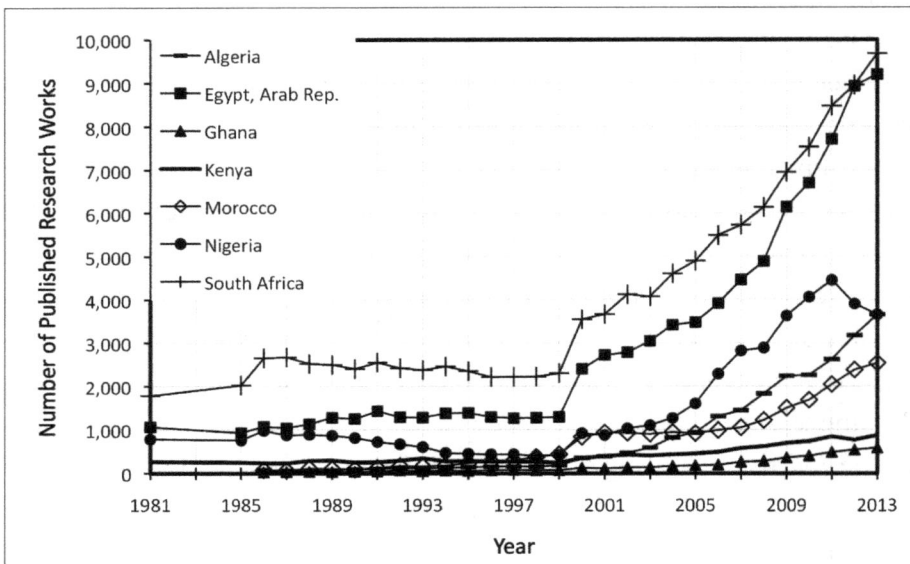

Figure 6:4 Scientific and technical publications in certified journals

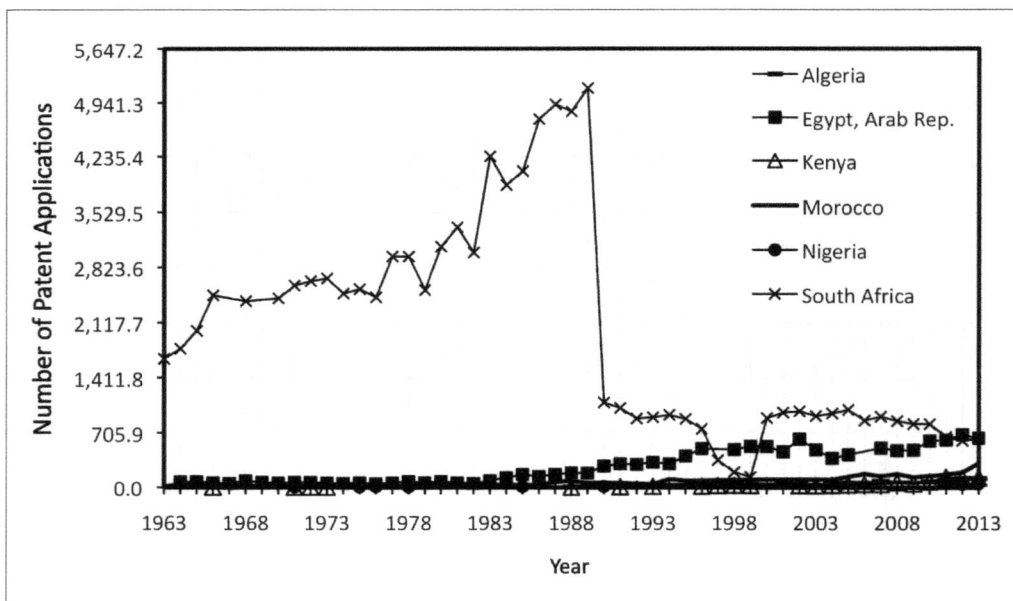

Figure 6:5 Number of patents awarded

73 Credit: World Economic Forum, Executive Opinion Survey, 2013 and 2014 editions

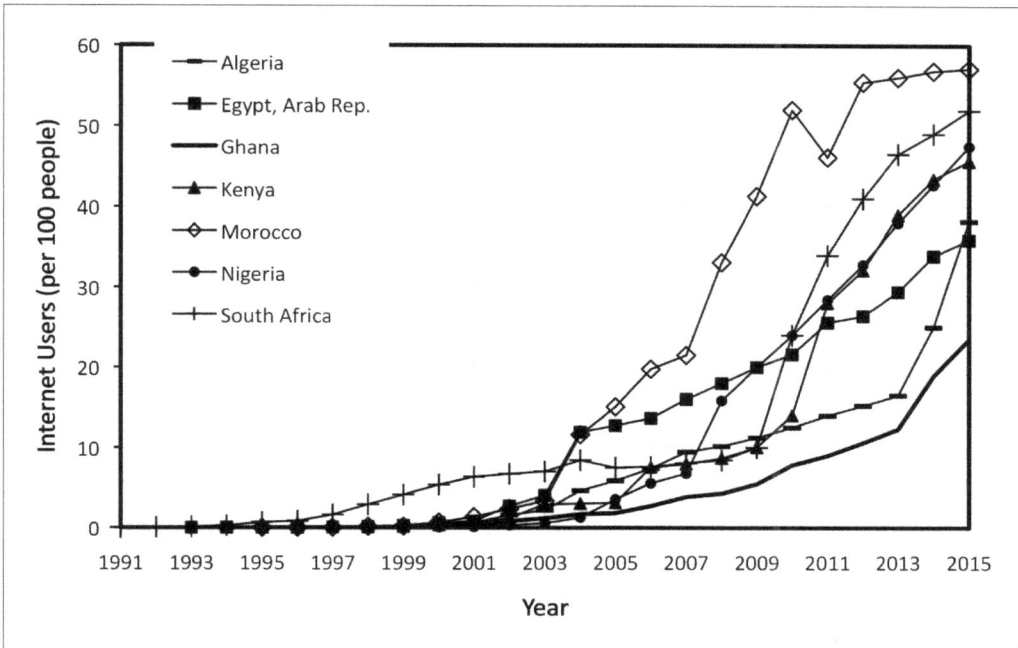

Figure 6:6 Number of Internet Users (per 100 people)

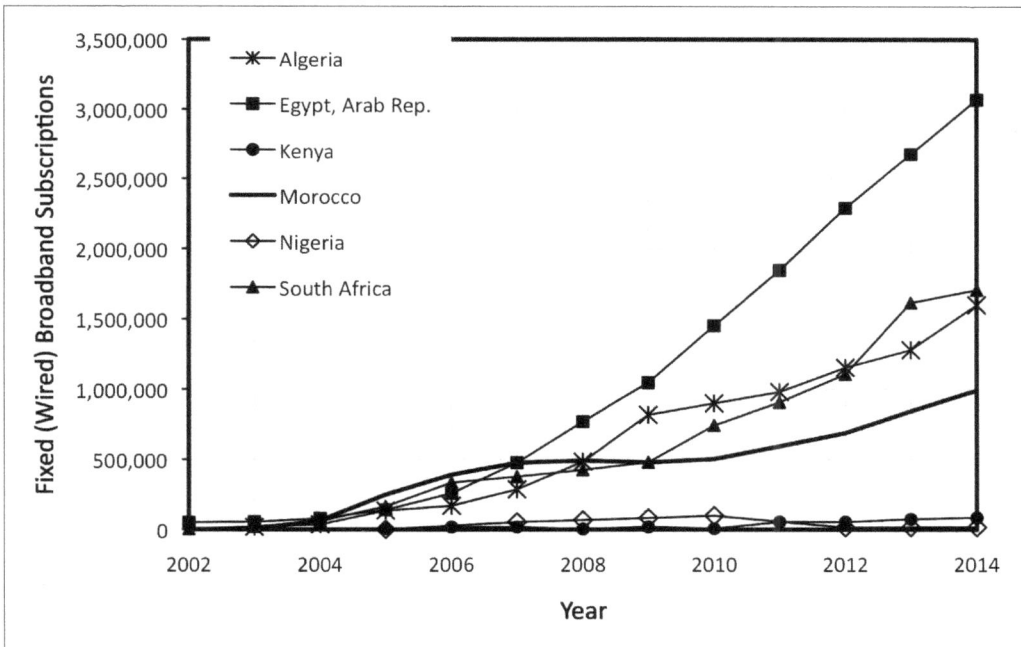

Figure 6:7 Number of Fixed Broadband subscriptions

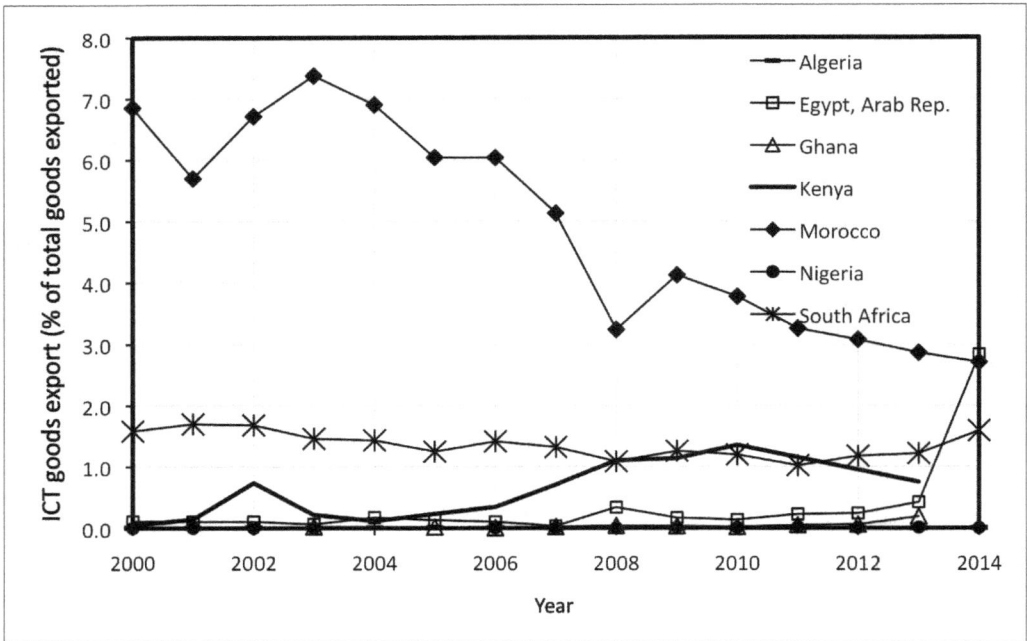

Figure 6:8 ICT Export as percentage of total goods exported

Chapter VII

THE STATE OF NIGERIA'S SPACE JOURNEY

Any journey in life requires some amount of preparation, namely, the destination, the proper tools needed for the journey, and the leadership with requisite knowledge and experience that will safely guide the journey to its destination.

In the case of Nigeria's Space Journey, the country's destination was defined in its National Space Policy document as follows:

To make Nigeria build indigenous competence in developing, designing and building appropriate hardware and software in space technology as an essential tool for its socio-economic development and enhancement of the quality of life of its people.

As indicated in Chapter V, on May 5, 1999, the Federal Executive Council (FEC) established the National Space Research and Development Agency (NASRDA) as the space organisation for Nigeria and appointed its first Director-General. On July 4, 2001, the same FEC adopted the National Space Policy and Programme document, the first tool of the journey. Other tools subsequently came on line; these included (i) the 25-Year Road Map for Nigeria's Space Mission (2005-2030) that was developed by FMST upon the directive of the President, (ii) the NASRDA Act 2010 that was adopted by the National Assembly, (iii) the training of NASRDA staff at academic institutions in the United Kingdom, the USA and China, (iv) the establishment, by NASRDA, of NigComSat Ltd in 2006 and GeoAppsPlus Ltd in 2007, and (vi) the disbursement of annual budgetary allocations to NASRDA by the Federal Government of Nigeria. With the goal, the leadership and the tools in place, what is the status of the Nigerian space journey to-date?

The Purchased Satellites Plus One

Once the government succumbed to the pressure, as stated in Chapter V, that it *"must treat this project with some urgency so as to position itself properly, as there are signs that an international movement is underway to restrict access to satellite imagery by developing countries,"*[1] the coast became clear for FMST/NASRDA to implement the terms of the Agreement it signed with SSTL. Thereafter, SSTL took charge of the development, building and launching of Nigeria's first and second Earth observation satellites, NigeriaSat-1 and -2; it also provided guidance in the development and building of NigeriaSat-X and arranged for its launch along with NigeriaSat-2.[2]

The Earth Observation Satellites

On September 27, 2003, Russia launched Nigeria's first Earth observation satellite, NigeriaSat-1, into a polar orbit, in space, approximately 600 kilometres (km) above the Earth's surface. It had a revisit period to the same spot on Earth every 3 to 5 days, collecting earth resources data of Nigeria and the global community, with its 32 metre(m) medium spatial resolution sensor system.[3] A follow-up satellite, NigeriaSat-2, also purchased from SSTL, was also launched into a polar orbit from Russia, along with NigeriaSat-X, on August 17, 2011. NigeriaSat-2 carried onboard three sensor systems which included a 2.5 metre panchromatic, a 5metre multi-spectral high resolution sensor and a 32 metre multi-spectral scanner, the latter, to ensure continuity with NigeriaSat-1, as well as enhance the quality of data that are available for use, through NASRDA, in Nigeria and elsewhere.

Two issues most likely paved the way for the development of NigeriaSat-X satellite in parallel with NigeriaSat-2. First were the concerns raised by the report of the 2000 official visit by the Presidency to SSTL headquarters, in the UK, that the scientific and user communities in Nigeria were left in the cold and had no input into the specifications of NigeriaSat-1, nor in the definition of the parameters of its sensors. The second was the clamour, within NASRDA, for an indigenous satellite system. Similar in configuration and specification with NigeriaSat-1, NigeriaSat-X,

---

1 Report of the Presidential Committee on the Satellite Project was completed on February 9, 2001.

2 Images of these satellites are available on www.nasrda.gov.ng.

3 Earth resources data include data acquired of the vegetation, land, and water, fisheries, agriculture and of natural and man-made changes on the surface of the Earth, such as all kinds of pollution such as oil spills, forest fires, floods, deforestation, volcanic eruptions, and the earth's immediate environment.

according to NASRDA, was developed and built by a team of Nigerian trainee engineers, under SSTL supervision and at SSTL's facilities in Surrey, England. The features of NigeriaSat-X included a 22 metre next generation imager with improved resolutions and optics over a swath area of 600 km.[4]

In order to gain access to these satellites as well as programme them for data acquisition and download, NASRDA built two ground receiving stations at its Abuja headquarters, respectively, for NigeriaSat-1 and NigeriaSat-2.

Primary Objectives of the Satellites

NigeriaSat-1, with a revisit period of 3-5 days, had the following objectives:

- To give early warning signals of environmental disaster;

- To help detect and control desertification in the northern part of Nigeria;

- To assist in demographic planning;

- To establish the relationship between vectors and the environment that breed malaria and to give early warning signals on future outbreaks of meningitis using remote sensing technology;

- To provide the technology needed to bring education to all parts of the country through distant learning; and

- To aid in conflict resolution and border disputes by mapping out state and International borders.[5]

NigeriaSat-2, with a revisit period of 2-4 days, was developed with the following objectives in mind:

- To support food supply security, agricultural and geology applications;

- To support mapping and security applications;

- To support development of the national GIS infrastructure; and

- To provide continuity and compatibility with the existing NigeriaSat-1 system.[6]

NigeriaSat-X was seen as a training model (TM); its goal was to build the capacity of the Nigerian engineers that were studying at SSTL. The TM was used to impart

---

4 Swath width refers to the strip or area of the Earth's surface that the sensors on board a moving satellite "see" and image or collect geographical data.

5 https://en.wikipedia.org/wiki/National_Space_Research_and_Development_Agency (Accessed, October 12, 2014).

6 https://directory.eoportal.org/web/eoportal/satellite-missions/n/nigeriasat-2 (Accessed, October 12, 2014).

the SSTL's Know-How Technology Training (KHTT) hands on experience in the requirements, specification, project management, system engineering, manufacture, test, assembly/ integration and final system testing of a spacecraft;"[7]

At the handover of the two satellites to the Minister of FMST on July 10, 2012, it was noted that "the immediate deployment of the satellites to crisis areas to observe and proffer lasting solutions to nagging issues such as security in the northern region, flooding and degradation in the south of the country will be the next line of action."[8]

Use of Acquired Data

Data from these satellites can be used to address a variety of needs in Nigeria and anywhere else around the world. Such needs include the monitoring of the nation's life-support systems, environmental problems (such as gulley erosion in Enugu state and oil pollution in the Niger Delta), Nigeria's natural resources as well as the management (relief and rescue) of the natural disasters that Nigeria and its people experience from time-to-time, such as the extreme floods of 2012. Table 7:1 shows the results of a number of demonstration projects carried out, nationally, by NASRDA.

However, for an effective application of Earth observation data in development programmes, the challenge is to know what to extract and how to extract it. To achieve such a goal, there should be adequate investment in the fundamental ground infrastructure, such as in Earth observation ground receiving station and in key capability development and utilisation (people, organisation, training, equipment and documentation), and their sustainability. The availability of and timely access to the data would also enable entities such as Nigeria's mapping authorities to develop baseline maps that can be readily up-dated in the future. In most societies of the world, such maps are used routinely to plan numerous daily human activities such as agricultural development and management to ensure food security, health management, laying of power lines, in search of ground-water locations, and transportation planning and management. For the average Nigerian, however, what is most important is the practical use of the data to positively impact Nigeria's development and growth and enhance the quality of life of its people.

---

7 https://directory.eoportal.org/web/eoportal/satellite-missions/n/nigeriasat-x (Accessed, October 12, 2014).

8 http://www.channelstv.com/2012/07/10/nigeria-receives-two-earth-observation-satellites/ (Accessed, Accessed, October 18, 2014).

Table 7:1 Space applications as an economic driver. NASRDA'S Pilot Project *(Source: NASRDA)*

Year	Project	Government Agency	Project Status
2007	Tele-Health	Federal Ministry of Health	No longer operational because of the failure of NigComSat-1
2007	Tele-Education	Open University	No longer operational because of the failure of NigComSat-1
2008	Assessment of environmental change and its social and economic implications in the Niger Delta	Federal Ministry of Niger Delta	22,000 hectares of mangrove already lost. Niger Delta govt. expected to replicate studies.
2011	Land use, land cover map of Nigeria – Atlas, at a scale of 1/300,000 is for planning and to be used as a base map.	Office of the Surveyor-General of the Federation	Budget plan for a national update unknown.

To facilitate the use of the acquired data, nationally, NASRDA established a Department of Space Applications at its headquarters in Abuja, and Space Application Laboratories at Ile-Ife (Obafemi Awolowo University), Kano (Bayero University), and Uyo (University of Uyo). It also established GeoAppsPlus Ltd, in Abuja, as its commercial arm, with the mandate to promote the applications of space technology for sustainable development of Nigeria and Africa and to serve as the exclusive commercial distributor of satellite imageries from all NASRDA satellites, as well as for other value-added geo-spatial data, consultancy and training services for the African markets.[9] On February 10, 2011, it also signed an exclusive agreement with the DMCii, a remote sensing solution provider in the United Kingdom, to distribute, outside Africa, imagery from both NigeriaSat-2 and NigeriaSat-X.[10]

What have the Earth Observation Satellites done for Nigeria and Its Citizens?

With all of the above efforts and related investments, one would expect that by now, space would have become "*…an essential tool for Nigeria's socio-economic*

---

9 http://www.geoappsplus.com (Accessed, October 18, 2014).

10 http://www.dmcii.com/?p=7030 (Accessed, October 18, 2014).

development and would have begun to enhance the quality of life of its people," as articulated in the 2001 National Space Policy and Space Programme document. But that is not the case. The reasons are not far-fetched; some factors are the results of Nigeria's actions and in-actions, while others are external to Nigeria.

No Satellite Sensor Simulation Campaign

As a nation with active Earth observation satellites, Nigeria's actions and in-actions are many, and most of them are inexplicable. For example, in 1973, Nigeria specifically requested the USA to use its *Landsat -1 and -2* satellites to acquire data of its territory.[11] But it is very hard to pinpoint what Nigeria did with these data or the experience it gained in using them for national development before it embarked on purchasing its own Earth observation satellites.

As one of the first customers that bought into the SSTL-led Disaster Monitoring Constellation (DMC1) consortium, Nigeria could also have demanded the development of sensors that could meet its own specific needs and address its own peculiarities, but its representatives did not. There was no fall-back experience to guide them as indicated above. They were satisfied with the off-the-shelf sensors that SSTL prescribed and with the technical presentations made by SSTL, on the results of data acquired from such sensors in prior simulations, outside Nigeria. Specifically, the managers of the Nigeria satellite project did not demand a simulation campaign in Nigeria, for the sensor systems that finally became the eyes of the nation's first and second Earth observation satellites.

No Data Interpretation and Utilisation Campaign

The needed campaign for data interpretation and utilisation is also another critical step that Nigeria by-passed. We should note that both the Landsat and Spot programmes have been successful, in part, because of the huge national and global satellite data user campaigns mounted by both the USA and France in advance of the launch of their respective first Earth observation satellite. Consequently, data

---

11 Prof. H. A. Oluwasanmi, the then Vice-Chancellor (VC) of the University of Ife (UNIFE) (and since renamed Obafemi Awolowo University (OAU)) constituted, in the first half of 1973, an 8-man Remote Sensing Committee that was chaired by Prof O. S. Adegoke of the University's Department of Geology. The VC charged the committee to develop a proposal for Nigeria's participation in the Landsat programme and in particular for the acquisition of data of Nigeria by the next American satellite, Landsat-2. This author was then a member of the academic staff of the University's Faculty of Technology; he served as a member of this committee.

from these two series of satellites dominated the geo-spatial data market, worldwide, in the 1970s and 1980s. The United Nations (Space Applications Programme) also used the data from these two series of satellites to respond to the demands of the developing countries for a greater understanding and application of space acquired data in national development efforts. Indeed, the developing countries, led by a General Assembly Resolution proposed by Sierra-Leone, spearheaded the establishment of the United Nations Space Applications Programme,[12] through which the first generation of remote sensing scientists from these same countries received their first education and training in the analysis, interpretation and use of remote sensing data in development activities.

But the managers of Nigeria's first satellite project did not organise any national or regional satellite data user campaign, training programmes or applications workshops to familiarise and prepare prospective customers and user communities on the understanding and use of its satellite data, prior to the launch of NigeriaSat-1 on September 27, 2003. Thus, from late 2003 when the Abuja station began to download the data being acquired by NigeriaSat-1, the nation had not attained any appreciable knowledge nor gained any collective experience on how to process, analyse, interpret and use Earth observation data in its archives, including those it formally requested from the USA, for its national development programmes.

For example, most of the projects illustrated in Table 7:1 did not move beyond the pilot phase to become operational or institutionalised. Lack of the fundamental knowledge needed by the staff of the user agencies to apply the analysed data in their day-to-day activities was part of the problem. There was also the unwillingness of these same agencies to take ownership of these projects and provide for their execution in their budgetary requests. These problems led to the demise of the projects.

The secrecy that bedevilled the satellite project also screened off the essential institutions and the science and engineering experts that possessed the requisite knowledge and competences needed for such an undertaking. It was a total civil service affair, hatched and managed by Nigeria's civil servants, with limited input from outside the service. Those in charge of the project purposely did not draw on the broad mind-set needed for such an endeavour—a mind-set that is committed to meeting the needs of the country and its people, over and above all other interests.

---

12 Abiodun, Adigun Ade (2000). *Development and Utilisation of Remote Sensing Technology in Africa*, American Journal of Photogrammetric Engineering & Remote Sensing, pp 674—686, June 2000.

Consequently, the Nigerian Satellite Project has not lived up to expectation. The lesson here for Nigeria rests on the steps it must take to avert a repetition of similar situation, over and over again.

Problems with Data Acquisition by the Ground Receiving Stations

There were and still are multiple problems with data acquisition and distribution. The inability of satellite data users in Nigeria and within the Africa region, in particular, to directly access NigeriaSat-1 and -2 data, and related products and services, from NASRDA, has become a major bottleneck in the use of such data for a variety of development purposes and has also made a mockery of how these satellites and their data will transform Nigeria and Africa.[13] Above all, NASRDA's Space Application Laboratories are equally not operating optimally due to lack of skilled staff, weak infrastructures and endless funding problems. The satellite ground receiving stations in Abuja that are used for downloading data as well as undertaking up-keep of NigeriaSat-1 and -2 continue to encounter a number of technical challenges and have not functioned optimally, beginning with the first one that went into operation in 2003 when NigeriaSat-1 was launched. Nigeria is not the only country suffering from data scarcity from its own satellites; other African countries have gone through the same pains and have settled for Earth observation products and services from other sources within the global geo-spatial market.

The first alarm on the state of NigeriaSat-1's ground receiving station (antenna) went un-noticed until September 9, 2005. That was when a number of the first fifteen Nigerian engineers that trained at SSTL issued a memorandum and declared that "…the Nigerian ground station in Abuja has not been operational for the past six months, leaving NigeriaSat-1 solely under the operational back-up of SSTL, UK."[14] A similar scenario came to light on August 19, 2011; the news at that time was that Nigeria also had no access to its NigeriaSat-X since no antenna was built in Abuja to gain access to the satellite by the time it was launched on August 17, 2011. "A ministry (FMST) source had informed the Guardian Newspaper that only the ground

---

13 Akinyede, Joseph, *et al.* (2013). *Nigeria' Space programme implementation: Lessons learnt to-date,* The fifth Congress of the African Leadership Conference on Space Science and technology for Sustainable Development, Accra, Ghana, December 3-5, 2013.

14 http://www.gamji.com/article5000/NEWS5045.htm and reproduced in https://groups.yahoo.com/neo/groups/AlukoArchives/conversations/messages/408 {(both accessed November 10, 2014).

station at Surrey Space Technology Limited, United Kingdom (UK) has the facilities to receive images from NigeriaSat-X and monitor its operations."[15]

The problem of acquiring data to meet the nation's needs surfaced again in September 2013; that was when the government, through the Office of the National Security Adviser, asked NASRDA for images of NigeriaSat-2 for use in gaining a better understanding of the security condition in the north-East section of Nigeria. "The satellite images the government wanted were meant to help counter-terrorism agencies identify possible bases used by terrorist groups to plan and launch attacks."[16] But no satellite image was forthcoming from the ground station in Abuja because, according to the information provided to former President Goodluck Jonathan, "the ground station located in Abuja for the control of the satellites was not equipped with the laboratory to process images downloaded from the satellites," although "…in the original contract between Britain's Surrey Satellite Technologies Limited (SSTL) and the Nigerian Government, provision was made for the laboratory." And in the absence of any available images locally, "NASRDA bought the same [images] from SSTL, the UK firm that constructed the Nigerian earth observation satellites."

But that was not the only time Nigeria has bought satellite data from abroad, data that its own satellites could have provided at home. A case in point was the March 2013 FEC approved N3.7 billion contract with Infoterra of the UK, for the acquisition of multispectral images of Nigeria for use in developing needed maps, in part, "for the construction and rehabilitation of 13 roads, spread across the six-geopolitical zones of the country."[17] Could NigeriaSat-2 have provided the same images? If not, why not? What was the quality of the data bought from abroad in comparison with the highly acclaimed data that NigeriaSat-2 could have provided? As a multi-purpose data set, are the Infoterra—purchased data being used for other purposes beyond those stated in the original contract? We should note that the monetary value of the one-time data acquisition contract by the Federal Ministry of Works amounted to 43% of the actual value (N 8.6 billion) of the satellite (NigeriaSat-2) that Nigeria

15 Nigeria's ground station 'can't receive signals from launched satellite'http://www.nigeria70.com/nigerian_news_paper/archives/guardian/2011/08 (Accessed, November 18, 2014).

16 *Nigerian N8.6Billion Remote Sensing Satellite is a Failed Project,* Nigerian Geographic Intelligence Awareness, September 4, 2013, (Accessed, June 12, 2014) http://lagosstreetmap.blogspot.com/2013/09 (Accessed, November 18, 2014).

17 Could NigeriaSat-2 Supply the Satellite Imageries Delivered by Infoterra of UK to Nigerian Government for N 3.7Billion? BusinessDay NewsPaper, March 28, 2013 (Accessed, November 18, 2014).

could have used to acquire the same set of images. And with proper station keeping, the satellite could exceed its life span of 7.5 years and acquire tens of thousands of images of Nigeria and other parts of the world—data that can be put to good-use for a variety of purposes for the benefit of humanity.

Moving forward, Nigeria and Nigerians need to know why the nation's Earth observation satellites, associated ground receiving stations and data processing facilities could not consistently deliver the services they were designed to provide. Similarly, what has been the subsequent impact of the non-availability of such data on the nation's development and growth? Equally puzzling is the fact that Nigeria, an equatorial country, agreed to have its three Earth observation satellites launched into polar orbits.

Orbiting the Poles instead of the Equator

Figure 7:1 shows NigeriaSat-2 (on its polar orbit) and LAPAN-A2 (on its equatorial orbit) in operation, in space, as of June 15, 2016.[18] If NigeriaSat-2, with its 2.5 metre panchromatic sensor and a 5 metre multi-spectral high resolution sensor,

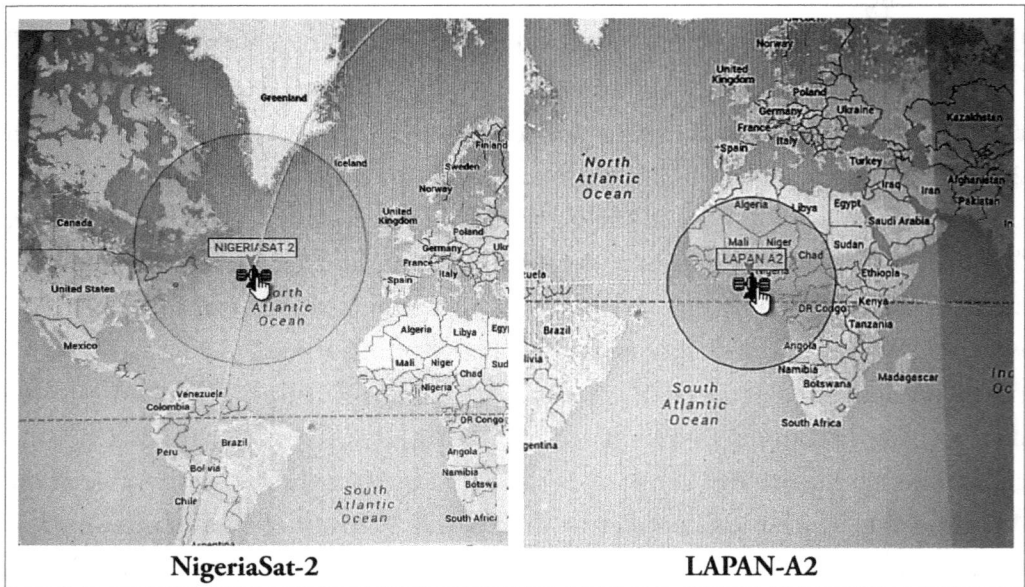

NigeriaSat-2   **LAPAN-A2**

Figure 7:1 Earth Observation satellites of Nigeria (on Polar Orbit) and Indonesia (on Near Equatorial Orbit) *(Sources: N2YO and Google Map)*

18 LAPAN-A2 is Indonesia's first indigenous satellite, built and tested at LAPAN's facilities in Indonesia. It was launched on September 28, 2015, into Near Equatorial Low Earth Orbit by ISRO from India's The satellite, with its 24 metre resolution 3-band multispectral imaging camera supports disaster management by earth observation (video surveillance), land use, natural resources, and envi-

(Continued on next page.)

had been launched into the equatorial orbit, it would have had up to 16 revisits, each day, to the same location on Earth, as LAPAN-A2 is doing, and it would have better served the economic, safety and security interests of Nigeria and all the other countries within the equatorial belt. There lies the wisdom and benefit of launching all Nigeria's Earth observation satellites into the equatorial orbit, as was the case with LAPAN–A2, instead of into the polar orbit.

Does the nation need Radar Satellite?

Meanwhile, on June 25, 2015, Nigerians woke-up to the realisation of their nation's plan to purchase a new satellite—a radar satellite - to be built by Menasat of the UK.[19] The justifications given for the acquisition were as follows: "It would help in tracking oil leakages and pipelines vandalism in the Niger Delta region, while it would also be useful for the military in the north east (of the country)."[20] As laudable as these justifications might be, there are a number of compelling reasons why Nigeria should not take this step on its space journey, at least, for now. There are other proven approaches that can lead to the attainment of these same objectives.

We should not forget that NigeriaSat-2 was supposed to "support mapping and security applications."[21] How effective, to-date, has it served as a security tool? And if it has not been effective, is it because of the machine or because of human failure?

---

ronment monitoring, as well as moon observation. Its high resolution video camera provides colour video observation of 11x6 km frame at 6 metre resolution is used particularly for ship-monitoring along the equator. LAPAN-A2 also provides digital and voice radio communication for search and rescue operations during disasters.

19 http://menasatgulfgroup.com/nigeria/ (Accessed August 14, 2015).

20 Adugbo, Daniel (2015).*FG, UK firm signs $250m satellite deal*, Daily Trust, Lagos, Nigeria, June 27, 2015; See also http://africanremotesensing.org/page-1846750?pg=7,7 (Accessed, October 3, 2015).

21 Curiel, Alex da Silva, Andrew Carrel, Andrew Cawthorne, Luis Gomes, Marting Sweeting, and Francis Chizea (2012). *Commissioning of the NigeriaSat-2 high resolution imaging mission* 26th Annual AIAA/USU Conference on Small Satellites, Utah State University, Logan Utah, August 13-16, 2012. 26 LAPAN-A2 is Indonesia's first indigenous satellite, built and tested at LAPAN's facilities in Indonesia. It was launched on September 28, 2015, into Near Equatorial Low Earth Orbit by ISRO from India's The satellite, with its 24 metre resolution 3-band multispectral imaging camera supports disaster management by earth observation (video surveillance), land use, natural resources, and environment monitoring, as well as moon observation. Its high resolution video camera provides colour video observation of 11x6 km frame at 6 metre resolution is used particularly for ship-monitoring along the equator. LAPAN-A2 also provides digital and voice radio communication for search and rescue operations during disasters.

Today, the nation-wide knowledge needed to analyse, interpret and apply the data already acquired by Nigeria's satellites that are orbiting in space is still at its infancy. Investing in a radar satellite, as was announced on June 25, 2015, without a good understanding and appreciation of the data from NigeriaSat-1, -2 and -X that the nation already paid for, will compound our problem. Why? Radar data require greater understanding and proficiency. We should also recall that, in the 1976-78 time period, the Federal Ministry of Agriculture commissioned the acquisition of radar data of Nigeria, under its Nigeria Radar Project (NIRAD), specifically for agriculture and forestry applications. The national impact of that effort, if any, has dissipated.

And even if we were to need radar satellite, *which for now, we do not*, why was Nigeria among the first countries that signed-up with Menasat for one? What is the experience of Menasat in space technology development, including the development, manufacture, launch and operation of satellites? Here again, the nation is faced with the same situation as in 2000 when its pre-occupation, at that time, was with the Nigeria Satellite Project and not with a Nigerian Space Programme.

Uncontrollable External Factors

There are factors that are external to the operation and management of a national space programme; many of these are beyond the control of Nigeria. In many developing countries, including those with Earth observation satellites of their own, such as Nigeria, government ministries, private establishments and donors often opt for Earth observation data from many sources, particularly the commercial ones. The latter respond to customer requests for their products and services, more efficiently, on demand, on a timely basis and without much hullaballoo. In addition, low resolution data that are being acquired by government-owned satellites cannot compete with data from commercial satellites that are equipped with high resolution (1m or less) sensor systems. Above all, the commercial entities, with established global networks, will continue to expand their international markets, even in the developing countries.[22]

---

22 Satellites such as *Ikonos, OrbView-3,* and *Quick-Bird* respectively provide high resolution images of 0.60 to 2.80 metres spatial resolution at nadir, to their customers world-wide. *WorldView-II* the successor to *QuickBird* (owned by DigitalGlobe), was launched on September 18, 2007, with a spatial resolution of 0.5 metre with three-meter geo-location. *WorldView-II*, similar to *WorldView-I* in specifications, will join its predecessors in space shortly. WorldView-3 satellite with a maximum resolution of 25 cm (9.8 inches) was launched on August 13, 2014. Meanwhile, Google's Terra Bella already placed seven satellites in orbit as of September 2016 to be joined by additional Earth observation satellites, by 2017. All of these are part of a constellation of satellites that are being dedicated to collect high-resolution sub-metre satellite imagery and high-definition video every day.

Strangely, while Nigeria was campaigning for the sale of its NigeriaSat-2 data, (2.5m panchromatic and 5m multi-spectral resolution), a few other satellite operators were providing, on request and free of charge, mid-to high-resolution data acquired by their own satellites. Meanwhile, China and Brazil have agreed to provide free images from their joint optical remote sensing satellite, CBERS-2B (2.36m panchromatic and 20 multi-spectral resolution) to African and Asian countries. The offer is intended to help the beneficiaries respond to such threats as deforestation, desertification, and drought. The USA is also making available, on request, free data and a guide table of its Landsat-7 satellite (30m spectral resolution).[23] The European Space Agency (ESA) is doing the same, and is providing free data from its Envisat MERIS-10 archives. India has also decided to help its neighbours in the Asia-Pacific region with free data acquired by its Earth observation satellites for disaster management. The Indian plan would also provide data analysis and training to countries without independent access. And to ensure that all in need are able to gain access to free data, a number of websites also provide an array of sources of free satellite imagery.[24, 25, 26]

There are also a few commercial satellite operators that can provide satellite images for use in counter-terrorism and vandalism activities, such as in surveillance and intelligence gathering, as well as in enhanced Earth observation services. The three USA-based companies that are in the fore-front of such efforts are: Digital-Globe with its WorldView-3 satellite and new WorldView-4—the latter launched on November 11, 2016,[27] Terra Bella (formerly Skybox Imaging), a Google subsidiary, with its SkySat series of satellites,[28] and Satellogic, whose focus is in meeting the needs of its customers with its Earth observation images and video clips services.[29]

With a 25 cm (9.8 in) resolution sensor on board its WorldView-3 satellite, DigitalGlobe is able to offer "an analytics service that could help track illegal human activity such as that of poachers and terrorists and predict locations where potential offences will take place next."

---

23 Gabrynowicz, Joanne Irene (2009). *President Orders Sweeping U.S. Policy Review*, Res Communis Blog, The University of Mississippi School of Law, University, Mississippi, July 5, 2009

24 http://www.ppgis.net/imagery.htm

25 http://www.sat.dundee.ac.uk/

26 http://gisgeography.com/free-satellite-imagery-data-list/ (15 sites of free images).

27 https://www.digitalglobe.com

28 https://terrabella.google.com

29 http://www.satellogic.com

Terra Bella is offering a comparable service with its planned 24-satellite constellation (seven already launched as of September 2016) that will occupy four different polar-orbit planes and provide high-resolution imagery, at one metre resolution and full-motion video that will "capture up to 90-second video clips at 30 frames per second" for commercial sale. The significant value of the data (imagery and video) acquired by a Terra Bela satellite "comes from comparing the data across time or location, looking for change. Individual customers can gain access to the Terra Bella constellation of satellites through a SkyNode; the process begins with the reservation of the needed satellite pass(es) "on a first-come, first-served basis."[30]

Similarly, as of mid-2016, six of Satellogic's nano satellites were in orbit. These satellites, which are part of its first service constellation, will provide 2-hour revisit times with 1 metre high resolution images. The sensors will provide Earth observation images as well as video. The company envisions the provision of real-time imaging of the planet, as well as offering its customers a platform where they can use the data from the constellation's sensors to generate data stream and related specific information needed for decision-making. While its initial demonstrations are "in agriculture and oil and gas," the capabilities offered by these sensors will also make them invaluable in a variety of security situations.

DigitalGlobe partners in Nigeria are already using WorldView-3 data to update the nation's outdated analog maps. In the meantime and if necessary, the radar data of Nigeria can be acquired and made available to Nigeria, at minimal cost, by countries with proven radar capability, particularly Canada, Germany, Japan, United Kingdom and the United States. Such radar data can be used to complement the data from LAPAN-A2, DigitalGlobe, Terra Bella and/or Satellogic.

Applying such data, and similar ones, to address the nation's development programmes and security interests, is the main and real issue. By so doing, Nigeria will achieve much more, at a shorter time, and at a considerably lower cost, than it has achieved from 2003 to the present time, with its three Earth observation satellites

---

30 A SkyNode is a compact system comprised of a 2.4 meter communications antenna and two racks of supporting equipment that gives you control of the Terra Bella constellation for your mission critical needs. From scheduling, tasking, imaging, and downlink, to image and video processing, a SkyNode includes a complete suite of easy-to-use applications that helps customer and his/her organisation get the right image at the right time. "We package this small-footprint system with our premier platform applications, which allow you to create orders, schedule collections, monitor the constellation, and produce and archive imagery. You'll go from downlink to accurate, relevant imagery and video in as little as 20 minutes." https://terrabella.google.com.

and others in NASRDA's pipeline. Chapters XI and XII provide a number of steps on how to accomplish these goals including the self-applied tests[31] Nigeria should scale through before it invests in any other Earth observation satellite.

Today, the availability of satellite acquired data is growing and the user has more choices of data suppliers than ever before. The challenge for Nigeria and other countries that intend to generate revenue from the sale of data acquired by their satellites is how to marshal their capabilities and resources to compete successfully with such evolving global developments that are here to stay.

With the successful launch of NigeriaSat-1 in 2003, the government decided in favour of a communication satellite as its next satellite project; the announcement was greeted with much cheer, nationally. The Nigerian Communication Commission (NCC) would be responsible for managing the communication landscape within which the new communication satellite would operate.

Nigeria's Communication Satellites

The enthusiasm of the Nigerian population for a national communication satellite was not a surprise. Over the years, a number of Nigerians have revelled at the fact that the Prime Minister of Nigeria and the President of the USA were the first leaders in human history to exchange communication signals via America's Syncom II satellite on August 23, 1963. It happened that the experience was just that, a one-time opportunity for Nigeria; thereafter, the nation moved to and stayed on the communication side-line while America and the rest of the world moved on.

Although, Nigeria had the opportunity to capitalise on the success of its Intelsat satellite communication stations in Lanlate (Oyo State), instead, as shown in Chapter IV, the nation's communications authorities of that era opted for wrong choices, such as tethered Aerostat Balloons, to meet the nation's operational communication needs. As indicated earlier, the project failed; indeed, it was a bottomless-money-swallowing pit and a source of major embarrassment for the nation. Subsequently, the nation descended into the communication abyss that was captured in Figure 2:1 of Chapter II.

By November 2003, the pulse of the Nigerian government in Abuja had been revved up by the on-going debate on the next move of the Regional African Satellite Communication Organization (RASCOM). In December 2004, the government

---

31 How well does the nation understand the space-related hard- and software it is acquiring from abroad? Has the nation been very successful in how it deploys, operates, uses, and manages these assets? Is Nigeria able to apply the array of Earth Observation data in its archives to make a significant impact in the nation's economic development and in the lives of its citizens?

responded to all these challenges by directing NASRDA's supervising ministry (FMST) to sign a contract, not with SSTL as specified in the November 7, 2000 Nigeria/SSTL Agreement, but, this time, with the China Great Wall Industry Corporation (CGWIC). The new contract called for CGWIC to build and launch the *first* Pan-African Communication Satellite, also known as the *first* Nigerian Communication Satellite (NigComSat-1). Nigeria chose Telstar Canada to provide needed technical and management consultancy services to NASRDA for the NigComSat-1 project. NASRDA claimed that China outbid 21 other contractors, mostly from the West, for the US$311 million contract to build and launch the satellite, a figure that climbed to US$450 million by the time the satellite was launched. This fund was expected to cover the cost of the satellite itself, the ground station in Abuja, the training of engineers in China and other related expenses.

NigComSat-1 was launched at 16:01 GMT on May 13, 2007, aboard a Chinese Long March 3B carrier rocket, from the Xichang Satellite Launch Centre in China. The satellite was a super hybrid geostationary satellite with a launch mass of 5,150 kg, and an expected service life of at least 15 years. It was located at 42.5000'E at the geostationary orbit, the same location that the ITU allotted Nigeria in 1988. On the Earth's surface, the satellite's location translates to just north-east of Mombasa (04°02'S, 39°43'E) in Kenya. The satellite was equipped to provide communication services for Africa, parts of the Middle East and southern Europe. The managers of NigComSat Ltd[32] also portrayed the launch of the satellite as part of Nigeria's drive to enhance rural access to technology and the Internet in the country and to boost Nigeria's and Africa's knowledge economy.

NigComSat-1 in a Graveyard Orbit

The expectations were high that the satellite would accomplish the above objectives and much more. But eighteen months after its launch, NigComSat-1 lost power on November 10, 2008 and stopped functioning; it is now in a *graveyard* orbit in space, approximately 36, 000 km above the mean sea level, at the Earth's equator.

Unanswered Questions on the loss of NigComSat-1

Artificial satellites, like any other man-made machine, do fail; NigComSat-1 will neither be the first nor the last one. According to CGWIC authorities, the satellite

---

32 NigComSat Ltd, an off-shoot of NASRDA, is the entity established by the Nigerian government, in 2006, to manage and operate the nation's communications satellite(s).

packed-up on November 10, 2008 because its solar panels did not deploy properly as a result of an anomaly in its solar array. Today, we know that a chain of events followed. The solar panel failed and could not recharge the satellite battery, hence the battery gave up. The subsequent chain reaction signalled and hastened the end of NigComSat-1's useful life.

To-date, a number of questions remain unanswered on NigComSat-1 and its demise. For example:

- When did the CGWIC authorities first know that the solar panels did not deploy properly and that the disabled solar panel would lead to the failure of the satellite? Was it within the first few days of launch or only just before NigComSat-1's demise, 18 months later, on November 10, 2008, when CGWIC notified Nigeria and the world of the satellite's demise?
- What corrective steps did CGWIC undertake to repair the solar panel?
- Did anyone keep a log of such efforts?
- When did NigComSat Ltd, in Nigeria, become aware of the dangers its satellite was experiencing in space?
- What steps did NigComSat Ltd take to seek requisite information from CGWIC?
- Throughout the 18-month active life of the satellite, is there a log of the time(s) when, during that period, NigComSat Ltd was in control of the satellite?

Only NigComSat Ltd and the CGWIC authorities know the answers to these questions. Nigeria has the capacity and should establish, on its own, the real cause(s) of why the NigComSat-1 failed, otherwise, it would be difficult to guide against similar failures in the future.

NigComSat Ltd and the Nigeria's Communication Satellite

The failure of NigComSat-1 notwithstanding, Nigerians know that the communication sector is indispensable to a nation's development, economy and security, including the delivery of its goods and services. It is also very essential for the safety, global interactions, information sharing, as well as the welfare and mobility of its people. Accordingly, the nation opted for a replacement satellite, NigComSat-1R.[33]

NigComSat-1R was launched in China on December 19, 2011. According to the then Managing Director (MD) of NigComSat Ltd, "*the satellite is expected to serve*

---

33 Images of NigComSat-1 and NigComSat-1R are available on www.nigcomsat.gov.ng.

no fewer than 40 countries in Africa, six in Europe and two in Asia; the real benefits of the satellite to ordinary Nigerians, apart from government and security agencies, as well as the media, would be in the education and health sectors."[34] On March 18, 2012, the MD of NigComSat expanded on the capabilities of NigComSat-1R and the services it would deliver as follows:

> *"In the upstream sector, the company is concerned with leasing of bandwidth, trunking services and the provision of satellite networking, operation and management services. In the downstream, the company's business objectives include telecommunications, broadcasting, provision of broadband Internet, tele-education, tele-medicine, e-government, e-commerce, real time monitoring services, navigation and global positioning services."*[35]

Shortly thereafter, the same MD assured the country that NigComSat-1R will also "enable intelligence gathering and strengthen security in Nigeria and beyond; drive the National ICT revolution in providing revenue diversification for the nation and offer cost effective solution and affordable access to meet Nigeria's telecommunications, broadcast, aviation, maritime, defence and security needs."

But the nation is still to realise any of the above economic predictions. Meanwhile, the percentage contribution of telecommunications to the Nigeria's gross domestic products (GDP) went from 0.62% in 2001, the year the Global System for Mobile Communications (GSM) system was introduced in the country to 9.8% by June 2016.[36] These gains were achieved without any contribution either from the nation's premier communication satellite, NigComSat-1, or from its successor, NigComSat-1R. And instead of properly managing its assets in order to generate revenue for the nation, just as other revenue-generating sectors of the nation's communication industry are doing, the managers of NigComSat Ltd continue to ask for financial commitments from government in order for it to stay afloat.

As of mid-August 2013, the landscape in which NigComSat-1R stepped into had acquired a number of distinct communication characteristics of its own. That same landscape is currently overwhelmed by the footprints of many foreign communications satellites, such as illustrated in Figure 7:2; they are serving, mostly, the urban communications needs of Nigeria and Africa today. Indeed, both Nigeria and South

---

34 *Nigeria: Real Benefits of NigComSat-1R,* ThisDay Newspaper, Lagos, Nigeria, January 4, 2012.

35 *NigComSat 1R Commercial Launch: Matters Arising,* Thisday Live, Lagos, Nigeria, March 8, 2012.

36 http://www.ncc.gov.ng

Africa are the destinations of choice of over 30 of these satellites that are located at the geostationary orbit (GSO) and are operated from different on-shore and off-shore locations around the world.

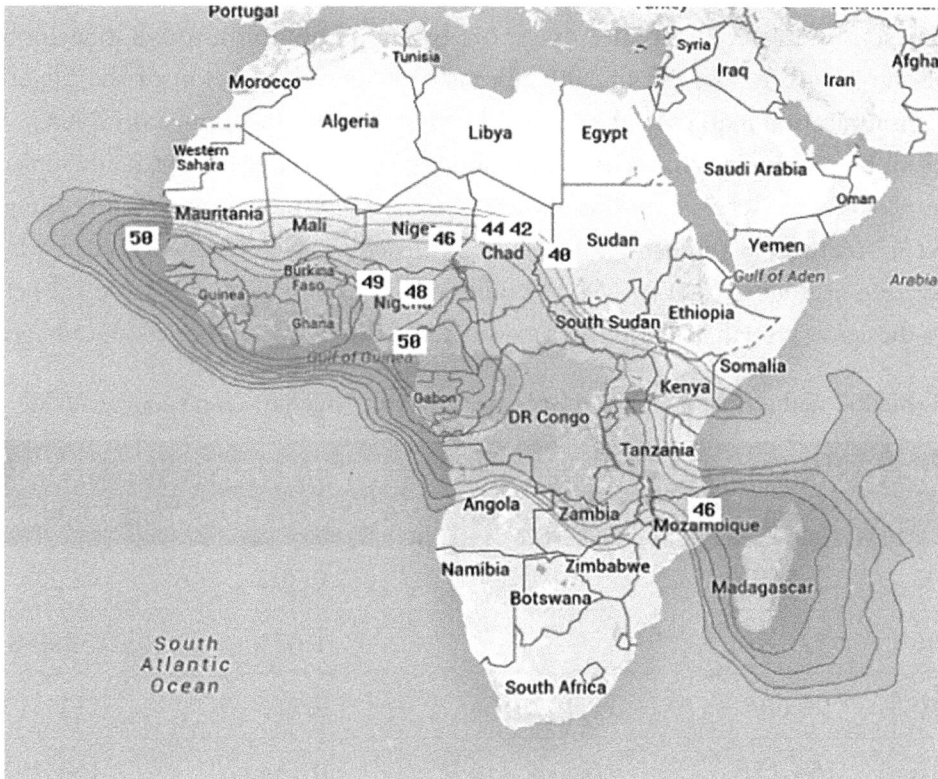

Figure 7:2 The footprint of Eutelsat 16 A communication satellite over Sub-Sahara Africa—Ku band (Source: LyngSat Maps)

The major satellite operators whose space assets are providing needed satellite bandwidths for satellite communication services in Nigeria and most of Africa today include Intelsat (USA), Eutelsat (Europe), New Skies Satellites (SES of Luxemburg), Inmarsat (Europe), GlobalTT.com (Belgium), ArabSat (Saudi Arabia) and SpaceCom (Israel)—the latter lost its Amos 5 on November 21, 2015 and its replacement was lost on its SpaceX launching pad on September 1, 2016. A number of these satellite operators also offer multiple communications service options such as a suite of Internet access services. There are also several new communication satellites on the drawing boards of these same and other satellite operators. The other operators, which include O3b (Netherlands and UK), Iridium (USA) and GlobalStar (USA), are prospective owners of multiple constellation satellites of the future that will operate from the middle Earth orbit (MEO). These future satellites are being targeted to meet the ever-expanding urban communication needs of Nigeria and its over 170 million people as well as that of Africa, the second-largest and second most populous continent on Earth with an estimated 2013 population of 1.033 billion people.[37]

These same foreign satellite operators brought the ubiquitous GSM mobile telephone service to Nigeria in 2001. Because of the deregulation of the Nigerian telecommunication market by the National Communication Commission (NCC), the use of mobile telephones has soared and mobile phone services have virtually replaced the unreliable and archaic land-line services of the NITEL monopoly era. According to the NCC records, the number of mobile phone subscribers (active) in Nigeria grew from almost zero in August 2001 to 154.125 million by December 2016.[38] Table 7:2 below shows the active growth in telephone services from 2002 to December 2015, in Nigeria.

Table 7:2 Subscriber Data for telephone service in Nigeria[39] *(Data Source: NCC)*

Year	GSM (Mobile) in millions	CDMA (mobile) in millions	Fixed Wired/ Wireless in millions	TOTAL in millions
2002	1.569	--	0.702	2.271
2006	32.185	--	1.673	33.858
2010	96.684	12.1323	2.736	111.517
2016	154.125	0.218	0.154	154.497

As of December 2016, the four main licensed GSM phone operators in the country included MTN (South Africa) with 40.12% of market share; Globacom Ltd (Nigeria) with 24.24%; Airtel (India) with 21%; and Etisalat (United Arab Emirate) with 15%.[40] Except for Etisalat, the same mobile GSM operators are also in the fore-front of the fibre-cable deployment effort in Nigeria, an action that is enhancing the country's ICT capability. As of December 31, 2014, the phone operators have installed a total of 80,938 km (land – 64,433 km; marine – 16,506 km) of fibre cables in the country. The five

---

37 http://worldpopulationreview.com/continents/africa-population/ (Accessed, May 10, 2015).

38 Because of erratic network services, many Nigerians own more than one mobile telephone device while many, many others, particularly the poor and those who live in rural and remote areas, have none.

39 http://www.ncc.gov.ng/

40 ibid.

leading deployments included those by MTN (19,200 km – land), MTN-FXD (12,518 km – land; 6,682 km – marine), GLO (10,869 km – land; 9,800 km – submarine), Airtel (6,314 km – land; 23 km – submarine), Multilinks (5,789 km – land).[41]

Initially, the availability of satellite bandwidth resources from a variety of communication satellites also resulted in the emergence of new services and businesses, including related employment opportunities. But incessant power failure and the increasing opportunities to access the Internet have slowed down the growth of *listed and licensed* cyber-cafes in the country which, as of November 2015 were 745 in number. Internet telephony is now ubiquitous—popularised, in part, by Skype and Google. The Internet has also opened the door to a new way of learning at the nation's universities; it is enhancing both academic and research work for teachers and students alike;[42] Google offered such an Internet helping hand to the University of Ibadan.[43] Many universities in the land are also embracing the Internet as a necessary component of the nation's educational infrastructure, the problem of service reliability due to incessant power failure, notwithstanding.

The NCC also licensed several indigenous companies to provide private communication network linkages, both nationally and internationally. Such companies are able to use Very Small Aperture Terminals (VSAT) to provide a variety of communication services to entities and establishments that include maritime services, government entities, *Internet service providers, international corporations, oil, gas and other mining companies*, banks, construction and transportation industries (shipping, trucking, railways and bus fleets), entertainment industry, emergency services at sea or in areas affected by natural disasters, video distribution and paid-TV services. The success of the providers of these services is judged by the integrity of the connectivity offered—specifically speed, reliability, security and cost.

Radio and Television (TV) Services

The nation's communication revolution has also witnessed the licensing and establishment of over 70 major radio stations in the country; most of them are FM

---

41 2014 Year End Subscriber/ Network Data Report for Telecommunications Operating Companies in Nigeria, Nigerian Communications Commission, Abuja.

42 Bamigboye, B. O., Nimbe Adedipe and Rueben Abiodun Ojo (2010). *Perception of the Internet as an Enabler of Scholarship among Postgraduate Students of the University of Ibadan, Oyo State, Nigeria, Library Philosophy and Practice 2010,* ISSN 1522-0222.

43 Aginam, Emeka (2012). Google extends support to University of Ibadan, VANGUARD Newspaper August 29, 2012.

stations and over twenty-two (22) of these stations are in Lagos State. There are also six (6) shortwave broadcasting stations in Nigeria; these are located in Abuja (Voice of Unity), Enugu Federal Radio Corporation of Nigeria (FRCN), Ibadan (FRCN), Kaduna (FRCN) and Lagos Voice of Nigeria (FRCN).

Satellite Television is another beneficiary of the communication revolution. At the time of this writing there are over 180 satellite TV stations in Nigeria. The major foreign satellite operators identified earlier are using their space assets to provide satellite communications services, including radio, TV and Internet services in Nigeria and to most of Africa. A number of these companies also offer multiple communications service options such as a suite of Internet access services. Without any input from its own indigenous communication satellite, Nigeria is gradually scaling one of the major hurdles in its development agenda. It is now up to its citizens and the governments (federal, state and local) to utilise this growing ICT capability to build a knowledge society—on the basis of what we do, invent, publish, manufacture and produce—to enhance our own quality of life. But because of the inability of NigComSat Ltd to be a leading contributor to this ICT growth in the country, it is thus not surprising that there was a national clamour for the sale of the government owned company and its assets, including the NigComSat-1R satellite.

The Future of NigComSat Ltd

The success of any communication satellite operator, including NigComSat Ltd, hangs on the quality of service and the integrity of the connectivity it offers to its clients—specifically speed, reliability, security and cost. According to such a client, Zinox Nigeria Ltd—"*We presently have huge bandwidth resources to meet our customers' short term and long term needs. Our focus is more on knowledge centers and under-served communities across the country.*"[44] Implicit in Zinox's announcement of February 27, 2013 was that it had not acquired any bandwidth from NigComSat-1R; it could also be perceived as Zinox's response to a critical January 31, 2013 newspaper article on the satellite and its operator, titled: *What is NigComSat-1R Bringing to the Table?*[45] Other factors could also have motivated Zinox to take such a decision and not utilise the services of NigComSat Ltd and its satellite, NigComSat-1R.

---

44 Zinox Relocates Tech Hub to Nigeria (2013). Ventures Africa, February 27, 2013.

45 ThisdayLive.com, January 31, 2013 (Accessed, August 9, 2014).

Economic Survival of those that could have depended on NigComSat-1R

Let us recall that the gap in time, between the failure NigComSat-1 on November 10, 2008 and the deployment, on December 19, 2011, of its replacement, NigComSat-1R, was very significant for the economic survival of those that could have depended on it. Other losses incurred are still unaccounted for—loss of confidence in the competence of NigComSat Ltd and its management team, loss of its customers and loss of revenue to the national coffers that funded the acquisition of NigComSat-1. Interestingly, the communication sector has experienced impressive growth, none of which is attributable to NigComSat Ltd and its NigComSat-1R. One should thus expect the communication service providers in Nigeria and elsewhere, that have made long–term bandwidth commitments with their foreign satellite operators, such as Zinox already did, not to jump ship in favour of NigComSat Ltd and its replacement satellite, NigComSat-1R, nor abandon their long time-tested, reliable, foreign satellite operators. In their long-term plans for an enlarged African market and its projected opportunities and benefits, the reality is that these foreign satellite operators would have locked-in the proven communication service providers, in Nigeria. How then can NigComSat Ltd survive without customers?

Is NigComSat Ltd up to the Competition?

Until NigComSat-2 is up there in space, functioning perfectly in partnership with a healthy NigComSat-1R, prudent satellite communication service providers, in Nigeria and abroad, would think very hard before investing in and leasing NigComSat-1R's bandwidths. The concern of such potential customers is a practical and an economic one; if for any reason(s) NigComSat-1R should pack-up, just as NigComSat-1 did, NigComSat Ltd has no immediate back-up communication satellite in space today which could take over the functions of NigComSat-1R without any significant service interruption. In the communication industry, secure and assured access, resilience and redundancy, exceptionally high availability and quality of service are all fundamentally important. It is thus inconceivable that any reliable Internet Service Provider (ISP), Mobile service or VSAT operator will opt for such a gamble. In the meantime, large and enshrined communication satellite operators that are servicing the Nigerian satellite communications market have multiple satellites in their arsenals. For example, SES of the Netherlands, in which the PTTs of African countries, including Nigeria, supposedly have shares, has over 50 communications satellites in space right now. Intelsat, Eumetsat, Inmarsat and Telesat, among many others, are equally or better endowed. And should any of their satellites that are serving the communication needs of Nigeria fail, the services on the failed satellite can be readily switched to another active satellite without much delay. NigComSat Ltd, with its one operating satellite, NigComSat-1R, cannot accomplish such a feat, today.

To respond to all these concerns, the nation was informed, on January 1, 2016, that its government had entered into an agreement with China to buy two more communication satellites. The latter, respectively code-named NigComSat-2 and -3 are to be stationed at longitudes 19ºE and 22ºE, as back-ups for NigComSat -1R, at a cost of US$701 million. While the nation was still digesting the significance of the new Nigeria-China agreement, the same management of NigComSat Ltd that could not successfully operate nor manage NigComSat-1R bragged, through one of its managers, that with NigComSat Ltd.'s three communication satellites, "it will be possible for the Nigerian telecommunications industry to *dominate the African market within a period of five years after the launch of the satellites.*"[46] A Punch Newspaper article titled: "*NigComSat-1R becoming white elephant four years after—Investigation,*" provides the reader with a sobering analysis of the ailments of NigComSat Ltd, as of April 2016.[47]

However, it is not enough just to purchase satellites and have them launched into space. Will the country have the capability to effectively operate these new communication satellites once they are launched? Or will they remain idle in space as NigComSat-1R is doing? A responsible planning programme demands that highly qualified communication engineers and other related staff should be recruited and trained as satellite operators at appropriate institutions, far in advance of the launch of the next communication satellite(s). Such individuals should gain essential practical experience with NigComSat-1R and use the latter to fully demonstrate the type of services the new satellite(s) will offer to potential customers (communication service providers).

Nigeria's Options: Partnership or Privatisation

Among the options opened to Nigeria on the future of NigComSat Ltd and its assets are public-private partnership and privatisation. In either of these two cases,

---

46 *Nigeria to acquire two satellites from China for $701m*, PUNCH Newspaper, Lagos, Nigeria, December 31, 2015.

47 *NigComSat-1R becoming white elephant four years after –Investigation,* Punch Newspaper, April 16, 2016.

the nation should ensure, with appropriate guarantees, that rural Nigeria will be a major beneficiary.

Partnership: Nigeria can enter into a public-private partnership with credible Nigerian investors of international repute who can attract a credible satellite operator with proven track records of reliable service. In such a relationship, what each party brings to the negotiating table is of critical importance. On a long-term basis, such a partnership will stimulate and grow a market of indigenous satellite operators in Nigeria. It is attainable and sustainable and is in the nation's interest. The Nigerian government as well as the State and local governments and the nation's academic institutions need a reliable national communication service provider to depend on and one that can meet their needs, nationally and globally. Because our public communication satellite operator, NigComSat Ltd, is currently not up to the task, private and public establishments that need dependable communication services have looked beyond NigComSat Ltd to meet their communication needs. A partnership between the government, on behalf of the Nigerian public, and these same Nigerian private establishments may be the answer. The other option is privatisation.

Privatisation: There are also many compelling reasons for privatising NigComSat Ltd, a parastatal (*company*) owned by the Nigerian government. Examples include the abysmal performance records and decades of mismanagement of many of the Nigerian government parastatals. The latter (extinct and existing) include NITEL (for telecommunication), NEPA (for energy supply-electricity), Nigerian Railways and Nigerian Airways, among many others. If NigComSat Ltd had fared any better, there would not have been mounted pressure for its sale.

There was a time when the Nigerian Airways, operated side-by-side with private airlines, M-TEL, the public mobile-phone carrier, did the same with private mobile-phone carriers, and today, NIPOST, the nation's public mail carrier, is in competition with private mail and parcel carriers. In each of these cases, the most efficient ones have and are surviving.

Examples of very successful privatization of publicly owned telecommunication companies abound both locally and globally. One such example is the privatisation of the former Korea Telecom, a public entity, was completed in 2002 and renamed KT; it is an integrated wired/wireless telecommunication service provider. In September 2012, KT was awarded the Global Supersector Leader for Telecommunications (World No. 1 Company in terms of sustainable business practices) by Dow Jones

Sustainability Indexes for two consecutive years.[48] KT offers a privatization example that Nigeria may wish to study and can adapt as appropriate.

ICT Manufacturing Capability and ICT Services in Rural Nigeria

Irrespective of the option the Nigerian government chooses, we must accept that a nation of over 170 million people who are blessed with the array of endowments described in Chapter X should be able to manufacture a significant amount of what it consumes. The same applies to our need to establish an ICT manufacturing capability that can meet the needs of Nigeria and its people and stimulate and grow a cadre of indigenous satellite operators. By taking these critical steps, Nigeria will become a major economic hub in West Africa, in particular, as well as in Africa. A critical contributor to that national effort is rural Nigeria. Thus, in order to enhance the quality of life in Nigeria's rural communities as well as strengthen the interdependence between rural and urban Nigeria, the government should consider ICT investment in rural Nigeria a national priority. How to move forward on these three ICT priorities is addressed in Chapter XI.

The 25-Year Road Map for Nigeria's Space Mission (2005-2030)

Since 2005, NASRDA's activities have focused on implementing the objectives of the 25-Year Road Map. Its two clear objectives are: "First, [it] will launch a Nigerian Astronaut by the year 2015 possibly in cooperation with China, and second, it will launch a Nigerian-made satellite by the year 2018, with a Nigerian rocket and satellite launched by 2025." En-route on the space journey and upon the directive of the FEC, NASRDA reviewed the implementation plan of the road map in 2007; in 2013, the National Space Council asked NASRDA to revise the road map. What follows is an analysis of a number of the key components of the road map.

The Cost Implications of the 25-year Road Map

At its inception in 2005, FMST requested $200 million (N 26.40 billion) as a take-off fund for the road map, with subsequent annual appropriation of $100 million (N13.20 billion) for the next fifteen years.[49] NASRDA estimated that the road map

---

48 Sustainability Indexes—Review". Sustainability-indexes.com.(Accessed, June 1, 2015).

49 *A Road Map for Nigerian Space Mission (2005-2030)*, Federal Ministry of Science and Technology, Abuja, July 2005

would cost N227.00 billion in 2007,[50] and N70.00 billion, $1.50 billion and £35.5 million, collectively, by 2014.[51] Meanwhile, NASRDA's efforts to implement the programme, as designed, have met with very limited success, in part, because of meager government funding, as shown in Chapter XV (Table 15:1, and Figures 15:1 and 15:3), unattainable and inflated goals/expectations, as well as unjustifiable justifications.

A Nigerian Astronaut Programme

The primary justifications, given by NASRDA, for sending a Nigerian astronaut into space are (i) to enable Nigeria to participate in the development and acquisition of new space technologies which will enable [the nation to] develop technologies which otherwise would have not been possible in Nigeria; and (ii) Some of the [space] equipment could provide services that can generate funds.

But what type(s) of technologies will Nigeria develop and acquire through its astronaut programme and what will these technologies do for Nigeria? For example, astronauts from Brazil, Malaysia and South Korea focused on agriculture (hybrid seeds), healthcare (growth of protein for pharmaceutical and health-related applications) on Earth and protein enzymes for industrial and medical applications. Similarly, what are the details of that particular "equipment that could provide services that can generate funds" for Nigeria? The amount of funds that will be generated and the time-frame when this will materialize remain unknown.

However, for now and in the immediate future, human access to the International Space Station (ISS) will be available only through the Russian *Soyuz* spacecraft, because the USA's Shuttle programme was discontinued in 2012. By 2017 or thereafter, commercial developers of space transportation systems such as *Space X* and the *Boeing Company*, two USA-based commercial companies, are expected to be able to provide similar transportation services to the ISS.[52] Additional launch opportunities will also become available when the Chinese *Tiangong* Space Station is launched in 2020 or thereafter. These developments should provide opportunities, in the future, for the developing countries, including Nigeria, to conduct experiments on board these space stations. But at what cost? And what are the benefits?

---

50 *Justification and cost implications for Road map for Nigerian Space Mission (2005—2030)*, NASRDA, Abuja, April, 2007

51 Personal communication with NASRDA Staff, in Vienna, Austria, in February 2014

52 http://www.nasa.gov/press/2014/september/nasa-chooses-american-companies-to-transport-us-astronauts-to-international/#.VBiuBPldURo (Accessed, June 4, 2015).

An alternative way for Nigeria to participate in human space flight and carry out space-based fruitful research is addressed in Chapter XII, under *Participation in the Human Space Technology Initiative (HSTI)*—an outcome of UNISPACE III recommendation.

The Development of Rocketry and Propulsion Systems

To achieve the second objective of the 25-year Road Map, in 2003, NASRDA established the Centre for Space Transport and Propulsion (CSTP) at Epe, Lagos State. Such a facility can be used to send a variety of payloads such as "satellites, astronauts and data gathering probes" into space, to attain specific strategic and economic development goals. But NASRDA went further in its review of the road map to lay out the benefits of setting up the above transport and propulsion facility at Epe, as follows:

> Setting up a launch facility in the Country would generate huge revenue to the Nation's economy more than what the oil/communication earns presently. Many space developed Countries would prefer equatorial launch window than any other site. This Technology has tremendous advantage since the nation is strategically located along the equatorial zone of the globe.

It would have been very useful if, in its review, NASRDA had illustrated the competitive advantages that Nigeria's equatorial launch site would offer viz-a-viz other existing equatorial launching sites. Here, there is need for a rethink and this should begin with a look, by NASRDA, at the current manifests of proven and established providers of space launching services. NASRDA would also need to clarify to the nation why the leading space countries would prefer Nigeria's equatorial launching facilities to the ones they have used, reliably, for years. If it is true that "*many space developed countries would prefer equatorial launch window than any other site,*" will such countries also prefer an untested, green-horn Nigerian satellite launching service, with no experience, or one that is being provided by those entities with proven years and decades of accomplishments, such as the European Space Agency from Korou, by Space-X, or at sea by Boeing *et al*? NASRDA should also have identified the existing launch facilities that are now available, world-wide, to all nations, including the cost, for each equatorial launch, from each of these facilities.[53]

---

53 Known equatorial launching services are available through (i) Brazil at its *Alcantara Launch Center*, Latitude 2.3⁰S Longitude 44.4⁰W; (ii) European Space Agency at *Kourou, French Guiana*, Latitude

(Continued On Next Page)

To also claim that Nigeria's earnings from providing satellite launch services would dwarf its current earnings from oil and communications is outrageously disingenuous; it is a deceitful act that has been deliberately planted in the review report to mislead the government and prod it to take a step that generations of Nigerians will forever regret. Such a fallacy was floated in 2001; by then, it worked, because it led the government to sign the Sale and Purchase Agreement for NigeriaSat-1 satellite, as shown in Chapter V.[54]

Chapter XII—Part Two on National Space Priorities, elaborates on the steps Nigeria should take to benefit from its geographical location near the equator, beginning with its gaining a scientific knowledge and understanding of the equatorial plane, its subsequent development of sounding rockets and the necessary acquisition of knowledge and use of the equatorial orbit.

Today, Nigeria and other African and developing countries must recognise that the transformation of sounding rockets into satellite launch vehicles faces a number of hurdles, including these three major obstacles: (i) high initial cost that continues to escalate as the programme progresses, (ii) launch vehicles are science and technology intensive, and gaining access to such knowledge requires cross-border knowledge sharing, and (iii) export control regimes make the sharing of knowledge, today, particularly between newcomers and the established entities in the space arena, almost impossible. The question then is: Who is going to give Nigeria and other African or developing countries access to rocket technology? Because of the dual-use of this particular technology—there is virtually no distinction between "peaceful" space launch vehicles and "military" ballistic missiles—the road to acquiring a satellite launch

---

5.2⁰N Longitude 52.8⁰W (iii) Italy at *San Marco Range off the Kenya coast*, Latitude 2.9⁰S Longitude 40.3⁰E (Dormant); (iv) Sea Launch Company (an international consortium) with a mobile ocean platform positioned along the equator in the Pacific Ocean. Sea Launch operates from a home port in Long Beach, California, USA and launches from a Pacific Ocean platform near the Christmas Islands on the equator, Latitude 0⁰ Longitude 154⁰W; (v) *Space X*, a USA commercial launcher offers space *launches from a variety of Launch facilities*. It launched Malaysia's RazakSat-1 into the equatorial orbit in 2011 from *Kwajalein Atoll, Marshall Islands, Pacific Ocean*, 9⁰2′52.30″N167⁰44′34.80″E, and is scheduled to do the same for RazakSat-2; and (vi) *Orbital Sciences*, a USA private company that provides launch services and has conducted 42 missions from six different launch sites worldwide since 1990. One of its launch sites is the *Kwajalein Atoll, Marshall Islands, Pacific Ocean*.

54 In 2001, part of the justifications used in winning government approval for the purchase of NigeriaSat-1 from SSTL of the UK, included the following: The government "must treat this project with some urgency so as to position itself properly, as there are signs that an international movement is underway to restrict access to satellite imagery by developing countries."

capability is a very tortuous one. Nigeria and other interested countries might want to study the circumstances that led to the demise of 'space launcher systems development' in such countries as Taiwan, South Africa and Argentina. The interested countries should also familiarise themselves with the international guidelines contained in the Missile Technology Control Regime (MTCR) which was established in 1987.[55] MTCR was not in place when India and China began their missile technology development, but South Korea began its own development after MTCR was established. South Africa is the only African member State that is signatory to the MTCR regime.

Justifications for Nigeria's Participation in Space

After over fifteen years of national space activities, it is both astonishing and alarming that NASRDA uses every opportunity to justify why Nigeria should be investing in a space programme. Elements of such justifications, renamed here as desirables, include "space race among nations, prowess, national integrity and geopolitical standing." The need for such justifications is a clear signal that much is amiss. At this stage of its existence, NASRDA should be presenting concrete examples of the impact of its activities on the socio-economic development of Nigeria and on the enhancement of the quality of life of its people. Unfortunately, these desirables will not advance Nigeria's economic development nor enhance the quality of life of Nigerians—the two critical products the nation's space programme must deliver.

Nigeria in a Space Race?

This author agrees with NASRDA that "the 21st century poses a challenge for the human race to live and work in space." That challenge is one that is not based on competition but one that is focused on collaboration and cooperation, among nations, in order to collectively solve the prevailing and future global problems. A case in point is human's limited but growing knowledge of the outer space environment, where Nigeria, twice, in 2010, almost lost its NigeriaSat-1 satellite to collisions with space junks; it was cooperation from the USA that saved the satellite on both occasions. Many other satellites that belong to other countries have also had similar encounters. As shown in Chapter III, a meteorite from Mars crash-landed in Katsina, Nigeria, in October 1962. In Chapter XIV, the reader will note that Nigeria and other countries along the equator went through the April 29, 2003 individual-national-panic over the possible crash-landing, near the equator, of the Italian Bepposax satellite. There

---

55 http://www.mtcr.info/english/FAQ-E.html (Accessed, June 15, 2015).

will be many more similar circumstances in the future. The regular threats of the impact of asteroids, comets and solar storms on all living things on planet Earth and to humans and human assets in space are here to stay.

These exigencies led the United Nations Committee on the Peaceful Uses of Outer Space (COPUOS) to invite all nations to partner with the international community to collectively seek and gain the necessary knowledge that would ensure sustainable human activities in outer space environment, beginning with appropriate scientific engagements in each country. In support of that same goal, the Asteroid Day, launched on June 30, 2015, called "for a 100-fold increase in the detection and tracking of asteroids or near Earth objects (NEOs) in our Solar System."[56] These are clarion calls that should be loudly drummed into the ears of Nigerian political leaders and its citizens as well as those in other countries, in order to ensure our collective safety and security. These are no agenda for any space race but bold plans for peace in our world.

Thus, NASRDA's exhortation that *"Nigeria, who is now in [the] space race, should not be left out in this endeavour,"* is misplaced. Who is in a space race with Nigeria and vice-versa? And for which prize? Such a counsel to the authorities in Abuja is contrary to the justifications Nigeria gave in 1982 and 1999 when it declared its interest to participate in peaceful space activities, namely:

*"for the purpose of development, **inter-alia**, in agriculture, mining, physical and geological surveys, forestry, fisheries, communications, and broadcasting, meteorology and research…".*[57]

True, there was a time, from the mid-1950s to the late 1980s, when space exploration was a race, particularly between two nations, the USA and the USSR. In the space race era, much money was spent, in a rush, to beat the opposition. In the process, mistakes were made and intended goals were not achieved on the first attempt of many activities.[58,59] More money had to be spent; and with caution and patience

---

56 Asteroidday.org (Accessed July 12, 2015).

57 *Nigeria's National Paper to UNISPACE-82,* UN-OOSA UNISPACE-82 Records: National Papers, 1982.

58 On December 6, 1957, USA's Vanguard TV3, USA's first attempt to launch a satellite into orbit was also its first failure. Two seconds after leaving the launch pad at Cape Canaveral, this rocket lost thrust and sank back down, rupturing and exploding its fuel tanks.

59 Soyuz 1, the first man space flight of the USSR, crash-landed 3 km west of Karabutak, on April 24, 1967. The one-day mission culminated with the parachute of the cosmonaut, Vladimir Komarov, not opening properly after his atmospheric re-entry. Komarov was killed when the capsule hit the ground at high speed.

in large supply, the goals were eventually achieved. That era of anything you can do, I can do better is gone, particularly for countries who believe in the development of their societies. Let us also not forget that, in August 1989, the space race era became history with the fall of the Berlin Wall. An era of collaboration and cooperation, born out of human common needs, rules the world today. The International Space Station (ISS), which Nigeria had hoped would host its first astronaut, is a unique symbol and a clear manifestation of this new era. Unfortunately, we are in a space race was the tune played by both Nigeria and the Regional African Satellite Communication Organisation (RASCOM) in the first decade of the 21st century. That was when Nigeria and RASCOM were in a race for the First Pan-African communications satellite trophy. In the end, as shown in Chapter VIII, both lost.

National Pride, Integrity, and Enhancement of Geopolitical Standing

Similarly, the need for Nigeria to attain integrity, continental leadership and national pride and to enhance its geopolitical standing are self-serving pronouncements as justifications for its space aspiration. Such declarations cannot influence nor win any friends and may indeed ward-off potential collaborators. Nigeria cannot buy "continental leadership" and "integrity" in S&T and related activities, such as space or in anything else. Nigeria and Nigerians must earn these desirables.

Nigeria will earn its integrity at home and within the global community and will attain its dream of continental leadership in Africa when, in reality, Nigeria succeeds in:

- Developing technologies that can power a vibrant national economy;
- Applying such knowledge to address the basic needs of the nation's teeming population, including the reduction of poverty and the enhancement of meaningful job opportunities in the land; and in
- Operating a national space programme that truly contributes to its development and growth as articulated in Chapters XI and XII, as well as to the solution of regional and global problems as presented in Chapters XIII and XIV.

Strategic Partners

In each activity category, NASRDA listed an array of "foreign Strategically chosen Partners that are well disposed to this [space] programme and will contribute meaningfully both financially, materially and technologically" in support of the Road Map. However, in each case, what assurance does the Nigerian government have that each of these strategic partners has signed on to this partnership arrangement, and is, indeed, prepared to make meaningful financial, material and technological commitments in support of Nigeria's 25-Year Road Map Space Mission? Specifically,

- What are the criteria for selecting these partners?
- What are the space history and related accomplishments of the chosen partners?
- Do these selected partners conform to the space policy of Nigeria?
- How "well disposed" are each of these chosen strategic partners?
- In concrete terms, what exactly, will these partners bring to the table in terms of "meaningful, financial, material and technological" support?
- Except for UN-OOSA and SSTL, other partners claimed by NASRDA, as of December 20, 2016, are not listed on NASRDA's website. And unlike the websites of other parastatals of the Nigerian government, such as the National Communication Commission (NCC), National Emergency Management Agency (NEMA), Nigeria Airspace Management Agency (NAMA) and National Information Technology Development Agency (NITDA), NASRDA's website is also very devoid of information and reports on its activities.

All of the above clearly show that the 25-year Road Map for Nigeria's Space Mission is very much in need of a revisit and a redesign by the Nigerian government, as recommended in Chapters XV and XVI.

Given the many steps Nigeria has taken on this space journey, particularly from the last decade of the 20th century till now, it is appropriate to step back and reflect on the lessons the nation has or should have learned in the process. That is the focus of the next chapter.

* * *

Chapter VIII

LESSONS LEARNED

Every experience in life, including a space journey, offers opportunities for reflection and learning. In this regard, and possibly as a result of his own personal experience, the Spanish philosopher, George Santayana, assured us that:

Those who fail to learn from history are bound to repeat its mistakes.[1]

As one reflects on Nigeria's on-going space journey, it might be equally appropriate to also ponder over the decisions that have shaped the nature of that journey and the lessons learned in the process. For example, Nigeria has acquired Earth observation and communication satellites, and its space-assisted meteorological services and search and rescue operations are also being directed from Abuja. Is the nation's space effort one that can lead to a robust technology development and subsequent enhancement of the quality of lives of Nigerians? A look at the lessons learned from this journey, to-date, should provide some clues.

Buy and Purchase of NigeriaSat-1 and NigeriaSat-2

In the 1999-2007 timeframe, the nation's Federal Ministry of Science and Technology (FMST) took significant steps and developed a number of policies and programmes including those needed for the management of its parastatals. Nigeria's space policy and programme, the 25-year Road Map on Nigeria's space mission, and Nigeria's science and technology policy documents emerged during this period. Some of these policies are working as designed; others need to be revamped.[2]

---

1 Santayana, George (1905). The Life of Reason, Volume 1, 1905,Originally published by Charles Scribner's Sons in 1905.

2 Abutu, Alex (2010). *Nigeria rethinks 'failed' science policy*, Science and Development Network, September 17, 2010.

Approaches to policy formulation and the execution of a given policy often affect the outcome of some of the intended programmes. As shown in Chapter V, in 1999, the Federal Executive Council (FEC) decided that *'Nigeria shall develop, design, build and launch its own satellite.'* At the time it took that decision, the FEC also established NASRDA, the nation's space organisation, and mandated it to pursue the attainment of the objectives embodied in its 1999 decisions. The space policy and programmes that the same FEC approved on July 4, 2001 also had the building of Nigeria's *indigenous competences* as its motto. As one reflects on the 2000—2001 decisions of the Nigerian government that led to the *'buy and purchase'* of NigeriaSat-1 and -2, it is apparent that the same government did not adhere to the directives of its own Federal Executive Council. Knowledgeable Nigerians at home and abroad, as well as *foreign friends of Nigeria,* cautioned against the *buy and purchase* approach as a way of developing space capabilities in Nigeria, but to no avail. The desire to *fast-track* a project, in order to showcase the accomplishment of the incumbent administration, often overrules the need to undertake a methodical process, without which the attainment of the intended goal is derailed.

What has Nigeria gained or lost by buying Satellites?
Mastery of the Enabling Technologies

Given that Nigeria went from the *Sardauna* towers (See Figure 3:2 in Chapter III) in the early 1960s to an altitude of 36,000 km in outer space by the year 2007 with NigComSat-1, one is inclined to believe that, in the process, Nigeria also retooled itself for the technological transformations and challenges of the future. But as shown in Chapter VI, that was not the case.

In the first decade of the 21st century, Nigeria and a number of developing countries acquired their first satellites, mostly in the micro-satellite range, in the open market, from those companies that marketed and are still building and marketing micro-satellites as indispensable tools of development and as the cheapest and fastest way for the developing countries, including Nigeria, to get into space. One message that was not preached by these satellite merchants is that a nation can actively participate in and benefit from space activities without going to space; many countries are doing so without any assets of their own in the outer space environment. For those countries that already swallowed the bait of *buy and purchase* of a microsatellite in the open market or are contemplating the same option, it is becoming apparent today that *the route to indigenous capability development in space is not by leap-frogging or fast-tracking;* indeed, it demands a different approach. It should begin with the retooling of a nation and its people, and with the mastery of the enabling technologies. That first step is an indispensable milestone in a nation's space aspiration, a preparatory step that would equip the nation to adequately assess its own capacities and capabilities and thereafter build upon them by investing in essential knowledge generation, development and utilisation.

The attainment of such a level of S&T maturity would lead the nation to initiate, develop, test and build a variety of hardware and software components, some of which may end up in a range of products including satellites. By buying or purchasing foreign-built satellites, as a short-cut or fast-track to technology development, Nigeria undermined its own long-term capability development. No space-capable country attained its space aspiration by buying or purchasing a satellite. Nigeria should never have taken such a short-cut; but by doing so, it jumped the critical fundamental steps of strengthening its own science and technology foundation that is needed to support a long-term national space endeavour. The short-cut Nigeria took has not earned it a place of honour at any true global S&T knowledge-sharing table nor won it any respect in any international space arena.

Why did Nigeria take these upstream steps instead of a down-stream approach? The downstream approach would have focussed the nation's attention on using space technology to help the Nigerian people meet their needs, either in the use of space-acquired data for natural resources management, in communications or in satellite navigation for precision agriculture, and for surveying and mapping.

In case we still do not understand the impact of our actions, in October 2011, South Africa drove the message home for us when it declared that: *"But unlike the others, we build our own satellites.*[3]*"* Earlier in the year, on April 24, 2011, Egypt cancelled the planned launch of EgyptSat-2, and announced that *"...60 percent of the components and software of the new EgyptSat-2 satellite will be Egyptian-made.*[4]*"* Going by the words of NASRDA, Nigerian engineers participated in the building of NigeriaSat-X that was successfully launched on August 17, 2011 along with NigeriaSat-2; by so doing, Nigeria demonstrated, in part, its cognizance of the need for a hands-on approach to self-development.

Nigeria also needs to learn how the British company, named Surrey Satellite Technology Ltd (SSTL) that built and sold both NigeriaSat-1 and NigeriaSat-2 to Nigeria, came into being and grew from a university research and development

---

3 http://www.southafrica.info/about/science/satellite-071011.htm#ixzz1hytnZoFU (June 15, 2014).

4 EgyptSat-2 Launch Excised (Satellite), SatNews Daily, April 24, 2011(Accessed, June 18, 2014).

project into a profitable commercial satellite company today. SSTL's first satellite project, a home-grown effort, began in 1979 and was concluded in 1981 when the University of Surrey successfully launched its first micro-satellite, known as UoSAT-1. That was the time when microprocessors and computer memory became affordable, resulting in the mass production of the new microcomputers of that era. The rapid advances in electronics in the late 1970s offered the University of Surrey and a number of dedicated staff, students, radio amateurs and their leader, today's Sir Martin Sweeting, the opportunity to *go* for the first modern micro-satellite. They developed the satellite and built it '*using donated materials,*' and '*commercially off-the-shelf* parts and '*a homemade clean-room.*'[5] UoSAT-1 was launched by Delta 2310 launcher from the Vandenberg AFB in the United States on October 6, 1981. It was that initial support and subsequent nurturing of that group by the University of Surrey that led to the transformation of the group into a company known today as SSTL. Thirty years after its debut, SSTL has become one of the major companies that build and sell micro-satellites.

In the process of developing, building and launching Nigeria's Earth observation and communication satellites, much of Nigeria's money went into foreign institutions and banks and was used to support research and development efforts in both China and the United Kingdom, but not in Nigeria. In fact and as shown in Chapter VI, Nigeria's research and development institutions have no funds to pursue their own research goals at home. Given the enormous cost of these satellites to Nigeria, fostering technology development in Nigeria, through bi-lateral cooperation, should have been key and conditional provisions of the satellites' *buy and purchase agreements.* One would have expected, justifiably, that Nigeria's aspiration to be a knowledge society should have led to collaborative research and development programmes in S&T and space science and technology between a number of Nigerian universities and counterpart tertiary institutions in both the UK and China. To-date, this has not happened; we are mostly undertaking "*business transactions.*" And since an informed consumer is a better customer, the question remains the same: *Was Nigeria a smart buyer?*

Support for Local Knowledge Development and Utilisation

A significant number of university educators and their students in the developing world, including Nigeria, can also deliver, if so challenged. Others have enough inner motivation to perform on their own. Such was the case with Prof. Ekundayo

---

5 http://www.spacedaily.com (Accessed, June 20, 2014).

E. Balogun and his physics students at the Obafemi Awolowo University, Ile-Ife, Nigeria. In 1986, Prof. Balogun and his students developed, fabricated and tested a functional satellite receiving station that was able to download data from operational meteorological satellites. If Prof. Balogun and his crew had been further encouraged and financially supported, they could have built on their success and would have developed and produced other tools and products as well as related spin-offs that could first satisfy the nation's market and possibly compete, successfully, with similar products for sale, both in Africa's market and beyond. In this connection, Nigerian leaders should reflect on the fact that the SSTL of today, that developed and built NigeriaSat-1 and NigeriaSat-2, was conceived and born as a rudimentary research and development project by students, staff and radio amateurs and nurtured at the University of Surrey, over thirty years ago, just as Balogun's meteorological satellite receiving station was in 1986. Local support in one case and lack of it in another made the difference. The opportunities are there, today, for Nigeria and other countries to replicate what happened at Surrey.

Earth Observation Satellites of the Future

Evolving technologies have also demonstrated that future transmission, reception, processing and archiving of large volumes of data will be far different from what we are used to, today. The emerging solution is the on-going research and development of an end-to-end user system, with one metre resolution imagery and high definition; this is being offered now, commercially, by Terra Bella and Satellogic as shown in Chapter VII.[6] Specifically, future Earth observation satellites would have capacities for simultaneous, global measurements and on-board timely analysis of the Earth's environmental data for real-time, mobile, professional and common users who can obtain their data needs, directly from the satellites, with the aid of their laptop computers. These developments could subsequently render existing ground receiving stations and associated image processing facilities obsolete. Should the latter become a reality, what would be its long-term implications for Nigeria and other countries with large investments in ground receiving stations?

With advances in science and technology development, it is now possible to go far less than the space altitude to carry out Earth observation activities and execute search and rescue missions. The revolution in robotics and the invention of new,

---

6 Zhou, G., O Baysal, J Kaye, S Habib, C Wang (2004). *Concept design of future intelligent Earth observing satellites*, International Journal Remote Sensing, 20 July, 2004, Volume 25, No. 14, pp. 2667-2685.

light-weight materials mean that all nations can afford to operate Earth observing sensing systems, with a variety of Unmanned Aerial Vehicles (UAVs) that they can own and control, inside their national borders, and in the service of their nation's interest and its people. These remotely operated systems, as shown in Chapters XI and XII, are reliable and are becoming increasingly robust. In fact, they are now being developed, built and operated by a number of universities around the world; they offer cheaper alternatives to satellites for acquiring, with the consent of the concerned state, the Earth observation data of any geographical territory on planet Earth. UAVs can also serve as surveillance and relay communications platforms.

Competition in the Global Market

Many Earth observation satellites, particularly the commercial ones owned and operated by such entities as GeoEye, Ikonos, Orbview, and World View arrived at their low Earth orbits many years ahead of those from the developing countries, such as Nigeria. New arrivals, including DigitalGlobe, Planet Inc (acquired Terra Bella on February 1, 2017) and Satellogic, are establishing a firm hold in the geo-information market with new products which include a guarantee of *real-time imaging of the planet made possible by constellations of very small high-resolution satellites, of platforms customers can use* to generate data stream and related specific information needed for decision-making, *and of high-definition video clips,* on demand. These commercial satellite operators have also weathered the storm of developing a new industry, and have mastered the art of responding, on time, to data and related requests on demand.

Collectively, they offer and provide products and services, such as imagery and value-added elements, that are on the high or up-scale end of whatever may be the norm within the industry and which their local competitors are not able to match. To them, the customer is the boss, and as such they often try their best to please and retain their customers. Because they offer such high quality products and are very prompt and efficient in their delivery of services, they have established their own communities of satisfied and loyal customers that cannot just be enticed away at any price. The end result is the continuing dominance of the commercial satellite operators in the global geospatial information landscape.

As shown in Chapter VII, there are also government-backed Earth observation satellite programmes from Brazil and China (CBERS-2), France (Spot), India, USA (Landsat series) and Europe (Sentinel) whose data are either free, are on-line, today, or will soon become free—a development that is being driven and accelerated by a number of websites that provide an array of linkages to sources of free satellite imagery.[7,8] It is thus understandable that customers, anywhere, that are in need of satellite data, will not, because of patriotism, rush to purchase imagery and related products they can obtain, hassle-free, and free-of-charge, on-line, or purchase at a comparative price, with higher quality and guaranteed timely delivery. Certainly, every customer expects value for his/her money.

The challenge is in the court of those who, as a justification for the purchase of their nation's satellites, fabricate unrealistic streams of huge income into the coffers of their governments from the sales of satellite data.

Missed opportunities—*Understanding and use of Earth observation data in development*

It is noteworthy that the Nigerian request to the United States in 1973, to use its satellites, Landsat -1 and -2, to acquire data of the Nigerian territory for use in Nigeria's economic development, paralleled the 1974 request of India's political leaders to the United States for NASA's assistance, in collaboration with India's science and technology institutions, to fight illiteracy[9]—a major socio-economic issue for the country. The effort, famously referred to as the Indian Satellite Instruction Television Experiment (SITE), focussed on educating the poor people of India on various day-to-day human issues and experiences, via satellite broadcasting to rural India, with the aid of a communication satellite, ATS-6, that the USA loaned to India, *gratis*. The one-year successful experiment, conducted from August 1, 1975 to 31 July 1976 demonstrated to India and to the rest of the world that India could use advanced technology to fulfil the socio-economic needs of the country and its people. It also set in motion the realisation of India's own space dream—the technical experience gained from SITE culminated in the indigenous development, construction and testing of India's first satellite, INSAT.[10]

SITE offered a significant lesson in knowledge acquisition for most countries and it was well publicised for many years throughout the United Nations system. The community centres established and used for SITE have now matured into *Village*

---

7 http://www.ppgis.net/imagery.htm (Accessed, June 22, 2015).

8 http://www.sat.dundee.ac.uk/ (Accessed, June 22, 2015).

9 The two leading space-related public institutions of that era, in India, were and still are, the Physical Research Laboratory (PRL), founded in 1947, with an initial focus on cosmic rays and the properties of the atmosphere, and the Tata Institute of Fundamental Research (TIFR), established in 1945,

10 Evaluation report on Satellite Instructional Television Experiment (SITE)—1981, Programme Evaluation Organisation (PEO) Study No.119, Planning Commission, Government of India, New Delhi, India.

Resource Centres (VRCs), which today number over 450 throughout India. The VRCs are connected to Knowledge/Expert Centres such as Agricultural Universities, skill development institutes, hospitals and reputable NGOs. Training programmes being conducted by the VRCs include those in agriculture/horticulture, fisheries, live-stock, water resources, tele-health, tele-education and weather.

But unlike India, Nigeria did not take the necessary steps to educate its people on how to process, analyse and use the data it requested and received from the United States. By the time NigeriaSat-1 was launched in September 2003, the nation had not acquired any collective experience on how to process, analyse, interpret and use, for its national development efforts, any of these data sets in its national archives. And despite its acquisition of similar data with its three Earth observation satellites, NigeriaSat-1, NigeriaSat-2 and NigeriaSat-X, from 2003 till today, the nation's understanding and use of Earth observation data for social and economic development, as shown in Chapter VII, is still at its infancy. The reasons are not far-fetched.

Those in charge of the Nigerian Satellite Project did not prepare the Nigerian population on how satellite applications and services could form part of the fabric of every sector of the Nigerian economy. Part of the problem was that the FMST negotiated with the commercial builder of NigeriaSat-1 satellite in total secrecy; the nation's intellectuals and inquiring minds including the nation's private sector were also not invited to participate in nor consulted on the project. If the reverse had been true, a number of critical questions would have been raised by the Nigerian scientists, political leaders and policy makers, prior to the launch of the satellite. For example, there would have been a drive to understand and use the Landsat satellite data at hand and from that experience, determine what should be the characteristics of the sensor systems for the planned Nigerian satellite.

And because Nigeria is close to the equator, Nigeria might have also chosen a different orbit for its satellites, just as Indonesia did in 2015, as shown in Chapter VII. NigeriaSat-1 and -2 were launched into the *default* polar orbit, with a revisit cycle of 3-5 days to the same location, instead of an equatorial orbit with a revisit cycle of about 16 times a day, to the same location; certainly, a Nigerian satellite in the equatorial orbit would have served Nigeria better. At the 2010 IAA-NASRDA *Symposium on the Attributes and Characteristics of the Equatorial Plane*,[11] this author raised the

---

11 International Academy of Astronautics (IAA)/National Space Research and Development Agency (NASRDA) International Symposium on the Attributes and Characteristics of the Equatorial Plane, 30th Nov.—2nd Dec., 2010, Abuja, Nigeria.

issue of the choice of orbit for NigeriaSat-1 and the then up-coming NigeriaSat-2, with the following question, that he addressed to the representative of SSTL at the symposium: With your knowledge of Nigeria as an equatorial country, why was its satellite, NigeriaSat-1, launched into the polar orbit instead of the equatorial orbit? In response, the SSTL representative revealed as follows: "The issue of NigeriaSat-1 and NigeriaSat-2 has to be looked into absolutely from a 'Business concept' angle." Certainly, that business has proven not to be in Nigeria's interest.

The wrong choice of orbit for the nation's existing satellites, notwithstanding, if Nigeria had undertaken a dedicated and intelligent use of Landsat data in its possession, the nation and its citizens would have gained the following:

- An enhanced ability to evaluate Earth-observation data and assess their value in various Earth science disciplines;
- The Capacity and capability to develop calibration techniques for Earth observation data including the development of software for image assessment;
- An enriched understanding of the type of sensor parameters that can meet the nation's development requirements as well as ones that could feature in its future custom-built-satellites; and
- The critical knowledge needed to undertake a project's proof-of-concept, and to develop as well as fabricate satellites and their payloads that could meet Nigeria's needs.

Nigeria's Communication Satellites— From NigComSat-1 to NigComSat-1R

Nigeria also bought NigComSat-1, its first communication satellite, from China. The satellite represented a milestone for China's satellite export business. For the very first time, the China Great Wall Industry Corporation (CGWIC) provided all the aspects of in-orbit delivery of a satellite to an international customer, Nigeria. Those familiar with the communication and tele-health services of NigcomSat-1 were satisfied; others concluded that if the satellite had operated for the duration of its life-expectancy of 15 years, it probably would have repaid itself. However, the satellite, launched on May 13, 2007, failed on November 10, 2008. Why?

Unanswered Questions

Artificial satellites, like any other man-made machine, do fail; NigComSat-1 was not the first or the last one. According to CGWIC authorities, the satellite packed-up on November 10, 2008 because its solar panels did not deploy properly. A chain of events followed. The satellite battery soon became a victim because the failed solar panel could not recharge it, hence the battery gave up. The chain reaction signalled and hastened the end of NigComSat-1's useful life. As indicated in Chapter VII, many unanswered questions remain.

A Learning Tool

In addition to NigComSat-1, other China-developed communication satellites that were built on the DFH-4 bus, such as Sinosat-2 of China (launched on October 28, 2006) and Venesat-1 of Venezuela (launched on October 29, 2008), also were lost or did not function properly. That notwithstanding, Nigeria put all its eggs in one basket on this satellite project, particularly in a circumstance where NigComSat-1 served as "a learning tool," "a test case," or "a guinea pig." The satellite manufacturer, the China Great Wall Industry Corporation (CGWIC), was the same one that also provided the launch services, the ground station construction, project financing, insurance and training for this communication satellite—NigComSat-1. Could Nigeria have been better-off with input from and participatory role by other players, particularly, more experienced entities? Or as noted by Lucy Corkin, could it be that *"African countries serve as a useful 'practicing ground' for Chinese firms that are trying to become competitive globally?"*[12]

In a Space Race with Whom and at What Prize?

As shown in Chapter VII, one should also not dismiss the impact of the pressure that could have been exerted on CGWIC, the manufacturer of NigComSat-1 by the Nigerian government and on Thales Alenia Space of France and Italy, the manufacturer of RASCOM QAF-1, by the Regional African Satellite Communication Organisation (RASCOM) authorities, in order to be able to claim the prize of being *"sub-Saharan Africa's first communication satellite in space."* In case there was such a "political push," the failure, on November 10, 2008, of NigComSat-1, a satellite once described in Nigeria as the *first Pan-African Communication Satellite*, can be linked to *a race* with RASCOM's QAF-1, also described by RASCOM as *Africa's first satellite;* the latter was launched from Kourou, French Guyana, by an Arianne launcher, on December 21, 2007. Surprisingly and ironically, both satellites only survived briefly in space and have since been replaced by the companies that insured them.

---

12 Corkin, Lucy (2007). The Strategic Entry of China's Emerging Multinationals into Africa. *China Report*, *43*, 309 ppf.

QAF-1R was launched as a replacement satellite for QAF-1 on August 4, 2010 and NigComSat –1R was launched on December 19, 2011 as a replacement satellite for NigComSat-1. Along with their respective manufacturing and launching partners, both RASCOM and Nigeria had to come up with additional funds and with heavy doses of caution and patience to match, in order to get it right the second time around with the replacement communications satellites. It was a price Nigeria and Africa did not need to pay; but they did because they were in a race of their own choosing. If indeed there was a race, the race certainly compromised all the talents, time and funds invested on both satellites. Sooner or later, the truth will emerge.

Becoming a Satellite Operator

The eagerness of Nigerians for efficient satellite communications services came to naught since NigComSat-1R could not deliver the ground-based applications it was designed to provide. National outcry soon followed, and the government has responded; and privatization is one of the options for consideration and subsequent action by the government. Perhaps we need to ask ourselves some critical questions: With only one satellite in orbit, alongside multiple satellites from over 50 international satellite operators, how can Nigeria be a competitive satellite operator? Have we invested in developing and building the critical competences that will enable Nigeria to provide functional communication services? Nigeria should first rectify the many loopholes that were exposed by the NigComSat Ltd saga, before it signs on for any additional communication satellites.

Points to Consider

Transparency and Honesty

In addition to all of the above, it should be noted that the true success of a nation's space programme can be built only on the foundation of transparency, honesty and competence. Entrenched vested interests could be lurking everywhere to derail that foundation through an oversell of unachievable expectations, goals and counterfeit ideas. As shown in Chapter V, in 2000, the Nigerian government of the day succumbed to the pressure, from highly educated Nigerians, that it *must treat this (Satellite—NigeriaSat-1) project with some urgency so as to position itself properly, as there are signs that an international movement is underway to restrict access to satellite imagery by developing countries.*[13] At that time, there was no proof to confirm this

---

13 Report of the Presidential Committee on the Satellite Project was completed on February 9, 2001.

counterfeit threat. Rather, the data that was to be denied Nigeria and other developing countries was and is now available, *gratis*, globally and can be accessed or down-loaded on-line by any one in need.

The government's own 2000 Special Review Committee also weighed in on the secrecy of the Nigerian satellite project and confirmed that:

> *"lack of detailed information to the space community and (the) general public has led to the initial unpleasant and serious objections for (the) acquisition of the satellite, ...and full participation by all relevant stakeholders has not been possible because the satellite project has been shrouded in official secrecy."*[14]

Ironically, none of these observations, concerns and alerts made any impact on the subsequent decisions of the government to purchase the satellite, decisions which have not served the best interest of the nation.

Because of vested interests, the secrecy that bedevilled the Nigerian satellite project screened off the essential institutions and the science and engineering experts that possessed the knowledge and competences needed for such an undertaking. Until Nigeria accepts that its future, as a knowledge society, is rooted in its recognition and reward of transparency and meritocracy, the series of painful national experiences, such as detailed in Chapters IV through VII of this book, will be repeated, over and over again. Where do we go from here?

Developing Nigeria's Capabilities

For a moment, let us reflect on all of the lessons cited above. Suppose Nigeria did not buy these satellites and all the associated accessories, launch services and monitoring infrastructure, but instead, over the past seventeen years of its space adventure, it had genuinely invested all of the same capital in the following:

- Upgrade the nation's S&T capabilities;
- Acquire requisite knowledge of earth observation data analysis, interpretation and utilisation;
- Develop and nurture vital fundamental elements of space-related sciences and technologies in Nigeria's schools and universities and at its science centres, including the development and nurturing of related human capacities; and
- Had doggedly followed such a plan-of-action, with undiluted commitment, where would Nigeria be today?

---

14 The Report of the Special (Review) Committee on (a) The Nigerian Space Programme, and on (b) The FMST (NASRDA)/SSTL Satellite Project.

Such a capital investment would have taken place if those charged with the development, implementation and management of the nation's space programme took to heart the admonitions of the Secretary to the Government of the Federation (See Annex I in Chapter V) and the nation's expectations he clearly articulated in his June 29, 2000 address. If that had been the case, by now, at least, many Nigerians would be able to attest to how space has improved their lives.

Similarly, Nigeria, as a nation, could have attained the following goals:

- Space science and technology would be advancing Nigeria's development process and contributing to the nation's growth;

- Today, Nigeria would be developing some technologies at home instead of importing virtually everything from abroad;

- Nigerian schools and universities would have become centres of excellence in a variety of disciplines; they would attract overseas students from different corners of the globe; and today, other universities and institutions, particularly from Africa, would be seeking collaboration with Nigeria instead of going to North America or Europe;

- Nigeria would not be running abroad for "consultants" and "experts," because by now, the nation would have grown and facilitated the long-term development and nurturing of its own experts;

- Nigeria would be nurturing its private sector and many Nigerians would be gainfully employed, not only in the space-related fields, but also in many other related areas that can improve the lives of all Nigerians;

- Nigeria's private sector would have matured and would be producing needed parts, even as the nation enforces a "local content" component policy; and

- Space could have matured into an operational tool of development for Nigeria.

But because the managers of the nation's space programme decided otherwise, the nation's space investment has not made much impact on the economy of Nigeria or on the lives of its people. How Nigeria should marshal its human and material endowments to attain the above objectives is the focus of the rest of this book.

* * *

REAWAKENING OF NIGERIA

Chapter IX

REBUILDING THE SCIENCE AND TECHNOLOGY FOUNDATION

In undertaking the reawakening of science and technology (S&T) in Nigeria, we should first remind ourselves of what UNESCO said of Nigeria in its 2010 report:

In the late 19th century, for instance, Nigeria was considered to be no more than a decade behind the United Kingdom in terms of technological development. The origin of this divergence in economic growth can be found in the disparate levels of investment in knowledge over long periods of time.[1]

Of the many possible questions that could follow a thorough assimilation of the above observation, here are three of them: How many decades ago was that? What went wrong? And how many decades behind the United Kingdom is Nigeria today in terms of technological development? The answer(s) to the second question can be found in Chapter VI of this book. In the case of the third question, Chapter X provides a ray of hope particularly in the richness of the nation's endowments and in the abundance of genuine Nigerian talents both at home and in the Diaspora. It is never too late; Nigeria has all it takes if Nigerians are singularly committed to rejuvenate and reawake S&T, in totality, in Nigeria, for the betterment of all Nigerians.

Reviving Science and Technology in Nigeria

The reviving of S&T in Nigeria demands more than imbibing an enhanced knowledge of the subject. The reawakening should begin with and can benefit from a number of small but significant steps. First, there must be a change of attitude towards

---

1 UNESCO Science Report, Executive Summary–2010, Paris, France (Accessed, December 7, 2014)

our own nation, a change that emphasises and re-emphasises "*in Nigeria's interest*," in our daily activities. The nation should also revisit those activities and programmes of the past that served Nigeria well and which the nation can and should build upon. Equally important are the lessons Nigeria needs to learn from the experiences in other lands. Nigerians of today and tomorrow should also learn from the examples of esteemed Nigerians, not only about what and how S&T works in development, but also how individuals with national responsibilities should discharge their duties with integrity, and in the interest of the nation and the people they serve. Such esteemed individuals initiated national programmes we should revisit and build upon.

Revisiting Programmes the Nation can build Upon

Putting people to work: There was a period in Nigeria's history when, through what was known then as *direct labour*, government departments, at all levels, used their own staff to execute government projects, nation-wide. That was when Nigerian civil servants, with professional education, actually performed, on the job, the tasks they were educated and trained to do. The ability to execute such tasks was the basis of their employment, in the first place. Such individuals included engineers, architects, surveyors (land, quantity and estate), geologists, planners, carpenters, and accountants, who executed all grades of technical and civil works that included electrical installations, railroad development, road and house construction and maintenance, as well as building and maintenance of water supply and sanitation systems, just to name a few. Nigerians of that era conceived, planned, developed, implemented and supervised, with distinction, a variety of projects that functioned satisfactorily and stood the test of time. Today, that is no longer the case.

The reality today is that many of the S&T professionals in government services are not doing anything related to their professional education and training, and as such have mostly forgotten the fundamentals of their profession. And if the professionals that are employed by government in Nigeria are actually not doing the jobs that are related to their professional education and training, how can the nation progress? At the federal, state and local levels, jobs that can easily be professionally executed in-house, with distinction, and which should be the contributions of these civil servants/professionals to the development and growth of the nation, are parcelled out as contracts to sundry contractors while the professionals-turned-civil servants are busy chasing, daily, one contract file after another. The concept of *direct labour* is very alien to today's generation of professionally educated Nigerian civil servants. Presently, the nation's graduates are sent into the fields, all over Nigeria, for a period of one year, as members of the *National Youth Service* Corps (NYSC), to put their knowledge into practice and in the process meet the needs of the common Nigerian in whatever part of the country they are posted to. For many of these young men and women, who thereafter, opt to join the Nigerian civil service, their one-year service in the NYSC might be the only opportunity many of these NYSC graduates would ever have in applying, with joy, the knowledge and professional skills they spent four or more years to acquire at the university. Certainly, there must be a better way. To progress, the nation must find solutions to the prevailing problems of under-utilisation, mis-utilisation and non-utilisation of many of Nigeria's professional talents that are undergoing decay in the various government ministries and parastatals, either by reviving the *direct-labour* programme or by developing an appropriate equivalent, at all levels of government.

Federal Emergency Science School: Nigeria should also revisit its approach to S&T education at the time of its independence. At that time, we understood and appreciated the role of science and technology in development. Because of the acute shortage of qualified indigenous S&T personnel needed to sustain the national development agenda in the country at that time, the Federal Government established the Federal Emergency Science School (FESS) at Onikan, Lagos. As the name implied, the school served as a tool to jump-start the science and technology capability Nigeria needed to propel its post-independent aspirations. Today, a very high percentage of Nigerian S&T individuals, such as medical doctors, pharmacists, scientists, engineers, architects and surveyors and who are 60 years and older, gained their entry into their chosen S&T careers through the FESS science immersion programmes. The federal and the state governments should resuscitate the Federal Emergency Science School concept or develop an equivalent, in order to inject some life into the ailing S&T condition in the country.

The Lessons of the Biafran War: There is one aspect of our national life many Nigerians do not wish to be reminded of. But in Nigeria's interest, Nigerians should not be bogged down with only the painful lessons of the Biafran War; there are also a number of positive lessons the nation can draw from that experience. Better still, Nigeria would be better-off if we chose not to close our eyes on those experiences but to learn a number of lessons from them. The centre piece of this argument is the illustrious life and contributions of Professor Gordian Obuneme Ezekwe (1929-1997).

Many have characterised him as being an inventive scientist and a technological whiz-kid, accolades that he and his colleagues won as they did all that was humanly possible, in the field of science and technology, to sustain the defunct Biafra during the civil war. They established mini petroleum refineries to ensure self-sufficiency in petroleum products, produced internationally acclaimed vaccines and developed and produced a variety of weapon technologies to counter the onslaught from the federal forces. After the civil war, Professor Ezekwe also led a number of S&T-related teams, variously called Research and Production Board, Science Group, PRODEV and PRODA.[2] The civil war, notwithstanding, his accomplishments in the war and thereafter did not go unnoticed by the Babangida administration which appointed him, in 1989, as the Federal Minister of Science and Technology. Thereafter, he ably served as the pioneer Vice-Chairman and Chief Executive Officer (CEO) of the National Agency for Science and Engineering Infrastructure (NASENI), at a time when Nigeria hosted a national symposium to mark the International Space Year.[3] The entrepreneurial vision that drove the Ezekwe-led inventive spirit is still there in Nigeria; it can and should be harnessed for the good and in the interest of a united Nigeria.

Learning from Experiences in Other Lands

A critical part of development is the sharing of knowledge amongst professionals as well as amongst nations. From observations made in other sections of this book, Nigeria is very much wanting on this score, particularly on sharing of knowledge across political boundaries. However, it is not too late. One such basic effort is to complement the revival of the Federal Emergency Science School initiative with a national science enrichment programme such as the 1998 *YEAST* initiative in the Republic of South Africa (RSA). Other learning experiences include the engagement of foreign scientists to contribute to local research effort, and the cultivation of entrepreneurial spirit within our universities. We can also learn from case studies of how S&T and R&D are positively transforming other countries. But first, let us get to know and learn from *YEAST*.

YEAST: 1998 was designated the Year of Science and Technology (*YEAST*), in the Republic of South Africa. *YEAST* focused on a whole range of programmes that were scheduled and presented, throughout the year, in a cooperative way, involving

---

2 Olunloyo, V.O.S (2009). *Ogbomosoland Heroes: Brig. Gen. Benjamin Adesanya Maja Adejunle (rtd), The Unsung Patriot,* Ogbomoso Development Forum, Lautech, Ogbomoso, April 11, 2009.

3 The International Space Year (ISY), 1992, was a year-long, worldwide celebration of space cooperation and discovery. The year 1992 marked the thirty-fifth anniversary of the birth of the space age: in 1957, the first man-made satellite—Sputnik I—was launched by the former Soviet Union during the International Geophysical Year, which highlighted scientific inquiry and international cooperation.

the public, students, pupils, academics, science councils, business and government departments. *YEAST* has had a significant impact on the development of South Africa more than most South Africans realise. It all started with an idea re-awakened in President Nelson Mandela by Nancy Fraser in 1993 on the traditional sources of power and authority that can help democratise science and technology.[4] Democratization of science and technology implies that the people, as non-experts, are an integral part of all deliberations on policy, regulation and control of science and technology such as in debates or controversies arising from such issues as science, communications and biotechnology and the need and right of the people to know. The Mandela government that came into power in 1994 made the democratization of science and technology a priority in post-apartheid South Africa. It subsequently used the authority and power it had to launch *YEAST* to accomplish that goal.

Through *YEAST*, the South African Astronomical Observatory (SAAO) started a Science Education Initiative as a contribution to address the problem "of an unequal educational structure, particularly in science, [where] there is a general lack of science literacy and awareness." To accomplish this goal, SAAO initiated the use of astronomy as a vehicle to stimulate an interest in science amongst teachers and students. This educational programme, which is still very much active, was given a boost by the South African government in 1998. The construction of the Southern African Large Telescope (SALT) has also provided additional stimulus to the Science Education Initiative.[5] The *YEAST* exposure and experience empowered the people of South Africa to debate the pros and cons of introducing genome maize in the country.[6] The impacts of *YEAST* on science education, agriculture and other aspects of S&T in national development in South Africa are worthy of study by Nigeria.

Attracting foreign-based research scientists: The adage that *no man is an island* is equally true for the countries of the world and their institutions. Institutions and nations seek collaboration, in part, because of the need to innovate jointly.

---

4 Fraser, N. (1993). Rethinking the public sphere: A contribution to the critique of actually existing democracy, in Calhoun', C. (ed.), Habermas and the Public Sphere, Cambridge, MA: Polity Press, pp. 109–142.

5 Rijsdijk, Case (2000). *Using Astronomy as a Vehicle for Science Education*, Publications of the Astronomical Society of Australia2000, 17, 156–161, CSIRO Publishing, East Melbourne, Australia.

6 Mwale, Pascal Newbourne (2008). *Democratization of Science and Biotechnological Development: Public Debate on GM Maize in South Africa*, Journal of *Africa Development*, Vol. XXXIII, No. 2, 2008, pp. 1–22, Codesria, Dakar, Senegal.

Development of technical and non-technical solutions to given problems can indeed result from and be more effective in a collaboration or partnership that promotes the exchange of formal and informal knowledge across and beyond national boundaries to include foreign specialists. By engaging in scientific activities locally with tangible and credible results, Nigeria can attract international support and collaboration. This is the way to ensure that not only problems in the industrialized world are addressed and solved, but that those of Africa, as a whole and in Nigeria in particular, also attract credible attention. And just as in Brazil, the first step must begin locally.

In order to enhance opportunities to tap into the global knowledge-rich market, in October 2006, Brazil opened the doors of one of its highly prized research institutions, Fiocruz, to foreign talents by granting research fellowships to foreign scientists.[7] Fiocruz maintains research institutions that specialise in health sciences, biological sciences and health-related social sciences. The fellowship programme, which is offered through the Oswaldo Cruz Foundation, allows Fiocruz to enhance international cooperation with both developed and developing countries. Initially, the fellowship was opened to researchers in the health sciences. The interaction between these foreign post–doctorate research fellows and their Brazilian counterparts is bringing much knowledge sharing to the advantage of Brazil and its citizens.[8] The fellowship programme at Fiocruz is contributing to the institution's ability to "be very active in the area of health research and development, and as a major manufacturer of vaccines, pharmaceuticals and biopharmaceuticals." Its Center for Technological Development in Health (CDTS), which became operational in 2015, is also home to the National Institute of Science and Technology for Innovation in Neglected Diseases (INCT-IDN); the latter is a major contributor to the understanding of tropical diseases which are mostly not on the radar of the world major pharmaceutical companies.

Brazilians have also developed other innovative ways of funding research activities. Research is taking a centre stage in Brazil's richest state of São Paulo, a situation made possible by its constitution, which guarantees the State of São Paulo Research Foundation (FAPESP) 1% of the state government's tax revenue. In 2010, the tax revenue translated into US $450 million, in addition to what the state received from

---

7 Esteves, Bernardo (2006). *Brazilian centre aims to attract foreign researchers*, SciDev.Net, London, UK, 3 July 2006 (Accessed, December 10, 2013).

8 http://www.cdts.fiocruz.br/inct-idn/downloads/p17-9_Carlos_Morel_Intl_Innovation_131Research_Media.pdf (Accessed, December 10, 2013).

the federal government in Brazilia.[9] Availability of such funds allows FAPESP to offer an international standard of remuneration to its research scientists and post-doctoral research fellowships. FAPESP's fellowships have included research work in aerospace engineering with emphasis on structural composites with carbon nanotubes; satellite image data mining; molecular and celular biology, immunology and plant genetics; bioenergy and biochemistry; global climate change and bioinformatics—the latter with a focus on signal transduction pathways in malaria parasite. Most of these fellowships are tenable at the various research centres and universities in the State of Sao Paulo.

Similar post-doctoral fellowship programmes abound world-wide. The National Research Councils of both Canada and the USA offer post-doctoral research fellowships in a variety of disciplines. In order to contribute to knowledge enrichment in the developing countries, the Third World Academy of Sciences (TWAS) in Trieste, Italy, promotes *scientific capacity and excellence in the South (Southern hemisphere) for science-based sustainable development* through the award of fellowships for post graduate research.[10] *The fellowship programmes operate under agreements with governments and national organisations in developing countries and offer a number of fellowships to young scientists from developing countries to carry out postgraduate research in developing countries other than their own.* Presently, TWAS postgraduate research fellowship programmes are tenable at the scientific research institutions or universities of the partner countries shown in Table 9.1 on the next page.

Nigeria and any of its 36 states can make a similar commitment either directly, as is done by Brazil and its State of São Paulo or indirectly through TWAS. The end result is the same—the cross-fertilisation of knowledge which includes the development, sharing and utilisation of knowledge in national development that would result in the enhancement of the quality of life of the citizenry.

Entrepreneurial spirit within the universities: Today, in most countries that have committed themselves to industrialise their economy, it is not enough to present a graduating student with a certificate or diploma as a manifestation of his/her academic accomplishments. What has become standard practice is involving the students in linking the innovative ideas of the university's research scientists and technologists with the entrepreneurial and management skills of the private sector with the funding support of the government to address the industrial production strategies of the nation.

---

9 Science in Brazil: An emerging power in research—Go south, young scientist, The Economist, London, UK, January 6, 2011 (Accessed, December 10, 2013).

10 http://twas.ictp.it/prog/exchange/fells/fells-pg(Accessed, December 11, 2013).

Table 9:1 Overview of research fields covered by each programme (+)
(*Source: Third World Academy of Sciences*) Accessed, January 14, 2014

Programme	Agric. scs.	Biology	Chemistry	Earth scs.	Engg. scs.	Mathem.	Medical scs.	Physics
CNPq (Brazil)	+	+	+	+	+	+	+	+
CAS (China)	+	+	+	+	+	+	+	+
CSIR (India)	+	+	+	+	+	+	+	+
DBT (India)	+	+	+	+			+	+
Bose (India)			+			+		+
IACS (India)		+	+		+	+		+
icipe (Kenya)	+	+	+			+		
USM (Malaysia)		+	+		+	+	+	+
CEMB (Pakistan)	+	+					+	
ICCBS (Pakistan)	+	+	+				+	
CIIT (Pakistan)		+	+	+	+	+		+

Through such a collaborative arrangement, the universities (professors, research scientists and students) acquire experience in executing a variety of research undertakings for government departments as well as low- and high-technology companies. The participating students and researchers also imbibe the principles of entrepreneurship and leadership; their activities contribute to economic development and jobs creation as well as pave the way for collaboration, nationally and across political boundaries, for applied research in many areas of human endeavour and national needs.

In the United Kingdom, for example, the University of Birmingham floated the Birmingham Investment Fund with the goal of attracting long-term capital that could support science and technology (S&T) efforts that would lead to the next generation of industrial consumer products.[11] The university is offering its breakthroughs in

11 http://www.birmingham.ac.uk/partners/rcs/index.aspx (Accessed, July 14, 2014).

medicine, engineering, energy and social science as its main entrepreneurial attractions. The university's Research and Commercial Services programme, the centre-piece of its entrepreneurial activities, has two key roles. The first role is to build industrial partnerships and collaboration by working with and providing the industry and the public sector with the knowledge and resources available through world-class expertise at the university. The second role is to support its academic staff in accessing funding for their research.

At the National University of Singapore (NUS), the university's Entrepreneurs Association (NUSEA) is a student managed organisation.[12] Within the framework of NUS' Overseas College Programme, students are sent to various entrepreneurial hubs in the world such as Stockholm, Silicon Valley (Stanford University), Beijing, Shanghai and Bangalore where they spend a year-long internship in a start-up company and take entrepreneurial related courses in NUS' partner universities. The aim of these activities is to promote the entrepreneurial spirit within NUSEA's members with the goal of cultivating and nurturing global entrepreneurs of tomorrow.

In Philadelphia, Pennsylvania, USA, the Drexel University's Baiada Center for Entrepreneurship bridges education and entrepreneurship.[13] The centre provides physical space to incubate ideas, entrepreneurs and companies in order to link research, coursework and entrepreneurial thinking of its students and researchers, with practical guidance and real-world thinking. This is accomplished through the Office of Technology Commercialization at Drexel. The Office brings "Drexel's resources and guidance in the commercialization of new technologies to the wide spectrum of talent at the university—from Drexel student teams with promising business ventures to Drexel researchers developing new technologies, to students and researchers that are creating new uses for advances in such fields as the biosciences." The university is thus able to develop entrepreneurial leaders and create successful technology ventures that benefit the community and beyond through new products and job creation.

Science and Technology as a Tool of Development
Examples from Brazil, Saudi Arabia and South Africa

It might also be instructive for Nigeria to examine what motivated other societies and how they changed the course of S&T in their respective lands and in the process,

---

12 http://www.slideshare.net/mengkiat/national-university-of-singapore-entrepreneurs-association(Accessed, July 14, 2014).

13 http://www.lebow.drexel.edu/Centers/Baiada/(Accessed, July 15, 2014).

altered, for the better, the history of their nation's development. The experiences of three countries, Brazil, Saudi Arabia and South Africa are illustrated here.

Brazil

Brazil's S&T successes are the results of two deliberate decisions of successive Brazilian administrations, namely, the recognition and acceptance of S&T as a tool of development and the subsequent national commitment to handsomely invest, on a consistent basis, in long-term S&T-based research and development activities. The end result has been a progressive development of the nation and related improvement in the well-being of the Brazilian people. The global community is not unaware of the S&T steps Brazil took and is taking to attain its expanding economic status, an economic success built on S&T that is better illustrated by Brazil's approach to ethanol production.

Brazil became an ethanol producing country in the 1930s. That was the time the country produced more sugar than it could use. The government subsequently directed the excess sugarcane into ethanol production and made the addition of ethanol to gasoline compulsory, in order to reduce its annual cost of imported petroleum products. The 1973 international oil crisis that originated in the Middle East doubled Brazil's expenditure on oil imports and everything changed thereafter. The government was forced to consider alternative sources of energy in order to decrease its dependency and spending on fossil fuels. It was not until 2008 that Brazil discovered crude oil deposit, in commercial quantity, within its territory; it has since become a significant net exporter of crude oil—in 2014, Brazil exported US$16.36 billion worth of crude oil.[14]

Prior to 1973, the Latin American countries, Brazil included, discussed and grappled with how science and technology should be integrated into the development process if the region were to overcome its underdevelopment. Three ideas on S&T as a tool of development surfaced in the late 1950s through the late 1960s. Subsequently, the Latin American countries debated how to follow these approaches in order to overcome their dependency and under-development by the year 2000, just as Nigeria has been doing with its 2020 Vision. Amongst the three contending scenarios was the proposal from the United Nations Economic Commission for Latin America and the Caribbean (UN-ECLAC), which encouraged domestic industrial

---

14 http://www.worldstopexports.com/highest-value-Brazilian-export-products/2959 (Accessed, July 29, 2014).

production that would create demand for local technology.[15] Let us recollect from Chapter VI that the former International Bank for Reconstruction and Development (IBRD) introduced a similar economic development concept in Nigeria at the time our nation obtained its political independence from Britain in 1960. To the Latin American countries, UNESCO espoused its own "linear supply push" approach which advocated advances in as well as the expansion of scientific, technological and educational undertakings as pre-requisites to economic development; China subsequently adopted this approach.[16] In 1968, Sabato and Botana, two thinkers from Argentina, offered their own idea, known as the *Sabato Triangle*, which called on the Latin American countries to overcome underdevelopment *"by integrating the three pillars of development including funding, namely research, private industry and the government."* In Latin America, Brazil stood out as the only country that really committed itself to imbibing much of the knowledge that emanated from a combination of these three economic development concepts.[17]

The 1973 oil shocks set the minds of Brazil's leadership and its research community ablaze. Brazil's leadership in ethanol production today is not an accident but a definite and clear manifestation of how the Brazilians doggedly committed themselves to the combined elements espoused by the above three S&T polices, a commitment that has ultimately transformed their nation for the better. The adage, *necessity is the mother of invention,* has also worked in Brazil's favour. In order to decrease its dependency and spending on fossil fuels from abroad, to power its economy, the government launched the National Alcohol Programme (Pro-Álcool) in 1975, with the prime aim of increasing ethanol production as a substitute for gasoline. But only a handful of Brazilians gave the research programme a chance to succeed. The government was undaunted; it was an investment challenge that was worth the price. Between 1975 and 1989, subsequent, Brazilian governments massively invested in infrastructure and research and in the process, successfully propelled Brazil into the forefront, as a leader,

---

15 Plonski, Guilherme Ary (2000). S&T Innovation and Cooperation in Latin America, Number One-2000.

16 Yeh, Stephen H.K. (1989). Understanding development: Modernization values in "Asia and the Pacific region, Paper prepared for UNESCO, Paris, August, 1989

17 Hatakeyama, Kazuo and Dirlene Ruppel (2004). *Sabato's Triangle and International Academic Cooperation: The Importance of Extra-Relations for the Latin American Enhancement, International Conference on Engineering Education and Research on "Progress Through Partnership."* Ostrava, Czech Republic, June 27-30, 2004

in the ethanol production market. Over the next 15 years, the production of ethanol soared from 0.6 billion litres in 1975 to 11 billion litres in 1990.[18]

Today, Brazil is the world's largest sugarcane ethanol producer and a pioneer in using ethanol as a motor fuel. According to the report of Brazil's Agricultural Trade Office, Brazil is expected to produce 26.57 billion litres (7.02 billion gallons) of ethanol in 2013, up from 23.51 billion litres in 2012.[19] In 2014, ethanol production increased to 25.6 billion litres.[20] The report also noted that Brazil currently has 399 ethanol refineries with a combined nameplate capacity of 40.7 billion litres. Most of this production is absorbed by the domestic market where it is sold as either pure ethanol fuel or blended with gasoline at levels between 18% to 25% ethanol for use in Flex-fuel vehicles. Motor vehicles that run on any mix of petrol and up to 100% hydrous ethanol are known as *flex-fuel vehicles (FFVs)*.

Currently, 15 automakers offer over 90 models of FFVs in Brazil. As of mid-2013, more than 90% of new cars sold in Brazil were FFVs because of consumer demand; these vehicles now make up about 60% of the country's entire light vehicle fleet—a remarkable accomplishment in less than a decade. Thus, at the gas-pump, in 2013, Brazilians have chosen to replace almost 40% of the country's gasoline needs with sugarcane ethanol. Because they emit fewer pollutants and reduce greenhouse gases, Brazilians believe that FFVs are the answers to global motoring in the future.[21]

Spin-off from the Ethanol Challenge

Brazil's S&T success story is not limited to its transformation of sugarcane into profitable ethanol. It was also a national challenge and a singular effort that brought an intensive cooperation and collaboration between public and private research institutions and funding from both the government and the private industry. It opened up a wide avenue for research opportunities into many agricultural problems that were once ignored, and it resulted in very significant scientific and technological advances in Brazil's agriculture and related industries.

---

18 Almeida, Carla (2011). *Sugarcane ethanol: Brazil's biofuel success,* SciDev.Net, London, UK,, 6 December 2007 (Accessed, August 15, 2014)

19 Voegele, Erin (2013). *Brazil predicts increased ethanol production, exports,* Ethanol Producer Magazine, September 27, 2013 (Accessed, August 15, 2014

20 http://gain.fas.usda.gov/Recent%20GAIN%20Publications/Biofuels%20Annual_Sao%20Paulo%20 ATO_Brazil_7-25-2014.pdf (Accessed, August 16, 2014)

21 http://sugarcane.org/the-Brazilian-experience/Brazilian-transportation-fleet (Accessed, August 18, 2014).

One such private institute that contributed to these advances is the Sugarcane Technology Centre in São Paulo; it invested $20 million annually into sugarcane research at the peak of the programme. *Vinasse*, at one time an environmentally unfriendly by-product of ethanol distillation, was successfully processed into fertiliser for the sugarcane and other agricultural fields. Similarly, combined efforts of researchers at this centre and at other institutions found use for *bagasse*, the left-over sugarcane fibre, 'to produce energy by building on existing methods of burning *bagasse* to power steam turbines for electricity generation.' The cost of ethanol production in Brazil is low because these same researchers developed *cauldrons* which could produce more energy from left-over sugarcane fibre, thus enabling the ethanol plants to become energy independent.

The ethanol research and development experience has rubbed-off positively on other aspects of Brazil's economic development agenda. Today, Brazil's major industries which include steel, commercial aircraft, chemicals, petrochemicals, footwear, machinery, motor-vehicles, auto parts, consumer durables account for 28% of the nation's GDP. Brazil is also a leader in science and technology in South America and a global leader in such fields as bio-fuels, agricultural research, deep-sea oil production, computers and remote sensing. The Brazilian government is also developing an environment that supports innovation with a focus on taking the output of scientific advances from the laboratory to the marketplace in order to promote economic growth.

Saudi Arabia

As most countries are doing today, Saudi Arabia is also conscious of the need to invest in S&T and related R&D and by so doing invest in its own future and in the future of its citizens. With its petro-dollars, it has invested in the old and standard industrial technologies that have put in place the necessary infrastructure for electricity, water supply and sanitation, basic transportation, health and communication for its citizens. The rulers of the Saudi Kingdom are also conscious of the fact that the petroleum reserve within their territory is a finite resource. What would become of the kingdom when the wells run dry would have been an engaging question among the country's leadership and their advisers; and what emanated from the on-going internal dialogue on this national concern was a variety of ideas on the country's future outlook. Not lost on the Saudis were the steps that Brazil, China, India, Malaysia and Singapore have been taking over the past twenty-five to forty years to consistently invest in scientific and technological research that would prepare their respective countries for eventual industrial competitiveness in the future. That future, they eventually agreed, will be dictated, in a significant manner, by the evolving new technologies.

Conscious of these global developments, King Abdullah bin Abdul Aziz Al Saud of Saudi Arabia asked his advisers a variety of questions on how Saudi Arabia could be an effective partner in that future. As was the case with space science and technology and nuclear science and engineering, the Saudis noted that, with the passage of time, new international rules and regulations will gradually surface as the global community tries to cope with the waves of the new technologies—such as nanotechnology, biotechnology and stem-cell research. To be able to contribute, on behalf of Saudi Arabia, to the formulation of those new rules and regulations would require some level of understanding and knowledge of these new technologies. Fast forward and the Saudis soon concluded that if they were to feature in such a future global dialogue with other countries—industrialised and developing—they should prepare their nation to make concrete and worthy input at such a conclave.

In 1986, and as a means of modernising his country, the King of Saudi Arabia cast his mind twenty-five years into the future and subsequently committed his country to establish a renowned 21st century S&T institution, now known as King Abdullah University of Science and Technology (KAUST).[22] KAUST was conceived, designed and established to lead many aspects of economic development in the Kingdom of Saudi Arabia. It was officially opened on September 23, 2009.

KAUST is Saudi Arabia's response and contribution to the global challenges of the future. With its nine research centres that specialise in such fields as water desalination, alternative energy, nanotechnology and stem cell research, KAUST is expected to galvanize the country and make it, not only a more robust oil *exporting country in the future*, but also a contender in global solar energy development and export. To gain respect, recognition and collaboration, and invariably access to knowledge from the international scientific community, KAUST has had two things in its favour, right at its inception. The first is the recognition and acceptance by the government to keep politics out of research or to keep research and politics separate. The second is the establishment of KAUST's Global Collaborative Research (GCR) programme, a mechanism put in place *to draw top researchers from around the world*, through the formation of partnerships with international research centres and universities. Today,

---

22 Yahia, Mohammed (2009).*Saudi science powerhouse opens its doors*, SciDev.Net, London, UK, 25 September 2009 (Accessed, August 18, 2014).

that partnership includes Imperial College, London, and the global Dow Chemical Company, USA. The KAUST type of academic freedom, wherein the government has accepted to keep politics out of research is, today, the envy of most of the academic and research institutions in Africa including Nigeria.

South Africa

South Africa joined the basic space science community with its Hartebeesthoek Radio Astronomy Observatory (HartRAO). The latter, built in 1961 by NASA, was one of USA's three *deep space* research stations (*Deep Space Station 51*) which tracked many unmanned USA's space missions, including Lunar Orbiter spacecraft (which landed on the Moon), the Mariner missions (which explored the planets Venus and Mars) and the Pioneer missions (which measured the Sun's winds). In 1975, the United States withdrew from the station and handed it over to South Africa which immediately converted it into a radio astronomy observatory, also in 1975; it has remained very active and functional ever since. The knowledge and skills acquired at HartRAO emboldened South Africa to aim for two significant astronomy programmes. The first was the construction and successful commission of the *Southern Africa Large Telescope (SALT)*, in Karoo, South Africa.

SALT was designed to unravel some of the mysteries of the universe in which we live. It was developed and built to record distant stars, galaxies and quasars a billion times too faint to be seen with the unaided eye—as faint as a candle flame at the distance of the moon.[23] The goals of SALT include the promotion of high technology investment and availability of opportunities for local scientists to participate in world-class science. SALT is also contributing to competence building in such areas as microwave and antenna systems; space applications technology, including satellite systems; radar and digital electronics, including computers, microprocessors and information technologies; and to the building of a vibrant industrial growth of RSA. The telescope is inspiring the youth of South Africa to study science and engineering; it is also being used to develop a more scientifically literate population in South Africa.[24]

---

23 Buckley David, Marguerite Lombard, Mike Lomberg, Kobus Meiring, Roelien Theron (2005). Africa's Giant Eye : Building the Southern African Large Telescope, The SALT Foundation (Pty) Ltd, November 2005 (Accessed, August 18, 2014).

24 Rijsdijk, Case (2000). *Using Astronomy as a Vehicle for Science Education*, Publications of the Astronomical Society of Australia 2000, 17, 156–161, CSIRO Publishing, East Melbourne, Australia.

In 1998, President Nelson Mandela's administration committed the first US$8 million required for the construction of the Southern Africa Large Telescope (SALT), with a hexagonal mirror array that is 11 metres across.[25] At the inauguration of SALT in 2005, the project scientist, David Buckley, remarked:

SALT was an initiative of South African astronomers that won support from the South African government, not simply because it was a leap forward in astronomical technology, but because of the host of spin-off benefits it could bring to the country,.........Indeed, the SALT project has become an iconic symbol for what can be achieved in science and technology in the new South Africa.[26]

On November 10, 2005, President Thabo Mbeki inaugurated SALT, the largest single optical telescope in the southern hemisphere. On that occasion, President Mbeki remarked that:

This giant eye in the Karoo would tell us as yet unknown and exciting things about ourselves. .....We have to have a scientifically literate work force if we are to make the advances we so desperately need.

The success of its SALT project emboldened South Africa to compete for the construction and hosting of the €1.5bn *Square Kilometre Array (SKA)* radio telescope project.[27] SKA was conceived in 1991 as a global project which aims to provide answers to fundamental questions about the origin and evolution of the Universe. South Africa and Australia were shortlisted in 2006 for the final site selection for the project, from the list of nations that initially bid for the project. The latter included Australia, Britain, China, Canada, India, Italy, New Zealand, the Netherlands, South Africa and the USA. On 25 May 2012 the SKA Organisation announced that the SKA project would be co-shared by both South Africa and Australia, with a majority share coming to South Africa. While the preliminaries of design, governance and funding campaign will dominate the 2014—2016 time period, with the prototype of

---

25 South Africa contributed about a third of the total of $36 million USD that financed SALT for its first 10 years ($20 million for the construction of the telescope, $6 million for instruments, $10 million for operations). The rest was contributed by the other partners—Germany, Poland, the United States, the United Kingdom and New Zealand.

26 http://www.southafrica.info/about/science/salt-telescope.htm (Accessed, August 19, 2014).

27 The Square Kilometre Array (SKA) promises to revolutionise science by answering some of the most fundamental questions that remain about the origin, nature and evolution of the universe. It will operate over a wide range of frequencies and its size will make it 50 times more sensitive than any other current radio instrument.

phase 1, SKA1, being completed by 2016, the actual construction of SKA1 is scheduled for 2018—2023; the detailed design of SKA2 is scheduled for 2018—2021.[28] The eight other African countries collaborating on SKA with South Africa include Botswana, Ghana, Kenya, Madagascar, Mauritius, Mozambique, Namibia and Zambia; these countries will host the remote stations of SKA.

Currently, Australia, China, Italy, the Netherlands, New Zealand, South Africa and the UK have signed an agreement to create and provide funding for the SKA Organisation.[29] More than 70 R&D institutes in 20 countries, together with industry partners, are participating in the scientific and technical design of the SKA telescope. The SKA project is expected to drive technology development in antennas, fibre networks, signal processing, and software and computing, with spin-off innovations, including those that would benefit other systems that process large volumes of data.[30]

By making these investments, including the indigenous development and construction of its Earth observation satellite, SumbandilaSat, South Africa has sent a signal, not only to Africa, but to the global community that it is more than prepared to contribute to knowledge generation, development and sharing in the global space efforts of the future.

The competences that are now driving Brazil's, Saudi Arabia's and South Africa's march to the future are also essential for the emergence of Nigeria as a knowledge-society, for improving the well-being of Nigerians, and for energising its industrial growth; they are also essential prerequisites for Nigeria's space journey. It is a challenge Nigeria must address. But first, let us also learn from the experience of two illustrious members of the Nigerian family, on how they addressed the challenges of their time.

Home-grown Experiences that are Worthy of Emulation

As Nigeria rebuilds its S&T foundation, such an effort should benefit from the experience of two of the country's esteemed professionals turned public servants; their records of selfless service and high integrity in executing the programmes they

---

28 http://www.ska.ac.za (Accessed, August 19, 2014.

29 International partners join forces and agree [to] funding for detailed design of Square Kilometre Array telescope". SKA Organisation, November 23, 2011. (Accessed, August 25, 2014).

30 http://www.southafrica.info/about/science/ska.htm#.UkxkDRDD9dg#ixzz2gaenOUL2 (Accessed, August 25, 2014).

initiated, and their subsequent exceptional performance on the job are worthy of emulation by all Nigerians. The legacies of Prof. Olikoye Ransome-Kuti and Prof. Gordian O. Ezekwe will remain enviable benchmarks for others to follow, particularly in Nigeria and Africa. Here are their records.

Legacies of Esteemed Professionals turned Selfless Public Servants

Professor Olikoye Ransome-Kuti: The Babangida Administration appointed Prof. Olikoye Ransome-Kuti as the Federal Minister of Health in 1985. In that capacity, he committed all his energies to the restoration of the nation's very weak health-care system. By the time he left his ministerial post in 1992, he had already institutionalised primary health care in Nigeria. His experience between 1968 and 1976, when he directed the Institute of Child Health of the College of Medicine at the University of Lagos and subsequently served as the first Director of the Nigerian National Services Scheme Implementation Agency, shaped most of his ministerial activities. These and other experiences ably prepared him to act on his beliefs in prevention and community health when the opportunity came.

Amongst his main health-care targets were poor child-immunization rates, lack of access to basic health services and AIDS, all of which he successfully tackled with significant measurable achievements, and with both local and international support. In 1994, his sterling achievements in Nigeria earned him a post-ministerial advisory position at the World Bank, in Washington, D. C., where he produced a blueprint for adopting and sustaining primary health care in Africa for years to come. Apparently and unfortunately, his successors as ministers of health did not follow up on his 1989-1992 primary health care efforts. President Obasanjo, desirous of resuscitating the concept of primary health care as the foundation of Nigeria's health care delivery system, sought him out in 1999, and appointed him Chairman of the National Primary Health Care Development Agency (NPHCDA), a post he held until he passed away on June 1, 2003.[31] All indications are that Nigeria is not medically healthy today; the primary health care condition in the country has gone back to the poor state that prevailed in the land just before Professor Ransome-Kuti was appointed to his last post in 1999.[32] Because the wealth of a nation is an integral

31 Okonofua, Friday (2006). Remembering Olikoye Ransome-Kuti, The GUARDIAN Newspaper, Lagos, Nigeria, May 29, 2006 (Accessed, October 2, 2014).

32 Improving Primary Health Care Delivery in Nigeria—*Evidence from Four States,* World Bank Working Paper No. 187, Africa Human Development Series, April 2010. (Accessed, October 3, 2014).

and an indivisible part of its health, it behoves the Nigerian leaders to re-institute the Primary Health Care Delivery model that Professor Olikoye Ransome-Kuti laboured, selflessly and with all his energy, to develop, establish and institutionalise throughout the country.

Prof. Gordian O. Ezekwe: As a renowned scientist and engineer, Prof. Gordian O.Ezekwe knew that irrespective of Nigeria's actions and inactions, the nature of global power structures would continue to be dictated by advances in science and technology development, understanding and appropriate applications. His highly acknowledged inventive mind during the civil war, albeit on the side of Biafra, earned him, in 1989, the post of Federal Minister of Science and Technology in the Babangida Administration, a position he held until 1991. He later found greater satisfaction in serving the nation in another capacity. As the pioneer Vice-Chairman and CEO of NASENI, from 1992-1997, he dutifully engaged the NASENI platform and successfully translated, into practical programmes, his strong belief that Nigeria must urgently embark on the production of adequate quantities of primary and intermediate capital goods required for major equipment development, fabrication and mass production. He was convinced that such an approach would provide the necessary environment for a sustainable industrialization of Nigeria and in the process reduce its import expenditure to about 80%. Under his leadership of NASENI, he established five technology centres that focused on scientific equipment, electronics, hydraulic equipment and engineering materials. Most of these centres are still active and functioning today.[33] Given its record of performance, the government should mandate NASENI to drive the development and nurturing of private industries that can sustain the nation's S&T revival efforts.

Knocking at the Doors of New Technologies

Before the country returned to civilian rule in 1999, Nigeria had already established the National Mathematical Centre and the Sheda Science and Technology

---

33 Abiodun, Adigun Ade (1997). *Research: The foremost prerequisite for sustainable development in Nigeria*, Keynote Address presented at the National Conference on Research as Backbone for Sustainable Development, Abuja, Nigeria, August 11-15, 1997. Published in Research Capacity Building for Sustainable Development in Nigeria, UNILAG Consult, University of Lagos, Nigeria, Peter O. ADENIYI, Editor, 1999.

Complex, both in Abuja.[34,35] On his assumption of office in 1999, President Olusegun Obasanjo also recognised how, globally, such advances in S&T were influencing national development in other countries as well as global power structures. He subsequently gave his strong support to the initiation and development of the nation's programmes in space science and technology, information and communication technologies, and bio-technology. Before he left office in 2007, he also laid the ground work for what has now become Nigeria's Vision 2020. The goal of Vision 2020 is for Nigeria to be one of the 20 most powerful economies in the world. UNESCO featured prominently in assisting Nigeria to attain this and other development goals.[36,37] The nine strategic targets of Vision 2020 are all S&T related and are expected to culminate in the establishment of a National Science Foundation with an initial endowment of US$5 billion that would be supplemented by donors.[38] The authors of the 2011 Science, Technology and Innovation (STI) policy also expected the policy to drive and guide all of the S&T

---

34 The National Mathematical Centre, Abuja, was established in 1988 to develop appropriate initiatives and resources of international standing for reawakening and sustaining interest in the mathematical sciences at all levels in Nigeria, and also as an adequate response to the dramatic decline in the production of teachers and specialists in the mathematical sciences.

35 The Sheda Science and Technology Complex (SHESTCO), Abuja, was established by the Federal Government of Nigeria in 1993 as a multidisciplinary research and development centre. There are three National Advanced Laboratories at SHESTCO—a Biotechnology Laboratory, a Chemistry Laboratory and a Physics Laboratory. A Nuclear Technology Centre, which houses a state-of-the-art gamma irradiation facility, is also located there.

36 UNESCO Science Report 2010 on Nigeria (Accessed, December 7, 2014).

37 Osotimehin, Folarin and Olufemi A. Bamiro (2006). *The Nigerian Science, Technology and Innovation System Review Project*, Colloquium on Research and Higher Education Policies, UNESCO, Paris, 29 November—1 December 2006, (Accessed, October 2, 2014).

38 The nine strategic targets of Vision 2020: (i) Carry out a technology foresight programme by the end of 2010; (ii) Invest a percentage of GDP in R&D that is comparable to the percentage invested by the 20 leading developed economies of the world; (iii) Establish three technology information centres and three R&D laboratories to foster the development of small and medium-sized enterprises; (iv) Increase the number of scientists, engineers and technicians and provide them with incentives to remain in Nigeria; (v) Support programmes designed by professional S&T bodies to build STI capacity; (vi) Develop an STI information management system for the acquisition, storage and dissemination of research results; (vii) Attain progressively 30% of local technology content by 2013, 50% by 2016 and 75% by 2020; (viii) Develop new and advanced materials as an alternative to the use of petroleum products; and (ix) Establish a National Science Foundation.

components of Vision 2020.[39] If the policy is to achieve this objective, a rigorous review and a far-sighted re-design of its contents are in order; its recommendation for the establishment of S&T ministries at each state level, in particular, is at variance with the realities in both Nigeria and the rest of the world.

Science and Technology Development at the State and Local Levels

In Brazil, India, Europe and in the United States, state governments are investing in research and development on renewable energy programmes, bio-engineering, electronics and in many other areas of need for their respective societies. Through such investments, these countries are able to develop and manufacture products as well as generate new knowledge. "However, as part of the nation's STI policy strategy, states are being encouraged to establish or strengthen S&T Ministries at the State level."[40] Nigeria and its states do not need any new bureaucracies; the Steven Oronsaye report which recommended the scrapping, dismantling and rationalization of many government departments, agencies and parastatals provides innumerable arguments in support of such a view.[41] Nigeria needs a civil service, at the federal and at each state level that will deliver services that would meet the need of the people. The annual disparities between capital budget and recurrent expenditures at all levels of government do not make room for such a service delivery.

Similarly, in other parts of the world, development activities are taking place in all corners of society and not only at the centre; that is the case in the State of São Paolo, Brazil, and in the State of Gujarat, in India. The same should be true of Nigeria. Each state and each local council of the federation should make its own contribution to science and technology revival in Nigeria through the establishment of state science parks, museums and planetariums. Not everyone can nor should be trooping to Abuja to experience human adventure when they can do the same in their own local environment. Each state should invest in enhancing its own environment, know and develop its natural endowments and identify and invest in those technologies that

---

39 Science, Technology and Innovation Policy—Draft 6, Federal Ministry of Science and Technology, Federal Government of Nigeria, Abuja, May 2011.

40 See page 3 of the STI Policy at http://www.fmst.gov.ng/uploadfil/sti_policy2011.pdf (Accessed, October 7, 2014).

41 Oronsaye, Steven (2012). Report of Steven Oronsaye's Presidential Committee on the Rationalisation and Restructuring of Federal Government Parastatals, Commissions etc. (Accessed, October 7, 2014).

could boost its sustained productivity. That is how to spread development in any forward looking society.

Nigeria's S&T Engagements in the 21st Century and Beyond

Nigeria's S&T engagements in the 21st century and beyond should rest on four firm pillars; these are the decision makers, the youth, the tertiary institutions and the private sector.

The Decision Makers

The first building block of Nigeria's S&T foundation consists of Nigeria's leaders, in decision-making positions, in the nation's parliament, as well as in government, the private sector and the NGOs. The understanding of the role of S&T in development by such individuals and their subsequent commitment to ensuring that it becomes the driver of the nation's development efforts just as is the case in Brazil, India, South Korea and South Africa, is what will make Nigeria a knowledge society. These leaders need to understand that Nigeria cannot grow and develop unless it invests in the building of S&T capacities and capabilities which translate into educated and skilled work-force, world-class functioning S&T institutions, R&D establishments, sound infrastructures, laboratories, mechanical workshops, clinics as well as industrial establishments that can transform or turn the results of research and development efforts into products and services for Nigeria and the global market.

By taking these steps, Nigeria will be able to build at home, most of what its citizens need; and Nigerian factories can process the nation's raw materials (cocoa, groundnuts, palm-oil, crude oil, iron and cotton, and forest products) into finished goods for sale at home and abroad. As shown earlier, we should not forget that in spite of its leading role as the largest producer of crude oil in the world, today, Saudi Arabia is investing in renewable energy research and development at KAIST. Nigeria should do the same, by taking bold steps in renewable energy investment because it is good for our environment, our health, our nation's growth and development, and for the pocket book. All of the above should provide opportunities for Nigeria's private sector to grow and contribute significantly to the nation's future, including the creation of a variety of meaningful jobs for the nation's teeming population. In the process, Nigerians will gain a variety of knowledge, skills and experience that can power other major initiatives of the nation, such as poverty reduction, food security, space programme and Vision 2020. Nigeria must also make the future of its children an integral part of its S&T future, since the Nigerian youth constitutes the nation's workforce of that future.

The Youth, the Nation's Workforce of the Future

Enhanced Basic Science Education and Improved Science Curriculum

Globally, the universal adage is: "*Our children are our future.*" This is equally true for Nigeria, and by implication, whatever Nigeria invests in its youth today is what it will reap in the future. That investment can take many forms. The goal is to positively affect the early development of the Nigerian youth in S&T, since whatever happens to them in the early stages of their life could affect their future development and orientation as well as the development of Nigeria and our world. Dedicated national education television programmes can help young people discover where they fit into their society, develop closer relationships with peers and family, teach them to understand complex social aspects of communication and the basics of science. An extended exposure to a national science enrichment programme is a good beginning. That was the goal of *YEAST* on S&T education in South Africa.

In a formal classroom setting, science immersion programmes, illustrated with everyday experience, at the elementary school level and at speciality (e.g. technical) high schools, would reinforce the enrichment programme young people have been exposed to and help to sow the seeds of S&T in the Nigerian youth. Since these children are living real lives in real societies, teaching Nigerian youth about economy and how to manage money, such as is done in Jamaica's educational system,[42] and the re-introduction of civics, home economics, hygiene, general nature study (with the aid of satellite images at the high school level) into the nation's education curricula, would adequately equip them to face a future that is full of many challenges.

In the mid-1950s, the Nigerian government established FESS as a way of addressing the challenge of the shortage of S&T personnel in the country. The FESS period witnessed a heightened interest in mathematics and the sciences, nation-wide. Unfortunately, today, mathematics and science subjects are not in the knowledge menu being sought by most of the Nigerian youth. A look at the annual results of the West African Examination Council (WAEC) should shed some light on the problem. The inculcation of mathematics in the education curricula of the Nigerian youth should be a basic requirement in both the elementary and the secondary schools, and its teaching should be fun and rich in everyday practical examples. Annual competitive exhibitions that demonstrate the scientific and practical ingenuities of our youth

---

42 Barnaby, Teneica, Elizabeth Gordon and Kasan Troupe (2015).*Making it count: Teaching financial literacy in Jamaica,* Vocational and transnational paths, Commonwealth Education Partnerships 2015/16, Commonwealth Secretariat, UK.

should be introduced and nurtured at the local, state and national levels in order to demystify and give meaning to mathematics and the sciences among the youth at each level of the nation's educational system. These steps are necessary because *Mathematics* is the over-riding tool in the mastery and application of sciences and technologies, including its diverse applications in such fields as computer science, finance and engineering. Mathematics is also the unifying field on which all other disciplines and human transactions depend. It is a major requirement of a knowledge society, which Nigeria aspires to be. The foundation of that future begins with a sound system of education that is anchored in science, technology, engineering and mathematics (STEM) for the Nigerian youth.

An in-depth understanding and appreciation of mathematics can pave the way for a rewarding engagement with basic scientific exploration, the catalyst for the evolution of a variety of new technologies that can drive national development and growth. Thus, Nigeria must prepare a solid foundation for its scientists and engineers of the future; it is all about sound education and the imperatives of being and remaining productive and competitive in the global economy. The process begins when Nigeria determinedly and conscientiously sows and nurtures such a STEM education programme that can produce a STEM proficient workforce of the future. The science, technology and engineering components of that effort consist of instructions—theory and practice—on basic astronomy, and atmospheric and Earth observation sciences, as appropriate, for each level of education—elementary and secondary school levels as well as at the initial two years of tertiary education.

When such a programme is built on foundational science skills curriculum, with essential instructional models, and is backed by adequate and encouraging incentives, it should excite the Nigerian youth, ignite their inquiring minds, revolutionise their learning and positively change their attitude towards STEM. At the tertiary level, the National Universities Commission (NUC) should encourage a number of well-equipped universities to expand STEM to include the understanding and use of the Earth observation data of Nigeria in their immediate communities. In this manner, students of today can be prepared for a responsible management of the Earth, its environment and its resources in their adult lives. Both RECTAS and ARCSSTE–E, co-located with Obafemi Awolowo University, should be recruited by the NUC to develop hands-on materials for use in the expanded STEM education programme.

Many qualified teachers are needed for these immersion programmes, and most of them will require appropriate training in each subject area. In addition, Nigeria's decision makers and education authorities, nation-wide, should re-examine the nation's short history and learn from those actions that worked, under similar circumstances, in the past. For example, there was a time when it became necessary to significantly upgrade the salaries of teachers of science (physics, chemistry and biology), mathematics, geography, agriculture, information technologies, technical drawing and similar subjects in Nigeria's elementary and secondary schools. The nation took this line of action because of its recognition and appreciation, at that time, of the contributions of these segments of its educational system to national development.

The future of Nigeria demands a similar incentive—the recognition and appreciation of the S&T contributions of its educators today. As it did in the past, such an action would witness a greater sense of dedication and performance by the teachers; it would also provide an incentive for attracting more qualified individuals to the profession and thereby contribute to a greater popularization and achievements in mathematics and the sciences, nation-wide. It will enhance the nation's capabilities and capacities in S&T and provide crops of students, from year-to-year, that are well equipped for S&T education at the nation's tertiary institutions, and subsequently as researchers at its R&D establishments—these are the ingredients needed to energise and advance the development and growth of the nation.

Education Reform at the Tertiary Level

At the tertiary level, there is much to fix in Nigeria's educational system. It ranges from dilapidated infrastructure, non-functional laboratories to empty shelves in libraries. All these symptoms and more were captured in the 2005 report, "Nigeria Education Sector Diagnosis," developed and produced by the Federal Ministry of Education.[43] The report is awash with every conceivable problem in the nation's elementary and secondary schools, and at its tertiary institutions; they are still present today, ten years later. How to ensure adequate and consistent funding in order that the nation's education system could be productive and competitive is among the major challenges. Job creation for Nigeria's university graduates, in a country with very limited industries, is another challenge, more so because such jobs should not only provide an income for the employee but should also be rewarding for him or her in all its aspects.

---

43 Federal Ministry of Education (2005).*Nigeria Education Sector Diagnosis—A Framework for Re-engineering the Education Sector*, Education Sector Analysis Unit, Federal Ministry of Education , Abuja, Nigeria, May 2005 (Project Coordinator: Dr. G.A.E. Makoju (Mrs.) (Accessed, January 7, 2015).

The above report also cited disaffection among students in polytechnic colleges. Their disaffection is rooted in the inability of most of the existing polytechnics to offer courses or programmes, such as ceramics, leather work and auto electricity, which could serve as the foundation of thriving industries for polytechnic graduates. The report recommended that such courses should be introduced in order to stimulate and encourage students' interests in polytechnic education. But there is more to it than the introduction of the right courses. Let us also recall that polytechnics were established in Nigeria to develop a variety of technical skills that could support the research work of the scientists and the professional practice of the engineers and technologists. As of 1999, there were 30 polytechnics in Nigeria; that number increased to 47 in 2005. As of 2015, there were 25 federal, 38 state and 24 private accredited polytechnics in Nigeria.[44]

To their credit, many of Nigeria's polytechnics have inculcated into their graduates much technical competence and skills, including those needed for the successful manufacture of a variety of apparatus, electronics parts and related equipment. Such polytechnics have saved themselves and other establishments in the society much needed funds that otherwise would have been used to purchase every replacement part from abroad. But because of the nation's obsession with paper qualifications, graduates of polytechnic colleges are generally denied the status and recognition that are accorded engineers and other technologists. To gain such recognition, a majority of the graduates are using their polytechnics diplomas as stepping stones to obtain higher degrees from engineering and related institutions. And Nigeria's decision makers are compounding the problem as they succumb to the persuasion of their constituencies to convert polytechnics into universities, to the detriment of developing the basic ability acquired through hands-on skills. The Nigerian reality is that the average individual, with a university degree, automatically develops a disdain for manual work. Who then will do the technicians' jobs as Nigeria tries to industrialise? The clamour for societal recognition, appreciation and commensurate material and financial rewards, by polytechnic graduates, an indispensable component of the work-force that is crucial for Nigeria's industrialization, is a legitimate demand that Nigerian authorities should promptly examine and redress.

---

44 http://www.myschoolgist.com.ng/ng/list-of-accredited-polytechnics-in-nigeria/ (Accessed, January 7, 2015).

Rationalisation and Revitalisation of Nigeria's S&T and Tertiary Institutions

Today, a hundred and forty-one (141) universities dot the Nigerian landscape,[45] with several new ones in the pipeline. However, not every university in the country needs to be a generalist that offers degree programmes in every conceivable discipline. Globally, there are many focused graduate-level research universities whose specialties are dictated by the needs of the host country. Federal Universities of Agriculture at Abeokuta and Makurdi and the Colleges of Agriculture in the nation offer clear examples. At this stage of Nigeria's development, among its critical educational needs are focused tertiary institutions that can address specific problems of the nation. By focusing on key areas of national interest, a university can develop a niche for itself that is unique and can have maximum impact on both its immediate environment and the nation at large. The focus could be in such fields as fundamental and applied sciences, health sciences, engineering (all branches), electronics and the computer, energy (finite and renewable), marine sciences and forestry, just to name a few.

The almost countless number of motor vehicles in Nigeria, both on-and-off the road, comes to mind. With a high resolution satellite image, one could count millions of abandoned and non-functioning motor-vehicles in Nigeria's villages, towns and cities, by the roadsides and in the open-fields. Similarly, frozen images of all the running vehicles on Nigeria's roads, at any given moment, would also yield an untold number. Where did all these vehicles come from? Nigeria has not manufactured any of them! Assembling cars is totally different from manufacturing cars. Either as a whole or as knocked-down parts, all these vehicles were imported; new ones are arriving, daily, by land and sea—a practice that began long before Nigeria became a politically independent country in 1960. And when these vehicles break down, the parts needed for their repair are equally imported. Trying to put a figure on how much money Nigerians have shelled out in our over 80 years of buying motor vehicles and spare parts from abroad is an almost impossible task; if it could be done, the sum would be staggering.

Learning from the above lesson, can anyone imagine what an economic wonderland Nigeria would become were it to establish and steadfastly run a genuine Nigerian automobile engineering institution that develops, models, tests and manufactures the best possible motor vehicles, including their spare parts and ancillaries, for Nigeria and the world market? The spin-off and impact of such a motor vehicle establishment

---

45 http://nuc.edu.ng/ (Accessed, September 22, 2016).

in Nigeria on Nigeria and its people would be incalculable. The economic impact of replicating such a feat, many times over, in other segments of the economy, would be overwhelming. It is achievable in a knowledge rich Nigerian society; it will offer countless jobs and enhance the quality of life of Nigerians, and it will also offer Nigeria a competitive edge in the global economy. One or two of the Federal universities in the nation can be re-oriented and revitalised to accomplish such a goal.

Given the state of Nigeria's S&T academic and research establishments as presented in Chapter VI as well as in the *2012 Needs Assessment Report on Nigerian Universities system*, it is apparent that a comprehensive re-orientation, revitalisation and rationalisation of these institutions is very much overdue.[46] Nigeria will not be the first country to undertake such a reform exercise; within the last decade, similar exercises have taken place in Ethiopia,[47] Kenya,[48] Morocco,[49] South Africa[50] and Tanzania[51].

The justification for such an overhaul of the general university system and research establishments is dictated by home-grown needs and the inability of established institutions to be responsive to the aspirations of the people they were established to serve. Given the conditions here in Nigeria and in particular with its close to 65% rural population, it is imperative that our educational programmes and research orientation must be relevant to our own needs. Village and community problems should be brought to view early in our educational system and in our research

---

46 Nigerian Needs Assessment Report on the Nigerian Universities System; A presentation made to the National Economic Council (NEC) of Nigeria in November 2012 (Accessed, January 7, 2015).

47 Yizengaw, Teshome (2003). Transformations in Higher Education: Experiences with Reform and Expansion in Ethiopian Higher Education System, Regional Training Conference on *Improving Tertiary Education in Sub-Saharan Africa: Things That Work!,* Accra, Ghana, September 23-25, 2003 (Accessed, January 8, 2015).

48 Kenya's 2008-2012 Strategic Plan—Quality Higher Education, Science, Technology, MHEST, Nairobi, 2008 (Accessed, January 10, 2015).

49 The Academic Reform in Moroccan universities and its relations to the Bologna Process, 2000. (Accessed, January 10, 2015).

50 Howie, Sarah J. (1999).*Challenges to Reforming Science Education in South Africa: What Do the Third International Mathematics and Science Study Results Mean?* Science and Environment Education: Views from Developing Countries, The World Bank, Washington, DC, 1999 (Accessed, January 12, 2015).

51 Luhanga, Matthew L. (2003). The Tanzanian Experience in Initiating And Sustaining Tertiary Education Reforms, Regional Training Conference on *Improving Tertiary Education in Sub-Saharan Africa: Things That Work!,* Accra, Ghana, September 23-25, 2003 (Accessed, January 12, 2015).

orientation in order to educate the student on the issues that his/her profession must address in the society. The communities within the immediate vicinity of Nigeria's universities justifiably expect these institutions to make a great difference in their lives. This can be accomplished through a variety of out-reach programmes and physical interactions between each university and its immediate community, depending on the needs of the community. For many of these communities interactions between the university and the educational establishments in the immediate community can result in knowledge enrichment.

The universities should also be innovative in order to achieve a degree of financial self-sustenance as practised in other comparable institutions around the world. Such goals are achievable in an environment where it is possible to marry knowledge, skills and industrial know-how to economic development. The Nigerian government and the Nigerian university system can borrow a leaf from the Innovation Centre at Coventry University in the United Kingdom. This university prepares the graduates of its product-oriented courses to start up and run new companies that capitalize on the ideas these same students have developed while at the university. Other variants of this approach are described earlier, in this Chapter, in the section on *Entrepreneurial spirits within the universities.*

For Nigeria to advance technologically and industrially, its leadership must be visionary and courageous enough to initiate a bold policy that will transform a number of the nation's public universities into front-line speciality universities, with encouragement to the private institutions to follow suit. Such institutions abound globally. Amongst them are:

- The Earth University, Guacimo, Costa Rica which was established as a private, international, non-profit university—An agricultural university with a focus on investigating sustainable agriculture in tropical environments and seeking a balance between agricultural production and environmental preservation;
- Al Akhawayn University, Ifrane, Morocco which was established in January 1995 by the King of Morocco as a centre for academic excellence and inter-cultural tolerance;
- Indian Institute of Science, Bangalore, India—A private research university with a focus on fundamental and applied sciences and engineering. It was conceived in 1896 and established in 1909;
- Korea Advanced Institute of Science and Technology (KAIST), established in 1971in Daejeon, South Korea by the Government of South Korea as a research oriented science and engineering institution; and

- Pohang University of Science and Technology (POSTECH), Pohang, South Korea, was established in 1998 as a private research university dedicated to research and education in science and technology.

These five research and development institutions, including KAUST in Saudi Arabia, and similarly focused research institutions around the world, offer very encouraging examples that are worthy of emulation for Nigeria. Because of their individual focus, they are able to have maximum impact in their chosen specialities without diffusing their energies and resources to other fields that are already well served by other institutions in their local communities. In the process, these institutions are able to respond to the needs of their respective societies. Because they have also become very competitive globally, they have continued to attract intellectual and financial collaboration from the international community. Above all, the focus of each of these institutions and their state-of-the-art facilities and infrastructure are also causing a reverse brain drain.

Building the Inquiring Minds of the Future

The future of S&T in Nigeria and the resurgence of Nigeria as a buoyant country hinge on and demand the development and nurturing of Nigeria's own inquiring minds. Such minds are not born overnight. The building of such minds should begin with the nation's conscious and continuing effort to make science a fun activity from an early age, a process that should develop into enriched science and technology programmes at the secondary school and at the undergraduate levels at our universities. The building and nurturing of such minds should culminate in front-line research in appropriate disciplines at the post-graduate levels of Nigeria's universities and in its research laboratories, public and private. There is no other choice. The knowledge requirements of new and advanced technologies such as biotechnology, nanotechnology, material science, renewable energy, climate change, sustainable development and the space enterprise cannot survive on standard industrial technologies alone.

That was the opinion of the Indian Parliament when it enacted the Indian *Scientific Policy Resolution of 1958* that *also* proclaimed and declared, *inter alia*:

> ... *It is only through the scientific approach and method and the use of scientific knowledge that reasonable material and cultural amenities and services can be provided for every member of the community...*[52]

---

52 Scientific Policy Resolution 1958,Doc. No. 131/cf/57 of 4th March 1958/13th Phalguna, 1879, Department of Science and Technology, Ministry of Science and Technology, Government of India, New Delhi.

India's resolute and practical adherence to this declaration, for decades, has successfully propelled it into the status of a space power today, with all the opportunities, challenges and international recognition that come with that achievement.

The issues of basic and fundamental sciences and high technologies also featured in the work of the United Nations South Commission under the leadership of the late Hon. Julius Nyerere, the former President of Tanzania. In 1990, the Commission issued its report, titled, *The Challenge to the South*, which states as follows:

Unlike the standard industrial technologies...., mastery over new sciences and technologies requires high expertise in the relevant basic sciences. Experience has shown that high technologies cannot simply be transferred; the notion that it would be possible for the South [which includes Africa and the Diaspora] to obtain them from abroad without the development of an indigenous broad-based scientific and technological infrastructure is mistaken.[53]

In 1976, the Republic of Korea had earlier arrived at the same conclusion as the South Commission did in 1990; it established and generously funded a world-class Electronics and Telecommunications Research Institute (ETRI) to support its various industrial development programmes, particularly the space-related ones. Similarly, the City of Campinas, which is 88 km (55 miles) from São Paulo, offers a vibrant, high-tech university and research environment that is equally worthy of study by Nigeria. Providing and nurturing such an enabling environment led to the much heralded achievements and prosperity of Japan.[54] Today, millions of Toyota, Nissan, Honda and Mitsubishi cars are plying the global roads, while Seiko and Casio watches are adorning the wrists of millions of people globally.

Nigeria needs its own S&T front-line research institutions that are affiliated, associated or integrated with Nigeria's best universities, public and private, and that can command the respect of the international scientific community. This is a critical step the leaders of Nigeria must take if they are truly committed to develop Nigeria using Nigerian talents. Nigerian leaders must recognise and accept that a scientist/technologist is a thinker, a doer, who has to be constantly engaged and challenged. Furthermore, skilled labour is an internationally mobile commodity which responds only to appropriate economic and other suitable incentives. Among the industrialised

---

53 *The Challenge to the South (1990)*. The Report of the South Commission, Oxford University Press, New York.

54 Abiodun, Adigun Ade (1994). *21st Century Technologies: Opportunities or threats for Africa* (Futures 1994, 26 (9), pp 944-963, Butterworth-Heinemann Ltd.).

countries of the world, the back-and-forth flow of skilled labour is a continuing process. But in the case of Africa, particularly Nigeria, the flow of such skilled labour is a one way operation—out of Nigeria. When and where Nigeria's decision makers have failed to recognise and productively engage Nigerian talents and achievers, other countries, industrialised and developing ones alike, are offering them promising opportunities and nourishing environments to contribute to their own technological march towards the future.[55,56] Only Nigeria's decision makers can stem the outflow of Nigeria's students, university graduates and professors into the waiting arms of the global community; they are also the ones, who, through their actions, can attract a sizeable portion of Nigerian citizens in the Diaspora to return home. Such returnees and their counterparts in Nigeria require a motivating research environment that will empower them to freely contribute to competence building in different areas of S&T in the country. By taking these steps, Nigeria will advance its own development and win international collaboration in the process.

* * *

55 Abiodun, Adigun Ade (1998). *Human and institutional capacity building and utilisation in science and technology in Africa: An appraisal of our performance to-date and The way forward,* African Development Review Journal. Vol. 10, No. 1, June 1998, pp. 10-51, African Development Bank, Abidjan).

56 CARNESS, Kelly H. (1997).*Building the 21st Century Workforce: The Challenge of technological Change,* Address by the Deputy Assistant Secretary for Technology Policy, US Dept. Of Commerce to the Independent Schools Associations of the Central States of the United States, April 11, 1997.

Chapter X

HARNESSING NIGERIA'S POTENTIAL

The successful transformation of any society depends on its ability to harness its potential for the good of its people. Indispensable components of such a potential include, *but not limited to*, men and women of inquiring minds and adequate material/financial resources. The harnessing of this potential is deemed successful when a nation has the vision and the political will to commit its material/financial resources, on a long-term basis, to support the creativity of its citizens in translating scientific knowledge into economic productivity which includes adding value to raw materials.

Because of its human and material assets (agriculture, petroleum and gas, and solid minerals), Nigeria has been perceived, for over half a century, as a country with *great potential*. According to *Toyin Falola et al*, Nigeria was also dubbed the "Giant of Africa,'" and many people both inside and outside the country believed that Nigeria would soon rise to claim a leading position in African and world affairs. "Nigeria also saw itself as a beacon of hope and progress for other colonized peoples emerging from the yoke of alien rule."[1] But of what value is a potential that brings no development and has no impact on societal well-being?

While launching the *International Fund for the Technological Development of Africa* in Nairobi, Kenya, in February, 1994, UNESCO's Director-General responded to the relevance of such a potential, not only for the benefit of Nigerians but also for the attention of all Africans, particularly the African political and policy leaders, in these words:

---

1 Falola, Toyin and Matthew M. Heaton (2008). *A History of Nigeria*, Cambridge University Press, New York, USA.

"Africa must evolve its own blueprint for development and must no longer depend on external models. Your continent possesses the necessary talent and resources. What it needs is the knowledge and expertise to realise its human and natural potential."[2]

In 1995, the then Nigeria's Minister of Science and Technology officially weighed in on Nigeria's potential and how they should impact the Nigerian society, in these words:

It [Nigeria] is well endowed with abundant human and material resources yearning to be harnessed through science and technology. Nigeria is, indeed, an excellent case study in the evolution of social systems. She is classified as developing, as indeed she is, but analysts believe that with the requisite policy formulation and implementation, she could become a developed country by the opening decades of the 21st century.[3]

This chapter examines these key ingredients (human and material resources, as well as national vision, political-will and commitment) needed, not only for a vibrant space programme, but also for the overall transformation of the society; it also analyses how Nigeria has measured up, to-date.

The Nigerians

Today, skilled Nigerians, with expertise in many spheres of human endeavour, can be found in Nigeria as well as in several countries where they are contributing to the advancement of their host societies. For some time now, the doors of international education and research institutions have remained open to many of them, as students, research scientists and educators. According to Paul Cullen, International Director at the University of Essex, United Kingdom, the international perspective Nigerians bring to research and the uncommon brilliance they display are some of the reasons foreign universities desire to have them in their institutions.[4] Are these not the descendants of the same black Africans that the colonial mentality, at one time, judged to be *primitive and unsophisticated people?* Could there be an innate aspect to the *"uncommon brilliance"* of their descendants? What was the Nigerian ancestors'

2 *Africa Technology Fund,* West Africa, No. 3987, 28 February—6 March, 1994, p. 364.

3 Maduemezia, A., S. N. C. Okonkwo and E. E. Okon (1995). Editors of Science Today in Nigeria, The Nigerian Academy of Science, Lagos, 1995.

4 Kobo, Kingsley (2009). *Foreign universities woo Nigerian students, AfricaNews,* Amsterdam, The Netherlands, April 18, 2009.

knowledge and understanding of what we today define as science, technology and even space, particularly before the British explorers, turned colonialists, arrived on Nigerian shores?

Such knowledge existed before the Europeans landed in Australia, Latin America and in Africa. From Australia and Latin America came the following views:

"Ever before the Europeans landed in Australia, the indigenous people of Australia had their own particular unique views of the cosmological and spiritual creation of their World; they understood the importance of the sky and knew of the general movements of the Sun, Moon, planets and the stars."[5]

"Great indigenous cities of Latin America also used the pristine order evident in the sky to establish social order on earth."[6]

Nigeria's Ancestral Knowledge

Nigeria's ancestral knowledge of science and technology was captured in the British view that UNESCO later quoted as follows:

In the late 19th century, for instance, Nigeria was considered to be no more than a decade behind the United Kingdom in terms of technological development....[7]

From its pre-independence days till now, the land that we know and call Nigeria had and still has many ethnic groups. Each group bonded with nature and all its elements in a specific manner, just as other communities in the world did. Each group recognised the dominance of nature as well as the importance of the outer space environment in human activities. Each group also had and still has its own assigned names for them and established its own traditions for celebrating the Sun, the Moon, the Stars, as well as the Rainbow, Thunder and Lightning. Expectedly, the celestial bodies featured in many human activities—agriculture, religion and daily activities.

According to Ahmad Kani, by the 18th century, the Borno kingdom, which is part of Nigeria today, became the most important centre of learning of mathematics in the Central Sudan.[8] He noted the abundance of evidence to prove that scholars

---

5 James, Andrew (2012). The Dawn of Australian Astronomy, Sydney, Australia http://homepage. mac.com/andjames/Page031a.htm.

6 Pre-Columbian Astronomy and Latin American—Less Complex Societies, http://science.jrank.org/ pages/8407/Astronomy-Pre-Columbian-Latin-American-Less-Complex-Societies.html.

7 UNESCO Science Report, Executive Summary—2010, Paris, France (Accessed, December 7, 2014).

8 Kani, Ahmad (1992). *Arithmetic in the pre-colonial Central Sudan*, Science and Technology in African History, Gloria Emeagwali (Editor), Edwin Mellen, NY, 1992 (Accessed, April 12, 2014).

from Hausaland and Borno consulted Coptic Solar Calendars in determining their economic activities, including agriculture. Kani further noted that during the 19th century Jihad movement in Hausaland, Abd al-Quadir b. al-Mustafa, who *was* reported to have studied "medicine, astrology, arithmetic, logic and astronomy," had a collection of books on "arithmetic and related sciences in the syllabi of the schools in 19th century Hausaland."

A more recent study of pre-colonial sub-Saharan Africa also showed that the Yoruba (of *South-West* Nigeria) had a well-developed base-20 *vigesimal* number system (similar to the Mayan) and used mathematics for astronomy, astrology and other mystical arts prior to the arrival of the European explorers.[9] In her 1973 work, Claudia Zaslavsky corroborated these findings on the Yoruba number system.[10] In southern Nigeria, the Oyo Empire, the centre of the Yoruba land, at that time, was renowned for its naturalistic terra-cotta and brass sculpture.

Unravelling what happened to advances in the above-recorded scientific knowledge and skills in pre-colonial-Nigeria remains a challenge. The latter included the unending struggles and conflicts among the different pre-colonial Nigerian classes, struggles that also wiped out most, if not all, of the recorded history as well as the oral history.[11] These struggles included trade and territorial wars within and outside the Yoruba Empires and the Nok, Fulani and Borno Kingdoms. The European navigators-turned-traders eventually turned the table against the Nigerian natives and in favour of the Europeans. Trade in agricultural products, especially palm oil and spices, soon gave way to all manner of conflicts, incited in order to accomplish an ultimate goal—the capture of young and able-bodied men and women that were later purchased by the Europeans as slaves from Nigeria's coastal middlemen.[12] To understand what destroyed whatever was left of Nigeria's history, one only needs to recall the subsequent overthrow of indigenous sovereignties through large scale

---

9 Kelley, Loretta (2003). Multicultural History of Mathematics: A Project Based Middle School Course; Douglas Ruby, Leo Tometich, Melinda Willis Roland A. Gibson,—as contributors http://webpages.charter.net/druby/Papers/MultiCulturalMath.pdf (Accessed, April 12, 2014).

10 Zaslavsky, C. (1973a): *Africa Counts: number and pattern in African culture*, Prindle, Weber and Schmidt, Boston, 328 p. (out of print); paperback edition (available): Lawrence Hill Books. (Accessed, April 14, 2014).

11 Asiwaju, Anthony I. (2001). West African Transformation, Malthouse Press Ltd, Lagos, Nigeria.

12 http://www.infoplease.com/encyclopedia/world/nigeria-history.html#ixzz2Sl35mMja (Accessed, April 11, 2015).

British punitive expeditions. The end-result was territorial occupation which soon metamorphosed into full-scale colonization of today's Nigeria.

Though unable to match the British military might, there were those Nigerians of that period who could not bear surrendering everything to the *invading marauders*. They devised a protective method of burying, underground, a number of the indigenous non-perishable and invaluable products, including records that escaped the grasp of the colonial master. Amongst such products were the Nigerian terracotta and bronze heads, statues and other products that have demonstrated, convincingly, to the world that our ancestors that worked those *"perfect moulds,"* with *"incised lines …and subtle curves"* and with such "realism and technical sophistication," had some science and technology skills.

In November 1995, Barry Hillenbrand wrote about Nigerians' ancestors, as follows:

> *In 1897, when the British troops entered Benin City, in southern Nigeria, they found bronze statues, plaques and (terracotta) masks exquisitely designed and so perfectly fabricated that the invaders concluded the trove could not possibly be the product of what the colonial mentality judged to be an unsophisticated people. Inventive theories were devised to explain the works. Perhaps the ancient Egyptians—or even a lost tribe of Israel—had a hand in casting them. Back in London, art experts took one look at the clean, naturalistic lines in the African busts and declared that the works reflected the undeniable influence of classical Greek culture.[13]*

In February 2010, Mark Huddson corroborated the views expressed in the 1995 Barry Hillbrand publication. Excerpts from the article by Huddson on the terracotta and bronze heads, statues and plaques from Ile-Ife, in Western Nigeria, went as follows:

> *When the German archaeologist Leo Frobenius set eyes on the first of these heads (terracotta and bronze) to be unearthed in 1910, its bronze features "of perfect mould," he declared that he had seen the face of Poseidon, sea god of the ancient Greeks—proof, he claimed, of the existence of an African Atlantis…Made between the 13th and early 15th centuries, using the lost-wax process, they challenge ideas about the primitivism of black Africa that are widely held even today…While much of the terracotta work is extraordinarily refined, it is the bronze heads that tug most at the imagination. Does their realism and technical sophistication*

---

13 Hillenbrand, Barry (1995). *African Visions*, Time Magazine, New York, November 27, 1995.

provide evidence of links between tropical Africa and the ancient Mediterranean?
Or was medieval Africa far more advanced than was previously imagined?[14]

Nigeria's Human Resources Today

Today, we know that the Benin bronzes cited in Hillenbrand's writing of 1995 as well as the terracotta and bronze works sermonised in Hudson's article of 2010 have nothing to do with the Egyptians, the Greeks or the Jews; they have everything to do with an indigenous and brilliant artistic Nigerian tradition dating back to, at least, the 15th century when powerful kingdoms ruled West Africa.

That inborn skill that characterised the Nigerian ancestors is still there today. It also accounts for why many Nigerian citizens have competed successfully and won at the best and most prestigious academic and research institutions in the world, and have left their marks of accomplishments in many international entities and multi-national industrial establishments.[15,16] Britain recognised the thirst for scientific knowledge in their claimed territory of Nigeria. As far back as 1925, the colonial master *spoke in London* of the yearnings of the colonies for education in technological and scientific disciplines. These yearnings eventually resulted, in 1948, in the establishment of the University College, Ibadan (UCI), the first university in Nigeria, established by the local authorities and Britain.

As could be expected, the yearning for knowledge in science and technology (S&T), notwithstanding, UCI initially focused exclusively on arts and divinity and it trained administrative assistants and necessary staff for the colonial bosses as well as for the propagation of Christianity in Nigeria. Meeting the yearnings of Nigerians for S&T knowledge would have to await the establishment, in 1952, of the Nigerian Colleges of Arts, Science and Technology (NCAST), with a tripartite residential college system in Enugu, Ibadan and Zaria; the graduates of NCAST contributed

---

14 Hudson, Mark (2010). Bronzes of West Africa, Financial Times of London, February 19, 2010, http://www.ft.com/cms/s/2/da457416-1ce4-11df-aef7-00144feab49a,_i_email=y.html (Accessed, April 16, 2015).

15 Abiodun, Adigun Ade (1997). *Research: The foremost prerequisite for sustainable development in Nigeria*, Keynote Address presented at the National Conference on Research as Backbone for Sustainable Development, Abuja, Nigeria, August 11-15, 1997. Published in Research Capacity Building for Sustainable Development in Nigeria, Unilag Consult, University of Lagos, Nigeria, Peter O. Adeniyi, Editor, 1999.

16 Asuzu, Okechukwu C. J., (2005). The politics of being Nigerian, www.lulu.com (Accessed, April 16, 2015).

immeasurably to the introduction of S&T subjects in the national education curricula of that era.[17,18] By 2014, 1.7 million students were registered in Nigeria's 129 universities (Federal (40), State (39) and Private (50)) tertiary institutions[19,20]. Abroad, Nigerian students are poised to overtake Indian post-graduate students in the United Kingdom,[21] while in the USA, Nigeria is among the 25 leading countries in the graduate programmes of its universities.[22] These same environments groomed the geniuses we are familiar with today.

Albert Einstein and Isaac Newton are two names that often make the list when one seeks information on the modern-day geniuses of this world. Their mathematical prowess and those of others shaped the tools of our modern-day development. The universally agreed definition of *Mathematics* is that it is the study of quantity, structure, space and change. *Mathematics* is the over-riding tool in the mastery and application of sciences and technologies, including its diverse applications in such fields as computer science, finance and engineering. Indeed, it is the unifying field on which all other disciplines and human transactions depend. Timothy Gowers, Professor of Mathematics at the University of Cambridge (UK), re-affirmed all these attributes of mathematics in his 2000 Keynote Address at the Millennium Meeting of the Clay Mathematics Institute.[23]

Because of his prodigious knowledge of mathematics, **Professor Chike Obi** of Nigeria became Prof. Chike Obi of the whole world just as Albert Einstein and Isaac Newton were. He was universally acknowledged as a man of acute intelligence, a great thinker, philosopher and logician with an untiring quest for knowledge. He arrived on this planet Earth via Zaria, Nigeria, on April 7, 1921 and departed this planet from Onitsha, Nigeria, on March 13, 2008. In between his entry and exit, Chike Obi

---

17 Lewis, L. J. ((1959). Higher Education in the Overseas Territories 1948-58, British Journal of Educational Studies, Vol. 8, No 1 (Nov., 1959), pp. 3-21, Blackwell Publishing, London.

18 http://nigerianwiki.com/wiki/Nigerian_College_of_Arts,_Science_and_Technology (Accessed, April 18 2015).

19 http://wenr.wes.org/2013/07/an-overview-of-education-in-nigeria/ (Accessed, April 18 2015).

20 http://www.nuc.edu.ng/pages/universities.asp (Accessed, April 18 2015).

21 sp-nigerian-postgrads-set-to-outnumber-indian-students-in-uk-universities (Accessed, April 19 2015).

22 http://www.iie.org/Who-We-Are/News-and-Events/Press-Center/Press-releases/2013/2013-11-11-Open-Doors-Data (Accessed, April 19 2015).

23 http://www.claymath.org/annual_meeting/2000_Millennium_Event/Video/ (Accessed, April 19 2015).

obtained his B.Sc. degree in mathematics as an external candidate of the University of London and later became the first Nigerian to obtain a Ph.D. degree in mathematics in 1950 from the Massachusetts Institute of Technology (MIT). Though he was versed in all areas of Mathematics, Obi's primary area was *Non Linear Differential Equation*. He taught mathematics at the University of Ibadan (1951-1966) and the University of Lagos (1970-1985) where he retired as Professor of Mathematics in 1985. He was regarded as the foremost African Mathematical Genius of the 20th Century,[24] and was a winner of the Ecklund Prize from the International Centre for Theoretical Physics, Trieste, Italy, for original work in Differential Equations, and pioneering works in Mathematics in Africa. The Nigerian Government recognised him, *belatedly*, with the high honor of Commander of the Order of the Niger (CON) in 2000.

Prof. Chike Obi also served as a pathfinder for other like-minded Nigerians that are too many to count, not only in mathematics but also in other fields of human endeavour. In 2000, as the Nigerian space programme was beginning to take shape, the author provided the nation's administration of that era with the names and background information on ten Nigerians, at home and abroad, who were leaders in their respective fields of specialisation. Collectively, these Nigerians were an amalgamation of the expertise needed to drive and sustain Nigeria's budding space programme. The expertise in question included those in electronics and electrical engineering, nanotechnology, electronics/man-machine inter-phase, system analysis and robotics, computer command control and communications intelligence, remote sensing including radar technology, mechanical engineering and material sciences.

Prof. Adekunle Adeyeye epitomizes the new generations of such Nigerians today. He arrived on the world stage via Awe (Igbaye), Nigeria, in 1968, eight years after Nigeria obtained its political independence from Britain. Between that time and today, he has gone from his computer-programming job in Ibadan, Nigeria, after earning a first-class degree in physics from the University of Ilorin in 1990, to obtain his M.Sc. and Ph.D. degrees in microelectronics engineering and semiconductor physics, respectively, in 1993 and 1996, at the University of Cambridge in England. Today, Adeyeye is a Professor and founding researcher at the Information Storage Materials Laboratory at the National University of Singapore (NUS) where his research interests include spin-electronics and magneto-electronics devices, nanofabrication and nano-magnetism, data storage technology, half metals and novel materials. In 2002, the MIT award-winning magazine, TR100, named him as one of the top

---

24 http://arthurnwankwo.com/salute.html (Accessed, April 22, 2015).

100 innovators whose work and ideas will change the world; the award honoured his work on spin electronics. In 2010, he also won the NUS Research Horizons Award based on his work on "Novel Magnonic Crystal Based Structures and Devices.[25]"

There are many more Chike Obis and Adekunle Adeyeyes out there in Nigeria and abroad. For example, the December 2002 USA census report shed some light on the subject of the capacities and competences of Nigerian citizens. According to Prof. Ali A. Mazrui,

"...of the 400,000 African immigrant workers, ages 16 and older in the USA, 36.5% are in managerial and professional specialties. This figure compares with 30.9% for native-born Americans. Nigerian analysts have started analysing their country's skill transfer to countries such as the USA. The best-educated national group in the USA is probably the population of Nigerians. It is estimated that 64% of Nigerians, 18 and older living in the United States have one or more university degrees. Half the members of major Nigerian associations in the USA probably have master's degrees and doctorates. If these figures are correct, they indicate a skill transfer from Africa to the USA instead of the other way around."[26]

The transformation of Nigeria will begin when Nigerian talents, such as described above, are gainfully employed in translating Nigeria's natural endowments/resources into useful products for the benefit of Nigerians and for export to the global markets.

Nigeria's Natural Endowments

Among the well-known endowments of Nigeria is crude oil with a proven reserve of 37.2 billion barrels in 2012, the tenth largest in the world;[27] the country is also the 12th largest oil producer in the world, at an average daily production of 2.52 million barrels.[28] As of December 2012, Nigeria's proven estimated natural gas reserve, the ninth largest in the world, amounted to 182 trillion cubic feet (*tcf*), with a natural gas production of 1.525 tcf in that year,[29] thus making it the world's 25th

---

25 http://www.ece.nus.edu.sg/research/achieve_nus.html (Accessed, February 8, 2015).

26 Mazrui, Ali A. (2002). *Brain Drain between Counterterrorism and Globalization*, African Issues, 2002—JSTOR, Volume 30, No. 1(2002), pp. 86-89.

27 http://www.hydrocarbons-technology.com/features/feature-countries-with-the-biggest-oil-reserves/ (Accessed, April 23, 2015).

28 tain and innovation,today....personal experience, http://www.eia.gov/countries/country-data. cfm?fips =NI (Accessed, April 23, 2015).

29 tain and innovation,today....personal experience, http://www.hydrocarbons-technology.com/ features/feature-the-worlds-biggest-natural-gas-reserves/ (Accessed, April 23, 2015).

producer in 2012.[30] On October 20, 2014, the Governor of the Central Bank of Nigeria re-affirmed the natural endowment of Nigeria and noted that the country is also endowed with an abundant supply of natural resources, providing raw materials for a broad range of industries, yet it has an array of under-exploited solid minerals including coal, bauxite, tantalite, gold, tin, iron ore, limestone, lead, zinc, and a host of gemstones.[31] Other endowments include more than 80 million hectares of arable land, of which only 40 percent is currently cultivated and 267 million cubic metres of surface water that is available for irrigation.

Table 10:1 Mineral Deposits in Nigeria[32]

	Name of Mineral	Locations (States)	Proven reserve	Human uses
1.	Barite	Proven deposits have been found in Cross River, Plateau, Benue and Nasarawa States	111,000 tons proven, to 21+ million tons estimated	Barite is used as paint and plastic filler, in sound reduction in various engine compartments, as a smooth and corrosion resistant coat for motor vehicles, as a weighting agent in petroleum well and other fluids drilling and as a barium meal before doing a contrast CAT scan.

---

30 http://www.eia.gov/countries/country-data.cfm?fips=ni (Accessed, April 24, 2015).

31 businessdayonline.com/2014/10/nigeria-receives-67bn-fdi-in-13-years-with-opportunities-in-solid-minerals/#.VEhGKXl0zIU(Accessed, April 24, 2015).

32 Sources of data for this table include:

    (i) http://www.rmrdc.gov.ng/ (Accessed, August 5, 2015)

    (ii) http://nipc.gov.ng/miningnig/deposit.html; (Accessed, August 6, 2015)

    (iii) http://www.nigeria.gov.ng/2012-10-29-11-05-46/2012-11-05-09-52-15; (Accessed, August 5, 2015)

    (iv) http://www.foramfera.com;(Accessed, August 5, 2015); and

    (v) http://www.kpmg.com/ng/en/issuesandinsights/articlespublications/ (Accessed, August 6, 2015)

	Name of Mineral	Locations (States)	Proven reserve	Human uses
2.	Bitumen	Proven deposits have been found in Ondo, Ogun, Lagos and Edo States	Estimates have been as high as 42 billion tons. Oil-equivalent is estimated to be 27 billion barrels	The primary use (70%) of bitumen is in road construction, where it serves as a binder mixed with aggregate particles to produce asphalt concrete. Bitumen is also used in numerous other applications such as in flooring materials, as joint sealants, in pipe coating (both land and marine), in automotive applications, and as adhesives and for
3.	Coal	Enugu, Abia, Benue, Nasarawa, Plateau and Gombe States.	Nearly 3 billion tons of indicated reserves in 17 identified coal fields and over 600 million tons of proven reserves	Electricity generation, steel production, cement manufacturing and as a liquid fuel. Because of its high carbon content as well as its low sulphur and ash content, Nigeria's coal is adjudged to be best for use in electric power plants and for energy generation.
4.	Gold	Zamfara, Sokoto, Osun, Niger State, Kogi and Kaduna States.	Nigeria's gold deposits are adjudged to be comparable to that of Ghana as the two countries lie on the same gold belt.[33]	Gold is used in dentistry and medicine, jewellery and arts, medallions and coins, and in ingots. It is also used for scientific and electronic instruments, computer circuitry, as an electrolyte in the electroplating industry, and in many applications for the aerospace industry.

	Name of Mineral	Locations (States)	Proven reserve	Human uses
5.	Iron Ore	Kogi, Delta, Edo, Enugu and Niger States and at the Federal Capital Territory.	Over 3 billion metric tons	Iron ore is used to manufacture steel of various types and other metallurgical products, such as magnets, auto parts, and catalysts.
6.	Limestone	Abia, Akwa Ibom, Anambra, Bayelsa, Benue, Borno, Cross River, Edo, Ekiti, Enugu, Imo, Ogun, Ondo States	Very extensive in at least thirteen states	Limestone is used as a basic building material in the construction industry. It is also used to purify iron in blast furnaces, in the manufacture of glass, to neutralise excess acidity—which may be caused by acid rain—in lakes and in soils and a host of chemical processes.
7.	Columbite/ Tantalum	Niger, Nasarawa, Federal Capital Territory, Oyo, Gombe, Kaduna, Kwara, Kogi, Zamfara and Ekiti	By 1953, Nigeria was the largest exporter of Columbite in the world. Today's Columbite reserve is deemed extensive.	Applications include but not limited to electronic capacitor, metal cutting tools, important addition to super alloys such as high temperature alloys for air and land based turbines, camera lenses, chemical equipment and in mobile phones. Tantalum is also a valuable and rare element used in electronics manufacturing by electronics and telecommunications industries.

	Name of Mineral	Locations (States)	Proven reserve	Human uses
8.	Kaolin	Ogun, Ondo, Oyo, Ekiti, Borno, Plateau, Katsina, Sokoto States	Reserve of three billion tons of good kaolinitic clay	Kaolin is used extensively in the ceramic industry, where its high fusion temperature and white burning characteristics makes it particularly suitable for the manufacture of whiteware (china), porcelain, and refractories. It is also an important ingredient in ink, organic plastics, some cosmetics, and many other products where its very fine particle size, whiteness, chemical inertness, and absorption properties give it particular value.
9.	Talc	Niger, Cross River Osun, Oyo, Kogi, Niger, Nasarawa, and Kaduna states	Over 40 million tons deposits of talc have been identified	Talc was widely used in the manufacture of pottery in times past, it is applicable in the manufacture of soaps, lubricants, pigments and for talcum powder

But the failure to develop its natural endowments cannot be attributed to Nigeria's shortage of financial wealth. Without harnessing most of the material endowments listed in metres of surface water that is available for irrigation. Details on some of the minerals cited above and of which there are more than 34 different types, in 500 locations across the country, are shown in Table 10:1.

In addition to the above cited minerals, others, in various quantities, include tin, lead, zinc, titanium, rock salt, bauxite and gypsum. Nigeria is also rich in gemstones which can be found in different parts of the country, particularly Plateau,

---

33 Nigeria produced 600 kilogrammes of gold in 2010, while Ghana produced 82,000 kilogrammes (*Source: United States Geological Survey Mineral Resources Program*) (Accessed, August 7, 2015).

Kaduna and Bauchi states. The gemstones include sapphire, ruby, aquamarine, emerald, tourmaline, topaz, garnet, amethyst, zircon, and fluorspar. In addition to the endowments shown in Table 10.1, a look at some of Nigeria's financial statistics might be in order.

Nigeria's gross revenue from both oil and non-oil revenue sources, for the 3rd Quarter of 2014, amounted to ₦2,685.080 billion (US$16.782 billion), with an external reserve of US$38.29 at the end of the same period.[34] Its gross domestic product progression, between 2006 and 2014, is shown in Figure 10:1.[35] Where then is the bottleneck to Nigeria's transformation?

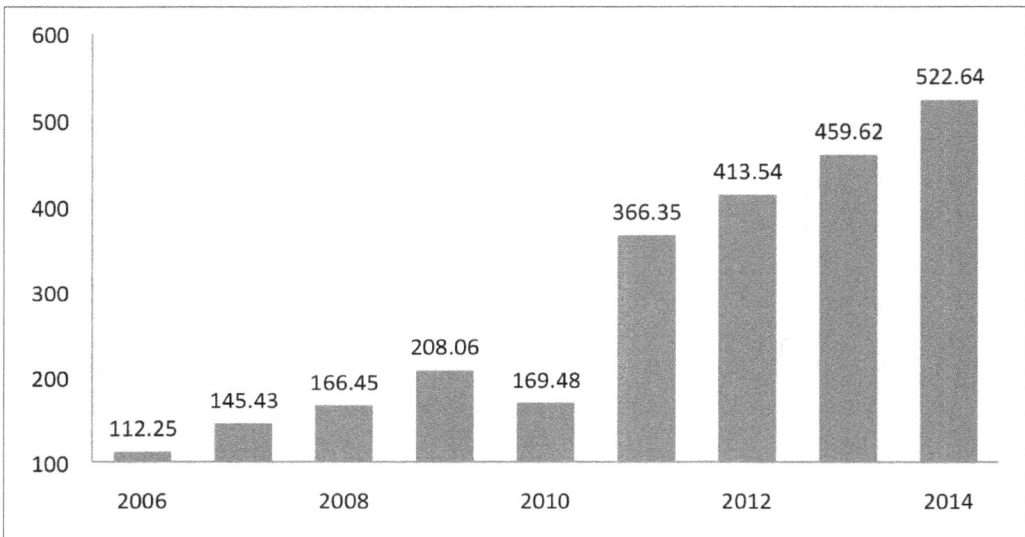

Figure 10:1 Nigeria's Gross Domestic Product (2006—2014) in Billions of US$ *(Source:www.tradingeconomics.com/world bank)*

Barriers to Harnessing Nigeria's Potential

Nigeria's human and material wealth notwithstanding, the nation is very far from becoming that envisioned "*…developed country by the opening decades of the 21st century.*" Part of the problem is that Nigeria, with its abundant petroleum resources, has long been plagued with what is known as *Resource Curse*, i.e., a paradoxical or illogical situation where, instead of resulting in a positive impact, such as enhancing the quality of life of its citizens, the natural resources endowment of a nation becomes

34 http://www.cenbank.org/ (Accessed, August 10, 2015).

35 http://www.tradingeconomics.com/nigeria/gdp (Accessed, August 10, 2015).

a setback to that country's growth and development—a condition attributable to the absence of strategic planning orientation, poor management decisions, lack of national commitment, and inimical acts committed, with impunity, against the interest of the nation and its people, for a long period of time. For over half a century, the desire for quick returns on investment in oil and gas projects took precedence over all other aspects of the economy. Agriculture, the bastion of the nation's economy before the discovery of oil, soon took a back-seat in the nation's planning and development agenda, while the long-term investment needed for developing the solid mineral sector got no further attention; and any investment idea not related to the enhancement of the petroleum sector was buried in the government cellar. Why?

Table 10:1 tells part of that story. A very significant aspect of this table is its fifth column which details some of the many useful products that are derivable from the many minerals that are within Nigeria's soil. But as a nation and a people, Nigerians have combed the global market for close to half a century, and still do so today, for the finished products made from these same raw materials that are right here, in abundance, in Nigeria. We know fully well that we can and we should bring man and machine together to process these raw materials into useful products, both for our own benefit and for export; but for a long time now, the needed political will and national commitment have been found wanting, in part, because some *vested interests* want to keep Nigeria that way.[36] Hence, Nigeria's mining sector, an engine of national development that should contribute handsomely to the transformation of the nation into an economically vibrant country, also remains a potential.

Ghani, *et. al,* certainly had the Nigerian condition in mind when they wrote as follows:

> *"While one half of the globe has created an almost seamless web of political, financial and technological connections that underpin democratic states and market-based economies, the other half is blocked, [partly by home-grown circumstances], from political stability and participation in global wealth.........they are locked into lives of misery without a stake in their countries or any certainty about or control over their futures."[37]*

---

36 Sanusi Lamido Sanusi (2013): Overcoming the fear of vested interest; https://www.youtube.com/watch?v=IjViGLJIU9g(August 17, 2013)—(Accessed, August 12, 2015).

37 Ghani, A. and C. Lockhart (2009). Fixing Failed States, *A Framework for Rebuilding A Fractured World*, Oxford University Press, United Kingdom.

This has been the case for the majority of Nigerians for a very long time, in part, because of the persistent failure of succeeding administrations to *sustain* Nigeria's S&T institutions and ensure the full contributions and long-term effectiveness of their staff and graduates in the productive sectors of the economy. This may not be unconnected to the lack of understanding of the dialectics of an S&T policy by the nation's decision-makers, i.e. their failure to link science and technology with development. While a policy statement on any issue is an articulation of the desired goal on that specific issue, the core of the policy is a clear delineation of the implementation strategies (means and process) for achieving such a goal. In the case of S&T, a nation's policy statement should include its objective (e.g. enhancing the quality of life of its citizens), and the means to accomplish it.[38] A nation's commitment to systemically translate such a policy into operational programmes will certainly transform the society.

In the absence of such a policy and its implementation plan, there is nothing to aim at, budget for or commit to. And when there is no national political will, with related commitment that can result in gainful employment, a multitude of educated and capable individuals will be idle as is now the case in Nigeria and in many other African countries. In several instances, the talents of these capable individuals remain unrecognised and therefore un-utilised, unchallenged and hence under-utilised, or misdirected and therefore misapplied.[39] Over the past several decades, highly skilled African graduates and professionals, particularly those with science and technology background, with Nigerians in the majority, have continued to emigrate, at an unprecedented rate and on a large scale across the entire African continent, into neighbouring countries and the international market. For many, a nurturing or an empowering environment, which is a very critical and non-negotiable factor in human progress, is missing in their home country.

Over forty-years ago, B.N. Bhattasali, the former Executive Director, National Productivity Council of India, examined this problem. According to him:

---

38 In the case of South Africa, its S&T policy is to pursue the public understanding of science and technology (PUSET) aimed especially at the rural and historically disadvantaged sectors of the economy through education, appropriate and judicious utilisation of human resources and phased industrialization. https://pmg.org.za/committee-meeting/4782/ (Accessed, August 13, 2015) http://www.saasta.ac.za/pdf/NSC_norms.pdf, (Accessed, August 13, 2015).

39 Abiodun, Adigun Ade (1981). *Technology development in Africa*, AFRICA, Magazine, No. 113, January 1981.

The quality and the character of a man's perceptions as well as his subsequent responses are determined in part by limitations imposed by, or opportunities available in his environment. If he is to manifest any real growth and reach his higher potential, his creativity would need nourishment from his environment.[40]

In pursuance of this goal, it should be remembered that life offers few pleasures more invigorating than the successful exercise of our faculties. Once in action, it unleashes energies for additional work. Nigeria's positive response to Bhattasali's challenge to provide a nourishing environment at home will offer a comparable opportunity for the Nigerian inborn skill and creativity, which are blossoming abroad to flourish at home for the nation's good.

The desire to mobilise the nation for such a response may have motivated the organisation of the *Conference on Bridging the Digital and Scientific Divide: Forging a Constructive Relationship with the Nigerian Diaspora*, in Abuja, on July 25, 2005. More than 250 Nigerian scientists from North America attended that conference where President Olusegun Obasanjo declared, as follows:

"…let this conference serve as the beginning of an enduring symbiotic relationship between our peoples in the science and technology sector abroad and those at home that would put in place the necessary structures for the technological transformation of our country."[41]

But the nation has not been able to sustain such a mobilization. To-date, it has not built a critical and indispensable foundation for the funding of research and development needed for its technological transformation. While other countries, such as Uganda, are lauding and considering emulating Nigeria's R&D funding initiatives, the Nigerian 2006 effort to establish a National Science Foundation with an endowment of US$5 billion and a similar effort in 2011 to establish a National Science Research, Technology and Innovation Fund are still yet to materialise or have fizzled. How long the potential of Nigeria will remain potential is any one's guess.

Can there be a Green Light at the End of the Tunnel?

The uncertainties in the crude oil market appear to be quickening a change of heart in the harnessing of Nigeria's potential. Presently, the nation's overdependence on its petroleum wealth is being shaken by the fluctuations in the price the world

---

40 Bhattasali, B. N. (1972). Transfer of Technology among the Developing Countries, Asian Productivity Organisation, Tokyo.

41 http://nigeriaworld.com/articles/2005/aug/303.html (Accessed, August 14, 2015).

is prepared to pay for crude oil. That price declined from US$107.26 per barrel on June 20, 2014, to US$45.59 on January 23, 2015; while it recovered to a US$62.00 by June, 2015, it went below US $39.00 by August 24 and in February 2016, it went below $30.00. The justifications for the downward spiral are not far-fetched. New crude oil producers are emerging and one-time avid buyers of Nigeria's crude oil, such as the USA, have also become producers and are scaling back their purchases or have not renewed their purchase agreements.

Above all, our world is witnessing intense competition in the development of alternative fuels that can replace the close to 30% of diesel and gasoline/petrol used in motor vehicles. Among the alternative fuels in commercial production, today, are biodiesel, electricity, ethanol, hydrogen, natural gas and propane. We must also take note that water-powered motor vehicles will be commercially available and will be plying the world's highways within the next seven to ten years, if not sooner.[42] What are we waiting for? Meanwhile, Nigeria is forced to compete in the Asian market for the sale of its crude oil, possibly at a reduced price. As an oil-dependent economy, Nigeria, at least in the short term, cannot reduce its daily crude-oil production because it has to pay its bills.

The net effect of these circumstances has been the current oil-glut with the corresponding fall in crude oil price, all leading to the undermining of the economies of oil-dependent countries. How to secure the economic future of such countries has become a dominant debate issue in such societies. In Nigeria, such debates have brought to the fore those celebrated days when the nation was the world's leading producer of columbite, groundnut (peanut), and cocoa, and its economy depended on agriculture, coal and tin, with coal being the main reason why Enugu became the capital of the former Eastern region. That was also the period when the nation was not only self-sufficient in food, but was also a net-exporter of food products.

In anticipation of a time when Nigeria would consider its material endowments, other than oil and gas, worthy of its attention, the on-going efforts of three entities of the Nigerian government deserve attention; these are (i) The Nigerian Geological Survey Agency (NGSA); (ii) The Raw Materials Research and Development Council (RMRDC); and (iii) The Nigerian Investment Promotion Commission (NIPC). They are indispensable entities in a successful harnessing of Nigeria's potential.

The Nigerian Geological Survey Agency (NGSA), a parastatal of the Federal Ministry of Mines and Steel Development (MMSD), has a vision "to evolve an open,

---

42 http://www.genepax.com/

transparent and flexible organisation that would provide geosciences information and knowledge for wealth creation and national development."[43] Its focus is to carry out basic geo-scientific research and advance the knowledge of geology and mineral deposits in Nigeria. It is also mandated to undertake research on behalf of and jointly with the Federal and State governments, any other government institutions, universities and other tertiary institutions and other persons. With the support of the Nigerian Government and the World Bank, the NGSA has successfully undertaken its largest airborne geophysical survey of Nigeria in its effort to position the country as an exciting destination for explorers. The products of this effort are the geophysical data sets of Nigeria. The latter, according to the NGSA, should help both local and international investors to take less risky and quick decisions on investment in Nigeria. The data sets, which are available on-line, for a fee, should boost mining activity in the country.[44]

The Raw Materials Research and Development Council (RMRDC) is a parastatal of the Federal Ministry of Science and Technology (FMST). Its mission is to promote the development and optimal utilisation of Nigeria's raw materials for sustainable industrial growth. In concert with its mission, it recently produced its 2014-2017 Strategic Plan.[45] The first key objective the plan seeks to accomplish is this: "To keep the Council focused and provide a guide for management in the next four (4) years to deliver on projects and programmes in a manner that will not only touch the lives of the citizens, but also enhance industrialisation and the overall socio-economic development of the nation through value added products from local raw materials." The document recognised the huge potential to create wealth and generate employment in Nigeria; it also noted that because of a number of constraints, the contribution of Nigeria's manufacturing sectors to the nation's gross domestic products (GDP) hovers around 4% in the last decade, as compared to the sectors' GDP average contribution of 46% in most emerging economies and developed countries. Among the constraints are the following:

(i)   Unfair competition from imported second-hand, counterfeit, smuggled and sub-standard products;

---

43 http://www.ngsa-nig.org (Accessed, August 15, 2015).

44 http://www.earthexplorer.com/2010-04/Putting_Nigeria_on_the_map.asp (Accessed, August 15, 2015).

45 Strategic Plan (2014–2017), Raw Materials Research and Development Council, Abuja, Nigeria, December 2013. (Accessed, August 15, 2015).

(ii) Weak business, financial and information management systems and practices;

(iii) Ineffective linkage between research institutions and industry;

(iv) Lack of engineering capacity to translate research and development results into finished products; and

(v) Low level of entrepreneurial capacity, amongst others.

The above constraints are beyond the mandate of RMRCD or its supervising ministry, FMST, to address; both the Federal Executive Council and the National Assembly, should, as a matter of urgency, attend to these constraints. Chapters X and XV of this book contain a number of proposals on items (iii), (iv) and (v) above.

The Nigerian Investment Promotion Commission (NIPC) was established by the Nigerian government, under the NIPC Act No. 16 of 1995, with a mandate to encourage, promote and co-ordinate investment in the Nigerian economy. Among the key responsibilities of the commission are (i) To co-ordinate, monitor, encourage and provide necessary assistance and guidance for the establishment and operation of enterprises in Nigeria; and (ii) To initiate and support measures which shall enhance the investment climate in Nigeria for both Nigerian and non-Nigerian investors.[46] According to the commission, "After years of dithering and being weighed down under squandered oil revenues and rising debt levels, the government finally demonstrated the political will to implement market friendly policies. Nigeria is richly endowed with a variety of solid minerals of various categories ranging from precious metals to various stones and also industrial minerals such as barytes, gypsum, kaolin and marble. Much of these are yet to be exploited. Statistically, the level of exploitation of these minerals is very low in relation to the extent of deposits found in the country." In promoting Nigeria as a place to invest in, the commission highlighted Companies Income Tax, Pioneer Status Tax Holiday and elimination of licensing fees as major investment incentives for prospective investors. While such incentives might attract foreign investors, what should be addressed is how to encourage Nigerians to play an active part in this segment of the nation's economy.

One of the challenges the Nigerian government faces is how to develop and nurture the mining sector as a credible contributor to Nigeria's future. To achieve that objective, the government should be an active angel- and co-investor in support of credible Nigerians that are genuinely interested in participating in the mining sector of the economy. This is critical in order to ensure that the effective control of Nigeria's mining operations is in the hands of Nigerian investors instead of expatriate

---

46 http://www.nipc.gov.ng/whyng.html (August 16, 2015).

representatives of multinational corporations. Necessary legislation should be put in place that will support such a plan and will ensure that the government recovers its investment as the mining operation becomes profitable. The bail-out of major banks, in the 2008-2010 period, by different governments around the world, following the 2008 world economic collapse, and the subsequent survival of these banks including their re-payment of the bail-out loans they took from their respective governments, offer a worthy lesson for Nigeria to study.

The current Nigerian Administration should complement the above steps by establishing the National Science Foundation proposed by former President Olusegun Obasanjo along with its US$5 billion endowment. Such a critical commitment, which should be secured with an appropriate legislation, will enable the nation to upgrade the S&T capabilities of the country and rebuild it into an enviable foundation of the nation's knowledge society of the future as proposed in Chapter IX of this book. Such a knowledge society shall build its future on the lessons Nigeria has learned from its more than 50 years of oil exploration; on a national diversified economy rooted in agriculture (food and cash crops, including processed foods); on the manufacture of a host of intermediate raw materials for local industries,including mining and mineral processing; on innovation, high technology consumer industries, exploration and processing of carbon deposits; and on the manufacture of petroleum by-products.

Thus, to attain its envisioned transformation, Nigeria must commit itself to a long-term investment in home grown industrial development that will result in the nation becoming a major producer of finished goods and services that meet the needs of its people while exporting the same to the global market. That is a sure way to curb the propensity of its citizens for foreign made products and services as shown in Figure 6:10 in Chapter VI. All of the above require that the nation imbibes a culture of change and adaptation. According to Charles Darwin, "*It is not the strongest that survive, nor the most intelligent, but the one most responsive to change.*" To make a difference and be most responsive to change means to be most adaptive.

John Stansell illustrated the concept of change and adaptation in his 1979 commentary on *Britain and innovation* when, in assessing the economic power and prosperity of his country, he proudly concluded that: "*Everyone knows that we are an industrial country, that our wealth is based on adding value to raw materials, and that our trained engineers are our lifeblood.*"[47] By implication, John Stansell is teaching Nigerians

---

47 John Stansell (1979). *Britain and innovation,* New Scientist, February 15, 1979, p. 458.

and many other resource-rich countries that although the geographical size of a nation may endow it with a given amount of natural resources and thus a manifestation of its potential power, the real power of a country is measured in terms of its economic prowess, that is, the proven capacity of that society to translate scientific knowledge into economic productivity, through its judicious and determined exploitation of technologies. That process also requires change and adaptation, including the need for Nigeria and other developing economies to invest in renewable energy development. Nothing prevents Nigeria, as well as other countries with abundant natural resources, from becoming industrialised and an economic power; the first step is a commitment to change and adapt, as appropriate, using the necessary knowledge to accomplish that goal. A well guided national science and technology commitment should make invaluable contributions in such a process.

A Nigerian leader who can give an unalloyed commitment to the implementation of the above recommended steps, and other commensurate ones, shall fit the mode of the leader described by Richard Dowden when, in 2011, he challenged Nigeria's leadership on the mobilisation of the nation's astonishing talents in these words:

"Nigeria's 140 million plus people are a quarter of sub-Saharan Africa's popula-tion and among them are astonishing talents. In business, law, science, art, literature, music, sport, Nigeria produces phenomenally talented individuals as if its superheated society throws up brighter, hotter human beings than anywhere else. The leader who manages to harness and direct all that human energy will create a formidable country that will change Nigeria, Africa and the world."[48]

Nigerians are hungry for such a leader. With the full support of the Nigerian citizens, such a leader should succeed in returning Nigeria to the level of a country which once fed itself and can feed itself, can withstand most turbulent economic conditions in the world, can offer a robust employment opportunities to its citizens, and can speedily climb out of the hole it has dug for itself in the past 50 years.

* * *

---

48 Dowden, Richard (2011). *Transforming Nigeria*, Speech delivered at the Nigerian Ministry of Foreign Affairs, Abuja on September 27, 2011 to mark Nigeria's 51st Independence anniversary. He is the Director of the Royal African Society, London, UK. www.youthhubafrica.org (August 18, 2015).

NIGERIA'S SPACE PRIORITIES

Chapter XI

NATIONAL SPACE PRIORITIES
—PART ONE

As the nation reassesses its space mission, part of that effort is the identification and pursuit of what should be its immediate and mid-term priorities. The recommended priorities in this book include the nation's need to learn from its past experiences; to re-build its science and technology foundation; to utilise its abundant human capacity to achieve its space goals; to know and understand the world in which we live; and to harness space to address the basic necessities of the Nigerian citizens including the bridging of urban and rural Nigerian communities. Other priorities include the need to meet the security and safety concerns of the nation; to gain a better understanding of the equatorial plane and to use that knowledge for the nation's future space effort; and to build and nurture the private sector that can support its space programme. All of these priorities are divided into two parts. Part One consists of the most fundamental priorities; these are addressed in this chapter and are listed in the first six bullets below. Part Two focused on the remaining six bullets, shown in the next page, and are addressed in Chapter XII. All these priorities, which are among the critical needs of the nation, are:

Part One
- Our approach to the future;
- Enhancement of the nation's science and technology capabilities;
- Understanding the world in which we live through Astronomy;
- Use of satellite data to power the nation's economy and development;
- ICT future in Nigeria; and
- Global Navigation Satellite System (GNSS).

Part Two

- Satellite services as tools of surveillance and intelligence;
- From Unmanned Aerial Vehicle(s)to Experimental Nano-Satellites;
- Sounding Rockets—Knowledge and use of the Equatorial Orbit;
- Participation in Human Space Flights Initiative;
- Building and nurturing the Private Sector; and
- A model contributor to the solution of regional and global problems.

Our Approach to the Future

To effectively plan for where Nigeria's space journey is heading, we should first examine and reflect on how we got to where we are today on the journey, as shown in Chapters V through VII, while Chapter VIII articulates the lessons we need to learn from. Nigeria also needs to grapple with what are universally recognised and accepted, in any new venture, as *the critical success factor*s. For Nigeria, space is such a new enterprise. To succeed, a nation's space programme must be built on the foundation of transparency, competence and meritocracy. Transparency shuns secrecy and it motivates the knowledgeable and skilled individuals, not just a select few, to contribute their vision on how to move the new national programme forward. Through regular and open national dialogue, as advocated in Chapter XV, the programme will benefit from constructive contributions and other related input from the community of stakeholders.[1] Such a regular forum should be a source of invaluable and accurate information on which to base informed national decisions. And to succeed in an era where space capabilities can effectively provide services that can address and meet the needs of Nigeria's teeming human population, the thought of embarking on a space programme for national pride and enhancement of geopolitical standing is out of place. Our focus should be on developing technologies that can power a vibrant national economy, and on operating a national space programme that can truly contribute to the nation's development and the wellbeing of its citizens.

---

1 As used in this this book, the word stakeholders refers to the representatives *at the decision making levels* of government, academia and private establishments that have responsibilities for the Nigerian space programme and vested interest in the use of its products and services to meet the nation's needs. These stakeholders are identified in the Section on "Partnering with all the stakeholders for a united space effort," in Chapter XV.

Enhancing the Nation's Science and Technology Capabilities

Nigeria's space journey and the overall economic development of the nation require *the rebuilding of a solid foundation in science and technology (S&T)* which, in the first two decades of its political independence, was the pride of Nigeria and the envy of Africa; indeed, it gained world recognition. Chapter IX provides some insight into how we can rebuild that foundation and thus empower Nigeria and its people. This will require our undertaking first-class fundamental and applied research and development activities that can fuel technological innovations. The latter will foster the emergence of science and technology enterprises that are critical to the nation's development and should sustain the growth of highly-skilled jobs in a variety of industries. All these are illustrated with examples from Brazil, China, India, Malaysia, Saudi Arabia, Singapore and South Africa.

In the aforementioned and other societies, *miniaturisation,*[2] *advanced manufacturing,*[3] *artificial intelligence*[4] *and big data analytics*[5] are combining to put a lot more power back into the hands of individual nations and their citizens. This is where the opportunity lies for Nigeria and Africa because African universities and researchers can get in on the ground floor and put these new tools to use in ways that will be economically and environmentally sustainable. We should remind ourselves of such examples as:

- The wind-up radio in those African communities without electric power or the funds needed to feed it with new batteries;

---

2 *Miniaturization* is the trend to manufacture ever smaller mechanical, optical and electronic products and devices, such as mobile phones, computers and vehicle engine downsizing.

3 *Advanced manufacturing* consists of the innovation in, and application of advanced materials and processes to manufacture new products using the emerging capabilities of new advanced technologies emerging capabilities enabled by the physical and biological sciences such as nanotechnology, chemistry, and biology.

4 *Artificial intelligence* consists of the theory and development of computer systems that are able to perform tasks that normally require human intelligence, The resulting programme or software acts as a helper by generating, investigating and recommending possible actions with justifications for such decisions. Such systems can be designed for a variety of applications including (a) failure diagnostics and maintenance of equipment, (b) handling of oil or chemical spills to avoid disaster, (c) management of administrative systems, and (d) helping farmers to produce better crops.

5 *Big data analytics* is the ability to collect, organise and analyse large sets of data in order to discover patterns and useful information.

- Micro financing in India as harbingers of rural community empowerment; and

- The possibility of an African farming community that buys and maintains its own UAV to track cattle and provide data on the health of each individual crop/economic tree in its community.

Add to all of these the power of positioning, navigation and timing (PNT) to enhance governance at the local level, and you will find comfort in the services of LocateIT in Nairobi,[6] a company that is successfully addressing, in Kenya, what it calls the "*Common inconveniences and Malfunctions*" within our rural and urban communities. All of the above and similar ones require knowledge and understanding that are built on a STEM system of education as shown in Chapter IX.

The emerging world is changing its patterns of governance, business practices, and methods of communication; it is also constantly undertaking research, and is finding new and improved ways of getting things done. In the process, a number of countries have become *knowledge societies*. All of the above and more, collectively known as innovation, continue to impact and enhance the quality of life of the citizens of those countries that are submitting and adapting the most to change. The computer, the Internet, fibre optics, the space and information age, the ubiquitous mobile telephones, and the continuous PNT tool known today as the Global Navigation Satellite Systems (GNSS), have been the drivers of these changes, particularly in the last two decades. Our knowledge and understanding of the sciences and technologies that are contributing to these and future innovations—particularly in quantum computing, nanotechnology, artificial intelligence, robotics, bio-engineering, advanced materials and others—will determine our commitment to build such a true *knowledge society* in Nigeria.

Our fulfilment of such a commitment begins with our acceptance that the days of doing business as usual are over. We must become active and real practical partners in innovation by committing ourselves to the following:

- Invest in basic scientific research—the critical foundation for new ideas, methods and products;

- Establish the right policies that would lead to business innovation and creativity, including facilitating partnerships between research institutions/ universities and related industries; and

---

6 www.locateit.co.ke (Accessed December 4, 2015).

- Instil the culture of academic freedom that will enable the universities and research institutions to collaborate and associate with the private sector with the goal of developing creative programmes that:

   ◊ Will enable students to receive hands-on education throughout their studies and thus help them build a career they can graduate into or be found as major assets by prospective employers, and

   ◊ Are relevant to the solution of local problems.

But first, let us assess our knowledge and understanding of the world in which we live.

Understanding the World in which We Live through Astronomy

Nicolaus Copernicus triggered a revolution when he claimed in 1514, in his hand-written unpublished 40-page script, 'Little Commentary', that the Earth was not the centre of the Universe.[7] The script was a summary of his naked eye observations and study of our world, the outer space, the planets and the stars and more. "For more than a thousand years before his time, astronomy[8] had been based on the Geocentric Model of the Universe, which stated that the Earth was the centre of all creation, with the Sun, the planets, and the stars all orbiting it."[9] Hans Lippershey of the Netherlands was credited for building the earliest telescopes which appeared in 1608. Subsequent generations of telescopes, beginning with Galileo Galilei's 1609 improvements of increasingly higher magnifying power than the earlier ones, 'opened the way to a deeper and more perfect understanding of nature.'[10]

Chapter I reminds us of how, from time immemorial, humans around the globe and in their varying degrees of sophistication have always looked into the heavens for clues that would guide their next steps as they journeyed through planet Earth. The need to understand the outer space environment has always engaged our untiring attention. Today, astronomy, as a tool of discovery, is being used in many parts of the world, including Africa, to stimulate interest in science among students, teachers and the general public as well as build the inquiring minds of the youth—the workforce

---

7 In 1543, Copernicus published *Little Commentary* as part of his book, titled: *On the Revolutions (De revolutionibus orbium coelestium).*

8 Astronomy is the observation and study of nature and the position and motion of the stars, planets, and other objects in the skies, including their relation to the Earth.

9 http://www.nmspacemuseum.org/halloffame/detail.php?id=123 (Accessed October 10, 2015).

10 https://www.aip.org/history/cosmology/tools/tools-first-telescopes.htm (Accessed October 10, 2015).

of the future. The reasons are not far-fetched. Most astronomical discoveries captivate the human imagination "by connecting to deep and long-standing questions about our origins and the nature of the universe, including planet Earth in which we live and call home." That inquisitiveness remains the driver of human response to Socrates' challenge that *"Man must rise above the surface of the Earth, to the top of the atmosphere and beyond, in order to understand the world in which we live."*

Benefits of Astronomy

By responding to Socrates' challenge, humans have continued to invent and use a variety of astronomical instruments, the most popular being the telescope, to satisfy our hunger for knowledge about our universe and everything within it. Astronomy is a discipline that continues to open our eyes and reshape how we see the world. With the knowledge acquired through astronomy, over many centuries, we are now aware of the world that exists beyond our own Blue Planet Earth. That world consists of a universe of planets, stars, galaxies and a variety of Near Earth Objects (NEOs).[11]

Until the advent of astronomical tools, the inhabitants of planet Earth mostly relied on the limited ability of the naked eye to monitor the threat of approaching space weather, asteroids and comets; they lived in fear of the consequences of their impact throughout the ages. Craters are their testimonies, and these abound in all the continents of the world—the largest one being the *Vredefort crater* in South Africa. We are not only able to monitor these threats to human existence on Earth, astronomy offers us the opportunity to work collectively to mitigate their impact, in the interest of the safety and security of our planet, Earth, and all its inhabitants.

Countries that are investing in knowledge generation in astronomy and related fields are also making gains in scientific discoveries and innovation, including job creation and economic growth. The fruits of their efforts in astronomy are finding applications in many areas of human needs that are driven by modern-day technologies. The latter include communications (Wireless Local Area Network—WLAN);[12] energy (use of a graphite composite material developed for an orbiting telescope

---

11 NEOs include near-Earth asteroids (NEAs), near-Earth comets (NECs), and meteorites, large enough to be tracked in space before striking the Earth. Asteroids and meteorites are extra-terrestrial rocks that are mostly composed of carbonaceous, stony and metallic (mainly iron) materials. Comets are fragile, irregularly shaped cosmic bodies composed of frozen gases, rock and dust.

12 Hamaker, J. P., O'Sullivan, J. D. & Noordam J. D. 1977, *Image sharpness, Fourier optics, and redundant-spacing interferometry*, J. Opt. Soc. Am., 67(8), 1122–1123 (Accessed October 12, 2015).

array, while solar radiation collectors have been developed to harness the power of the Sun for energy on Earth);[13] medical services (Magnetic Resonance Imaging scanners– MRI);[14] and Global Positioning System (GPS) satellites rely on astronomical objects — quasars and distant galaxies — to determine accurate positions and help millions of people, worldwide, find their destinations.[15]

Astronomy Programmes in African Countries

The STEM-based educational system is taking root in many developing countries. The science, technology and engineering components of that effort consist of instruction—theory and practice—on basic astronomy, and atmospheric and Earth observation sciences, as appropriate, for each level of education: elementary, secondary and tertiary levels. In several African countries, such efforts are being built around a variety of science centres some of which house a special theatre—*the planetarium*—built primarily for presenting educational and entertaining shows about astronomy and the night sky, or for training in celestial navigation. Variations of such centres exist today in a number of African countries, including Egypt, Ethiopia, Tanzania, Nigeria and South Africa.

Egypt: The driving force behind the development of the Egyptian astronomy was the annual flooding of the Nile which Egypt harnessed to build and sustain its civilization and agriculture. Hence, how to predict the flood's yearly occurrence, through the observation of stars, with a high degree of accuracy, became a national pre-occupation.[16] Among other developments that have been traced to astronomy in Egypt are the 5th Millennium BC stone circles at Nabta Playa that showed that the Egyptians had already developed a calendar and such national monuments as the pyramids and the temples, built to reflect the cardinal directions and important times of the year, such as when to start preparing the land for planting.

Fast forward, the oldest space research institute in Africa is Egypt's Helwan Institute of Astronomy and Geophysics built in 1903; subsequent ones include the Abu Simbel and Kottomia observatories.[17] These observatories have focused on four

---

13 www.ingenero.com (Accessed October 12, 2015).

14 https://www.iau.org/public/themes/astronomy_in_everyday_life/ (Accessed October 12, 2015).

15 http://www.universetoday.com/106302/how-astronomy-benefits-society-and-humankind/ (Accessed October 12, 2015).

16 https://explorable.com/egyptian-astronomy (Accessed October 13, 2015).

17 http://whc.unesco.org/en/tentativelists/5574/

major areas of study: geomagnetism, seismology, astronomy, and space research. Collectively, they shared in the search for planet Pluto in 1930, in the observation of Halley's comet in 1910, Kohotek comet in 1974, and Saturn's moon, 'Titan,' in 1980. The Kottamia's was one of a few world observatories that participated in the study of aerosols in the atmosphere.

Today, one of Egypt's modern science centres is the Planetarium Science Centre (PSC) located in Alexandria; it is dedicated to increasing the public's awareness, interest, and understanding of science and technology, including astronomy, through entertainment.[18] Activities at the three sections of the centre encourage curiosity, imagination, and creativity through a large number of diverse activities:

- *The Planetarium,* with its IMAX projection, aims to help establish a scientific culture in Egypt by offering the public a kaleidoscope of fascinating scientific shows that cover a diverse variety of scientific fields that are suitable for a wide range of groups.

- *The History of Science Museum* has permanent exhibitions that highlight the historical aspect of science in Egypt. It pays homage to scientists who have enriched scientific knowledge and aims to revive the scientific discoveries and great achievements of the ancient scholars; it also offers a variety of activities that targets school children in particular and the public in general.

- *The ALEXploratorium* is a hands-on science facility where young children and the youth are able to interact with its exhibits which cover various scientific topics, with emphasis on Physics and Astronomy.

Ethiopia: Ethiopia, with its new space observatory, the Entoto Observatory and Research Centre on Mount Entoto, sees its future in the stars. The country has concluded that if it is to develop, along with others in the East African region, it must strongly link its *economy to science.* That is why today, Ethiopia is the host of the East African regional node of the International Astronomical Union (IAU) Office of Astronomy for Development. At the national level, it commissioned, in 2011, two 1-meter class telescopes at the Entoto Observatory in the suburb of Addis Ababa. The observatory and the centre, both of which are opened for use by scientists from East Africa, are expected to accelerate the development of science and technology in the region.[19]

---

18 http://www.bibalex.org/en/centre/details/planetariumsciencecentrepsc (Accessed October 13, 2015).

19 http://www.iau.org/news/pressreleases/detail/iau1401/ (Accessed October 15, 2015).

Tanzania: The UNAWE-Tanzania[20], a branch of Universe Awareness (UNAWE), in partnership with *Astronomers-Without-Borders,*[21] has formed a community development non-governmental organisation, *Centre for Science Education and Observatory,* in northern Tanzania. Through this astronomy project, the youth of Tanzania are exposing very young children to the beauty of the universe and their place in it. In the process, they are broadening the minds of the children and are awakening their curiosity in science. Presently, with the aid of teaching models, the centre offers teacher workshops; it also donates telescopes and other science materials to schools in Tanzania, and sends *astro-science ambassadors* to provide training in the nation's schools.

Nigeria: In Lagos State of Nigeria, a week-long yearly Summer Science and Technology Camp for secondary school students has become an annual practice.[22] The camp, populated yearly by over 300 students who are chosen on merit, exposes the students to basic education in astronomy, celestial phenomena, weather patterns and concepts. As of 2013, plans were concluded to establish Astronomy clubs in schools and for members of the public through outreach programmes. These efforts were undertaken to ensure the success of the initiative of Lagos State Government to commence the study of astronomy and space science in its public secondary schools.

At the tertiary level, the atrocities associated with the Nigerian civil war of 1967 to 1970 wiped out the early gains in astronomy at the University of Nigeria at Nsukka and in other universities in the country. Since then, there has not been much financial support from succeeding governments to fund an energized national astronomy programme. Meanwhile, astronomers in Nigeria have reached out to their colleagues across the globe to respond to the science aspiration of West African youths. Subsequent discussions among interested scientists at the General Assembly of the International Astronomical Union (IAU), held in Beijing, in 2012, resulted in the launch of the West African International Summer School of Young Astronomers, first held in 2013 at the University of Nigeria, Nsukka, with 50 students in attendance.[23] The second summer school was also held at Nsukka, in July 2015, with programmes for undergraduate students and researchers. The Centre for Basic

---

20 http://unawetanzania.org/ (Accessed October 15, 2015).

21 http://astronomerswithoutborders.org/projects/telescopes (Accessed October 15, 2015).

22 http://www.nigeriaschoolsblog.com/secondary-schools/lagos-begins-study-astronomy-space-science-secondary-schools (Accessed October 15, 2015).

23 http://www.cita.utoronto.ca/sponsored-by-cita-the-west-african-international-summer-school-for-young-astronomers/ (Accessed October 18, 2015).

Space Science (CBSS) at the University of Nigeria, Nsukka which hosts this summer school, is also the host of the IAU West African Regional Office of Astronomy for Development (ROAD).[24]

Many African countries, including Nigeria, are looking forward to contributing to an astronomy project that should facilitate and enhance science and technology capability, especially in radio astronomy, in the continent. It will also enable Africa to contribute to a better understanding of our universe and the world in which we live. Nigeria cannot afford to be left behind in this endeavour.

Today, one of the cheapest vehicles for this capability development is the African VLBI[25] Network (AVN), an initiative of South Africa.[26, 27, 28] It is anticipated that AVN would initially work with its European counterpart, European VLBI Network (EVN), but it could operate independently if sufficient African antennas become available. Among the tools needed are the large redundant telecommunication antennas that litter the African landscape today, including those in Nigeria.[29, 30] In many countries around the world, satellite communications have been replaced by transoceanic fibre optics/cables. The latter provide communication bandwidths that are a thousand times greater than radio links via communications satellites, thus rendering a number of very large communication antennas redundant, particularly those in vogue when Intelsat ruled the communication world. In the case of Africa, such antennas similar to the Lanlate antenna shown in Figure 4:1 of Chapter IV, can be cheaply converted into stand-alone radio-telescopes or as part of a large array such as proposed for use in

---

24 http://westafrica.astro4dev.org/ (Accessed October 18, 2015).

25 VLBI stands for the African Very Long Baseline Interferometry.

26 Astronomical VLBI uses observed radio wave data to obtain information on space such as a star's position and distance from the Earth, or motion of a star. On the other hand, geodetic VLBI uses the data to obtain information on the Earth such as a position and a distance located on Earth, crustal movements, and Earth's rotation speed.

27 The African VLBI Network (AVN) is the initiative of South Africa's HartRAO, SKA Africa and the its Department of Science and Technology (DST).

28 Gaylard, M. J. et al (2014). An African VLBI network of radio telescopes—http://arxiv.org/pdf/1405.7214.pdf (Accessed October 19, 2015).

29 18 metre-antenna dish to 32 metre-dish were installed in Nigeria at Enugu, Kujama, Lanlate and Ikeja.

30 Opara, F. E., Omowa Edward and Esaenwi Sudum (2003). *Careers in Astronomy: Graduate School and Teaching,* The 2013 West African International Summer School of Young Astronomers, University of Nigeria, Nsukka. (Accessed October 18, 2015).

the AVN project. Both New Zealand and Peru successfully converted their redundant 30 metre and 32 metre antennas, respectively, into radio astronomy telescopes,[31] with follow-up successful experiments.[32]

These examples should motivate Nigeria to convert its Intelsat antennas in Lanlate and in other parts of the country into radio telescopes, and to subsequently join its other African partners in the establishment of and participation in the AVN programme. By taking these steps, Nigeria will be able to transform its dis-used antennas into useful space observational instruments that will foster fruitful scientific research both at the graduate and postgraduate levels of its universities. That decision is a political one with input from Nigeria's scientific community; such a decision may also lead to Nigeria's participation in the SKA project.

South Africa: In furtherance of the 1998 YEAST initiative on science and technology that was addressed in Chapter IX, today, astronomy has become a universal vehicle that is being used, nationally, in South Africa, to stimulate an interest in science amongst teachers and students. *The Southern Africa Large Telescope (SALT)* that President Thabo Mbeki inaugurated in Karoo, South Africa, on November 10, 2005 captured the essence of astronomy as a builder of South Africa's STEM-proficient work force of the future.[33]

Since then, South Africa has gone ahead and successfully competed and won the right to co-host, with Australia, the international €1.5bn *Square Kilometre Array (SKA)* radio telescope project.[34] In the past six years, South Africa has educated and trained over 700 astronomers that would be ready to man the SKA project when completed.[35] The eight other African countries collaborating on SKA with South Africa include Botswana, Ghana, Kenya, Madagascar, Mauritius, Mozambique, Namibia and Zambia; these countries will host the remote stations of SKA.

---

31 (a) http://arxiv.org/abs/1407.3346 (Accessed October 25, 2015)
    (b) http://adsabs.harvard.edu/abs/2006IAUSS...5E..55I Accessed October 18, 2015).

32 http://astrogeo.org/petrov/papers/wark.pdf (Accessed October 18, 2015).

33 http://www.southafrica.info/about/science/salt-telescope.htm (Accessed October 20, 2015).

34 The SKA project is an international effort to build the world's largest radio telescope, with a square kilometre (one million square metres) of collecting area. The SKA project is funded by an international consortium, including the United Kingdom, the Netherlands, Canada, South Africa and Australia. The SKA headquarters are in Manchester, UK.

35 Personal communication with Dr. Valanathan Munsami, Chief Executive Officer of the South African National Space Agency (SANSA), Pretoria, South Africa, November 15, 2016.

Africa's Contribution to Safety, Security, and Sustainability of Outer Space Activities

Back at the United Nations, the Committee on the Peaceful Uses of Outer Space (COPUOS) began, in 2010, its deliberation on the European Union initiative to develop the *International Code of Conduct for Outer Space Activities Agreement*; the latter seeks to enhance the safety, security, and sustainability of outer space activities. These objectives will require new space situation awareness (SSA) technologies and techniques that can protect satellite assets, detect and monitor orbiting debris fields, track the known and seek the unknown new space objects, including NEOs, and provide treaty verification.

Except South Africa, the rest of Africa is practically in an *observer mode* as the international community, through COPUOS, continues to deliberate on and define the political and legal steps, including the practical options it would take to address, with the aid of astronomy-led technologies, the safety, security and sustainability of outer space activities. Today, a number of African countries, particularly Algeria, Egypt, Nigeria and South Africa, have space assets; several other African countries have similar aspirations. But who is watching over the safety and security of these African assets in outer space? As advocated in Chapter XIV, there is an urgent need for the key African countries to make tangible contributions to the on-going and planned global efforts to enhance the safety, security, and sustainability of outer space activities. A functional world-class telescope in each of these countries (some with radio telescopes and others with optical telescopes), will give needed *support to the global mitigation of NEOs, space debris and space weather impact.*

Lessons for and Needed Action by Nigeria, Africa and Other Countries

The *Planetarium* and its IMAX projection at Alexander's Planetarium Science Centre are helping to establish a scientific culture in Egypt. Similarly, with their collective belief that science drives a nation's economy, the countries of the East Africa region are utilising the Entoto Observatory and Space Science Research Centre, in the outskirts of Addis Ababa, to accelerate the development of science and technology in the East African region. In Tanzania, very young children are being exposed to the beauty of the universe and their place in it; in the process, the exposure is broadening their minds and awakening their curiosity in science. And when he inaugurated the *Southern Africa Large Telescope (SALT)*, in Karoo, South Africa on November 10, 2005, President Thabo Mbeki gave the following as two of the justifications for the project: (i) *[SALT] would tell us as yet unknown and exciting things about ourselves; and (ii) We have to have a scientifically literate work force if we are to make the advances we so desperately need.*

Nigeria and other *astronomy-deficient* countries would need to make comparable investments, as indicated above, in both optical and radio telescopes in order for them to build their own scientifically literate work force in the knowledge–driven society of tomorrow. As shown in Chapters III and XIV, Nigeria and other African countries are not immune from the impact of space objects, both natural and human-made. As shown in Figure14:3, in Chapter XIV, we should not forget that on April 29-30, 2003, everything that was within the equatorial belt—latitudes 4^0 North and 4^0 South of the equator—was a sitting duck awaiting a possible hit from Bepposax fragments. We also need to remember that in the absence of any precise instrument to monitor the threat of approaching space weather, asteroids and comets, our ancestors lived in fear of such objects, on impact, throughout the ages. We must explore and act upon a collective mitigation effort; the technology is available and proven. What we need are appropriate collaborative activities at the diplomatic and S&T levels that will enable Africa's trained and competent scientists to add value to collective space security endeavours.

Nigeria, in particular, cannot afford to sit on the side-line and watch when its African partners are taking giant strides for the future of their nations and their people. For example, in the absence of a viable national astronomy programme, Nigeria will not be able to participate in the *Square Kilometre Array (SKA)*, nor will it be able to join other African countries to forge an African collective contribution to the United Nation's efforts on the safety, security and sustainability of outer space activities and environment. Certainly, Nigeria can and should do better.

Moving forward, in order to prepare and equip the nation for the national, regional and international activities outlined above, Nigeria should take the following steps:

- Invest in a world-class Astronomy Programme that is equipped with requisite tools and a modern national world-class observatory which should be appropriately sited.[36] Nigeria should also consider participating in the SKA

---

36 For optical telescopes, most ground-based observatories are located far from major centres of population, to avoid the effects of light pollution. The ideal locations for modern observatories are sites that have dark skies, a large percentage of clear nights per year, dry air, and are at high elevations. At high elevations, the Earth's atmosphere is thinner, thereby minimizing the effects of atmospheric turbulence and resulting in better astronomical "seeing."

project, a decision that will warrant its necessary investment in SKA-type radio telescopes.[37] The establishment of viable and sustained science centres in each of the nation's 36 states is a critical prerequisite to the building and bolstering of science and technology culture in the country. As shown above, such programmes will have untold positive multiplier effects on our nation, on its science and technology (S&T) development, on the stimulation of the intellect of Nigeria's youth, and on the fruitful engagement of Nigeria's inquiring minds, at home and abroad, in the development and growth of Nigeria. To achieve these objectives, such a programme should involve the participation of many of Nigeria's S&T capable universities and their research and academic staff;

- Seek an active participation of Nigerian scientists in (a) the international mission science team that is analysing the data from Canada's *NEOSSat; (b) in* the German (DLR) *Asteroid Finder* satellite programme, and (c) in Russia's International Scientific Optical Network (ISON) programme;

- Include basic astronomy in the science curricula of the nation's secondary schools and advanced astronomy courses as requirements for science and technology university education; and

- Convert unused or abandoned NITEL's satellite communication antennas into astronomical (VLBI) stations for use in the AVN project.

By taking these essential steps, Nigeria will acquire the critical capacity needed to solve its own problems at home, and would have earned its place at Africa's and international S&T negotiating tables.

Use of Satellite Data to Power the Nation's Economy and Development

Most Nigerians will never see a satellite or touch one; at best, they will see its pictures on their television screen or on the pages of the nation's newspapers. But it can positively impact their lives through effective ground-based applications.

---

37 Radio observatories are also preferentially located far from major centres of population to avoid electromagnetic interference (EMI) from radio, TV, radar, and other EMI emitting devices. This is similar to the locating of optical telescopes to avoid light pollution, with the difference being that radio observatories are often placed in valleys to further shield them from EMI as opposed to clear air mountain tops for optical observatories.

Impact on Nigeria's Economy

The value of a prudent Nigeria's space programme is in the productive ground-based applications, by Nigerians, of the data acquired by Nigeria's Earth observation satellites, to power the Nigerian economy and growth as well as enhance the quality of life of its people. When such data are efficiently and timely processed into information that can be used in the various aspects of the nation's development agenda, they will certainly have positive impact on its economy. In Australia, for example, through direct contribution and productivity impacts, that nation estimated that Earth observation systems contributed AU$496 million in 2015 to its economy.[38] The combined impact of the use of EOS services also resulted in the employment around 9,293 people in 2015. Interestingly, Australia gains these benefits without the risk and cost associated with owning and operating its own earth observation satellites. In 2012/2013, the contribution of Earth Observation to the UK economy amounted to £386 million.[39] At the global level, the total Earth Observation revenue from sales and value-added services by commercial operators was US$ 2.3 billion in 2012.[40] The remote sensing [Earth observation] market, world-wide, is expected to grow to US$ 6 billion by 2020 as nations use satellites for economic development. To gain a similar sense of the impact of Nigeria's space programme on the nation's development agenda and on the wellbeing of Nigerians, the nation needs an outside credible entity, authorized by the Nigerian government, to carry out such an annual economic impact assessment of the nation's space programme. That is the practice in other countries; it should also be the case in Nigeria in order that the nation can gauge the value of its space investment.

Specifically, Nigeria would need to energetically organise, manage and use its Earth observation data for national development in ways similar data have been used to achieve economic development in such countries as Australia, Brazil, Canada, India, South Africa, South Korea, Sweden, the United Kingdom and the United States. In these countries, the common trend in their acceptance of Earth observation from space/remote sensing as a tool of development has been its vigorous use to up-date

---

38 *The value of Earth observations from space to Australia in 2015*, Report to the Cooperative Research Centre for Spatial Information, Canberra, Australia (Accessed December 20, 2015).

39 *The impact of space on the UK economy 2015*, London Economics, London, UK (Accessed December 10, 2015).

40 *Bringing Space Down to Earth,* 2014 World Economic Forum, Davos Switzerland (Accessed December 15, 2015).

their base maps that were produced before the advent of satellite technology, and its day-to-day applications to address basic human needs. These needs include, but are not limited to, agriculture and food security, land-use and land and forestry management, water resources, fisheries and marine environment, mining, transportation, environmental management, disaster management, internal security. The aforementioned countries took different successful approaches to pursue the building of remote sensing capabilities at the state, provincial and local levels to ensure that space technology provides 'a variety of vital input for holistic and rapid development of rural areas.' Chapter VIII shows how, through appropriate user-education and training, both in theory and practice, the concept of Village Resource Centres, as practised in India, can be adapted to put essential space-derived information at the fingertips of Nigeria's farmers, fishermen and other users-in-need, particularly in rural Nigeria.

Data Availability and Knowledge of its Use to Power the Nigerian Economy

The economic case for investments in satellites is on the ground—with downstream applications and services that enhance the wellbeing of the citizenry and lead to many new jobs; the same should also be true for Nigeria. Chapter II provides examples of how an educated and effective use of the data acquired by Nigeria's Earth observation satellites can enhance Nigeria's productivity, particularly in such areas as agriculture and food security, land-use and land and forestry management, water resources, fisheries and marine environment, mining, transportation, environmental management, disaster management, and internal security, and in the process, it will enhance the nation's overall economic development. Presently, the nation's data archives are being enriched, daily, by its existing Earth observation satellites. These data sets are more than enough to meet Nigeria's needs for the next five to seven years or more. The data become valuable when they are properly managed and used; the latter includes data collection, storage, processing, and dissemination, analysis, interpretation and application. The nation benefits when the data are applied in the day-to-day national development activities that could result in an appreciable social and economic impact. The constraint today is that the use of the data that is at hand is very limited, in part, because the nation-wide understanding of the data is still at its infancy, and the user community, with a requisite knowledge of how to interpret and analyse the data, is equally limited. Until Nigeria is able to scale this hurdle, it is very pre-mature and not investment worthy for the nation to think of launching any new Earth observation satellite.

The nation's space effort will have a significant economic impact when its focus is on downstream applications, driven by a functional and sustainable data analysis, dissemination and utilisation plan[41] that entails the following:

- Development of essential ground segments that can cater to the processing, analysis and dissemination of data acquired by Nigeria's Earth observation satellites to end users;

- Development of user-assistance programmes that can be understood, assimilated, and readily applied, nation-wide; and

- Educating and training Nigerians, nation-wide, on how to apply the analysed data in their day-to-day activities, by taking the following steps:

  ◊ Develop necessary education and training curricula on the interpretation and utilisation of space acquired data for use in Nigeria's elementary and secondary schools, as well as at its tertiary institutions;

  ◊ Train, on a continuous basis, relevant and targeted cadre of staff members of user agencies and other establishments on the step-by-step application of Earth observation data in their day-to-day activities;[42]

  ◊ Accelerate the development and reach, nation-wide, of needed knowledge of space data utilisation. To attain these goals, space application tools (hardware and software), particularly for analysis and interpretation and use in educating the spatial-data user community would need to be readily available and accessible at the geospatial centres in Nigeria;[43]

  ◊ Understand the concept of *cloud-computing*, and introduce tools, such as *geospatial predictive analytics*. These are critical tools that will enable commercial users, government agencies and security establishments, such as the military, "to go beyond responding to change to anticipating

---

41 As advocated in Chapter XV, such a plan should be spearheaded by *The Think Tank, (an organ of the National Space Council)*, in collaboration with all the stakeholders.

42 Areas of global successful application of Earth observation data from space include: Land-use and agriculture, communication, environment, civil works, transportation, water resources, forestry, coastal management, mines and power, housing, population survey, emergency preparedness and disaster mitigation.

43 National Remote Sensing Centre in Jos, Federal Surveys of Nigeria, Geological survey of Nigeria, RECTAS and ARCSSTE-E, both at OAU Campus, Ile-Ife, and equivalent establishments at the State levels.

it."[44, 45, 46] With the availability of such tools and other analytical software in the geospatial market, users of Earth observation data, including those in Nigeria, should be able to address how to protect lives, manage risks and optimise resource allocations; and

◊ Encourage research at the university level. Nigeria's scientists are among the most active in the use of the nation's geospatial data for applied research; their contributions, based on their knowledge and rich experience, should be invaluable in the training and education of the nation's spatial data user communities.

By diligently following such a process, over a sustained period of time, the nation should witness a significant and measurable impact of its use of satellite data in its economic development and on the wellbeing of the Nigerian citizens. The consequent maturity of the nation's application of such data sets will guide the country's space future and related investment decisions, including a possible radar data ground acquisition system, in partnership with those that are knowledgeable about and familiar with radar technology. The USA developed, built and launched, in 1978, the first space-borne synthetic aperture radar (SAR) satellite, known as Seasat; its mission was the remote sensing of the Earth's oceans. Both Canada and Germany are also experienced operators of radar satellites, RadarSat-1 and -2, and TerraSAR-X respectively, with plans for future launchings of similar satellites in those same categories. The United Kingdom is equally developing a radar satellite, NovaSar. Nigeria can explore, with these countries, a collaborative arrangement that will enable it to build, in its own territory, a functional data reception station that can directly receive radar data of Nigeria.

But, for now, we should remain on *terra-firma* and learn, understand and be proficient in how to analyse, interpret and use such data to meet the nation's needs before investing in a hardware the nation rarely understands nor can effectively

---

44 Cloud computing refers to a variety of different types of computing concepts that provide customer access to application software and data bases, on-demand. The cloud infrastructure can be a very large computer or a series of large ones which customers can access, on a real-time communication network (typically the Internet), using their own desktop computers, laptops, tablets and smartphones.

45 Geospatial predictive analytics allows the combination of satellite imagery with other information that can be cross-referenced to a place, such as social media, history, cultural norms, and demographics, to help users analyse past events at a location to try to predict future events.

46 http://cdn.safe.com/resources/news/Geospatial-Today-January-2012.pdf (Accessed November 5, 2015).

utilise. The knowledge gained from such a broad and wide ranging experience in data interpretation, analysis and utilisation should guide the nation's decision makers with cogent advice from the Nigerian scientists, engineers, the nation's security entities and the data user community, on the necessity, in the future, of a direct acquisition of data of Nigeria by a Nigerian satellite.

Building such a direct data acquisition skill and competence should begin at a modest level, with a Nigerian developed and built *Unmanned Aerial Vehicle* (UAV) that can achieve, with less risk and at lower costs, the same goals that were once consigned only to satellites. Chapter XII addresses how such a mission can be undertaken in parallel with the above data processing efforts, and accomplished, on a competitive basis, by Nigerian universities. As time progresses, the UAV experience should mature into greater undertakings by the nation.

ICT Future in Nigeria— Bold Long-term ICT Vision and Investments

There are three ICT priority areas that deserve the immediate attention of the current Nigerian Administration, namely, the ICT needs of rural Nigeria, ICT manufacturing capability, and Indigenous satellite operators.

Meeting the ICT Needs of Rural Nigeria

Building up the ICT capability of rural Nigeria is essential because the breadbasket of the nation is rural Nigeria. Equipping rural Nigeria with necessary ICT features and infrastructure (such as computers, Internet, tele-centres, electronic conferencing and other networking facilities) will open up opportunities to rural dwellers for a variety of education and knowledge sharing possibilities. In 2014, Airtel made a contribution towards the attainment of such a goal when it launched its "Boost ICT Usage in Rural Areas" programme in order to:

- Enable its customers in the rural communities to experience data services that are aimed at enhancing their quality of work and living; and

- Bridge the gap between consumers in the rural settings and the world of opportunities before them; if given the right access, such an effort may bring their dreams to fulfilment.

As shown in Chapter II, through its Community Resource Centres (CRCs), the National Communication Commission (NCC) has initiated a number of ICT-supported programmes, in the rural areas of Nigeria. Because rural Nigeria is the breadbasket of the nation, this NCC worthy initiative should be strengthened to provide services that are comparable to those of the Indian Village Resource Centres (VRCs) that are described in Chapter VIII.

By empowering the rural communities with robust and functional ICT infrastructures, the nation's rural communities should begin to experience the benefits of ICT investments, not only in their agricultural and related economic productivity, including employment opportunities, but also in knowledge enrichment, health improvement and their overall wellbeing. The nation may also witness a significant level of urban to rural migration as the urban dwellers conclude that they might experience a higher quality of life in rural Nigeria. The nation's industries might equally experience the joy of relocating to more open areas in the rural communities and away from congested and polluted urban centres where monthly property rents are generally very high. Thus, as Nigeria strives to become a knowledge-based society, driven in-part by ICT, rural Nigeria will become an integral part of that effort.

ICT Component Manufacturing

Today, it is widely acknowledged that ICT is stimulating the development of home-grown solutions and driving entrepreneurships of many kinds in Nigeria and across Africa. But we must also acknowledge the sober reality of the long-term economic implications of our current ICT engagement which is mostly a service-oriented enterprise.

In Nigeria, the ICT industry is foreign supported and dominated, as well as monopolised by the importation and supply of huge quantities of various components and peripherals. We must also take stock of the colossal import cost and the absence of local content in the imported parts and components. Unfortunately, many of these components are incompatible, others are unreliable and below standard, and still others are being produced by *fly-by* companies. There are also the jobs, the gains and the benefits associated with their production, locally—all of which are eluding Nigeria today. While a limited number of local companies, with support from such electronic giants as Nokia (now partly owned by Microsoft), LG Electronics, Microsoft and Google, are engaged in software (applications) development, in Nigeria, a few others are assembling computers and electronic parts. But Nigeria's market is a very huge one—over 170 million people—and the creative abilities of its citizens are waiting to be productively engaged and fully utilised. Nigeria should take needed bold steps and begin to manufacture and produce a majority of what it consumes.

Given the current and future ICT needs of such a huge market, Nigeria should be bold on its long-term ICT vision and investments, with emphasis on public-private partnership in the development and nurturing of ICT manufacturing capability; the latter should provide the critical high-quality ICT parts, peripherals and components that can meet the nation's needs and that of other markets within the global community. The large array of mineral deposits in Nigeria, as shown in Chapter X, suggests that such made-in Nigerian ICT components should benefit from a significant amount of local content.

Indigenous Satellite Operators

Figure 7:2 of Chapter VII shows how the satellite communication landscape in Africa, particularly in Nigeria, is currently dominated by the footprints of multiple communication satellites that are foreign owned and foreign operated. The Nigerian press has informed the nation of the government's desire to acquire additional communication satellites in order to enhance the ability of NigComSat Ltd to deliver a robust communication service to the nation and its people. However, the nation must also note the inability of NigComSat Ltd, with its NigComSat-1R in space, for over four years, to successfully compete with these foreign satellite operators. Nigeria will need a cadre of competent satellite operators of its own to effectively utilise its satellite that is currently in space. This can only be addressed through essential training, education and the gaining of on-the-job experience by Nigerians. It is a step that Nigeria has to take if it intends to be in the satellite communication business, particularly through a public-private partnership; that is the route that can stimulate and grow a cadre of indigenous satellite operators in Nigeria. It is an effort that can contribute to Nigeria's aspiration as a major economic hub in West Africa, in particular, as well as in Africa. This is in Nigeria's best interest, and it is do-able.

Global Navigation Satellite Systems—
Why Nigeria and Africa should do a Rethink?

The Global Navigation Satellite Systems (GNSS),[47] especially the US Global Positioning System (GPS), which are used to pinpoint the geographic location of a user's receiver, anywhere in the world, have become ubiquitous global utilities that are

---

47 GNSS stands for Global Navigation Satellite System, and is the standard generic term for the collection of national/regional satellite navigation systems that provide autonomous geo-spatial positioning with global coverage. This term includes the Global Positioning System—GPS (USA), GLObal NAvigation Satellite System—GLONASS (Russia), Galileo (Europe), BeiDou Navigation Satellite System—Beidou (China), Indian Regional Navigation Satellite System—IRNSS (India), and Quasi Zenith Satellite *System*—QZSS (Japan).

accessible to virtually all people wherever they are on Earth. Initially developed for military uses, GNSS now provide services that are essential to the smooth functioning of most facets of the global economy.

Applications of GNSS

Today, GNSS applications include, among others, mapping by surveyors, for road navigation, management of fleet transportation, and for search and rescue operations particularly in natural disasters. Virtually all mobile telephones currently in use are GPS-equipped; it has also assuredly made getting lost *a thing of the past*. Major GPS applications also include aerial refuelling rendezvous, geodetic surveys, oceanography, law enforcement, fire-fighting, transportation (air traffic control and mass transit), space operations, management of road repairs, and off-shore oil exploration.[48] Farmers also use the GPS system in tractors, combine harvesters and in crop-dusters to map, plant, manage, and harvest their crops with efficiency and precision, a practice that has boosted US crop yields by almost $20 billion from 2007 to 2010.[49] Augmentation[50] of GPS services is also increasing the overall safety of air transportation systems, particularly in the reduction of accidents during airport approach and landing phase, including the ability for aircraft to perform their operations.

Given all of the above developmental challenges and related economic opportunities associated with an effective engagement with GNSS, first, a brief understanding of GNSS is important. Specifically, it is incumbent on the African leadership to acquire, through an African-led study, the necessary knowledge and understanding of GNSS, including its future direction, growth, development and Africa's level of participation and contribution; it is an undertaking that should be reinforced by at each national level.

Evolution and growth of GNSS

The American Global Position System (GPS)[51] can trace its origins to Sputnik 1, when scientists at Johns Hopkins University calculated *Sputnik-1*'s orbit by

---

48 http://www.gps.gov/systems/gps/modernization/sa/goldin/

49 http://www.explainthatstuff.com/howgpsworks.html

50 A satellite-based augmentation system (SBAS) is a system that supports wide-area or regional augmentation of GNSS signals through the use of additional satellite-broadcast messages. Such systems are commonly composed of multiple ground stations, located at accurately-surveyed points.

51 The GPS and other GNSS satellites provide free location, navigation and time information in all weather, anywhere, on or near the Earth, where there is an unobstructed line of sight to four or more such satellites.

measuring the induced changes in the frequency of the radio signal being transmitted from the satellite.[52] A subsequent enhancement of this approach, by the US Navy, made it possible for a nuclear submarine to accurately determine its own location. Fast forward, the first Global Positioning System (GPS) satellite, *Phase 1 Bird*, was launched on February22, 1978; the system became fully operational in 1995 with a constellation of 24 satellites.[53] The GPS has three major components, namely the network of navigation satellites in space, a control station on the ground used in managing the satellites, and GPS receivers in the possession of the users.

The former USSR (now Russia) began its own research and development programme on a *Global Orbiting Navigation Satellite System* (GLONASS) in the early 1970s, in the Cold War period, for much the same reason that the United States developed its GPS. To start GLONASS and match America's GPS/NAVSTAR, the USSR launched three satellites, Cosmos-1413, -1414 and -1415, on one rocket, on October 12, 1982. GLONASS became available for civilian use in 2007.[54] As of December 2014, the system had 24 operational satellites in space as well as back-up satellites.[55]

Selective Availability

But the new generation of GPS satellites launched from 1990 had an element called *Selective Availability* (SA) which was intentionally introduced into the GPS signals in order to degrade the civilian GPS navigation accuracy as well to protect USA's national defence. Within the United States, it was understood that the elimination of a "potential uncertainty in GPS performance for civil uses will make the system even more attractive to the world's users."[56] However, in the industrialised countries with the technical and financial means, the perception lingered on that the USA military "was dedicated to the development and deployment of regional denial capabilities in lieu of global degradation."[57] The latter became a recipe for the alternative GNSS systems we have in the world today, systems developed because of the need for individual sovereign capability.

---

52 http://www.jhuapl.edu/techdigest/td/td1901/guier.pdf

53 http://www.colorado.edu/geography/gcraft/notes/gps/foc.txt

54 http://sputniknews.com/science/20070518/65725503.html

55 https://www.glonass-iac.ru/en/guide/

56 Remarks by Hon. Mary Peters, Secretary of Transportation at the 36th ICAO Assembly, Montreal, Canada, September 18, 2007.

57 http://www.gps.gov/systems/gps/modernization/sa/faq/#time

Alternative GNSS Systems

In the past decade, subsequent developments of alternative GNSS systems began in Europe and Asia; these include Galileo (Europe); BeiDou (China), IRNSS (India) and Quasi Zenith Satellite *System*—QZSS (Japan).[58] While India had planned to participate in the Galileo programme with a €300 million investment, its perception that it would not have an *equal partnership status and role* in the project, most likely, dictated the development of its own satellite navigation system.[59] Indeed, on April 28, 2016, India launched the seventh and final satellite, IRNSS-1G, for its IRNSS system.

All of the above and many other successful applications, notwithstanding, as early as 1994, experience showed then that the GNSS signals being received from space were not perfect nor good enough for a number of applications here on Earth.[60] How to enhance and augment those signals to meet the higher standards required by users in various applications areas, particularly in high-risk safety industries, such as aviation, led to the development and growth of GNSS augmentation. The latter has become ubiquitous and it can impact farming practices, game park management and terrorist tracking everywhere, including in sub-Saharan Africa. GNSS has become an essential technology for the global economy. Our *knowledge-societies* of the future, with their driverless-cars, hyper-efficient parcel shipping, and automated air-traffic control, among others, will also require hyper-precise, efficient, reliable and continuous GNSS service.

GNSS Augmentation

A number of factors are responsible for the imperfection in the original stand-alone GPS signals. Such factors include the Earth's atmosphere: its upper region (ionosphere)

---

58 *The European Union Galileo Positioning System,* when completed, will consist of a constellation of 30 satellites (24 operational satellites plus six in-orbit spares); initial services will be made available by the end of 2016. As the constellation is built-up beyond that period, new services will be tested and made available, with system completion scheduled for 2020. *The Indian Regional Navigational Satellite System,* consisting of a constellation of seven satellites, is expected to be operational from 2016. *The Chinese BeiDou Satellite Navigation System,* which is operational, regionally, is being expanded as an eventual global service; the system will be a constellation of 35 satellites (5 geostationary orbit (GEO) satellites and 30 medium Earth orbit (MEO) satellites), that will offer global coverage by 2020. *Japan's regional Quasi-Zenith Satellite System* (a regional time transfer system and Satellite Based Augmentation System for the Global Positioning System, that would be receivable within Japan) will consist of four satellites that are scheduled for launch before the end of 2017.

59 *South Asia Monitor, Sept. 16, 2004.*

60 http://news.stanford.edu/pr/95/950613Arc5183.html

contains charged particles, and its lower region, the troposphere, is a turbulent and an uncharged region where Earth thunderstorms and all other Earth weather events occur. These inherent atmospheric disturbances contribute to the distortion and delay of satellite signals as they travel through space to the GPS receivers. Other major sources of errors that often affect GNSS signals include satellite orbit and atomic clock errors, receiver clock errors, receiver noise and resolution—these generally endanger the integrity, accuracy, continuity and availability of GNSS service.

To address a number of the above problems, particularly in the aviation industry, the International Civil Aviation Organisation (ICAO) standardised several GNSS augmentation systems including the Satellite Based Augmentation System (SBAS), in 2005.[61, 62] The need and urgency to compensate for these and other related errors so that a GPS receiver can be used for precise and accurate distance measurements gave birth, in 1994, to the process, known as *Differential GPS*, that led to the improved SBAS the world has today.[63] In October of that year, the system was installed and tested on a United Airlines 737 jetliner which flawlessly executed more than 100 blind landings.

In the operational mode, the SBAS calculates the range and integrity information for the GNSS satellites by a ground system and transmits these data to users within the coverage area through a satellite placed at the geostationary orbit (GEO). The main functions of an SBAS system consist of data collection from GNSS sources, ionospheric correction and satellite orbit determination, range and integrity calculations, independent data verification and SBAS message generation and broadcast.

Existing and Planned Expansion Satellite Based Augmentation Systems

Most of the regional SBAS programmes have been implemented to date, each complying with a common global standard. They are all compatible and interoperable

---

61 Existing and in-work SBAS include the Wide Area Augmentation System (WAAS—USA) which augments the signals of GPS, European Geostationary Navigation Overlay Service (EGNOS—Europe) augments the signals of Galileo position system, Multi-functional Satellite Augmentation System (MSAS—Japan) augments the signals of QZSS, and GPS-Aided Geo-Augmentation Navigation (GAGAN—India) augments the signals of GPS. Russia is developing its System for Differential Corrections and Monitoring (SDCM) to augment the signals from GLONASS. Both China and the Republic of Korea have signalled their intentions to develop their own respective SBAS facilities.

62 ICAO SARPS Annex 10 more related to SBAS standardisation. *Global Navigation Satellite System (GNSS) Manual*, Doc 9849 AN/457,International Civil Aviation Organisation, Montreal Canada, first Edition, 2005.

63 http://news.stanford.edu/pr/95/950613Arc5183.html

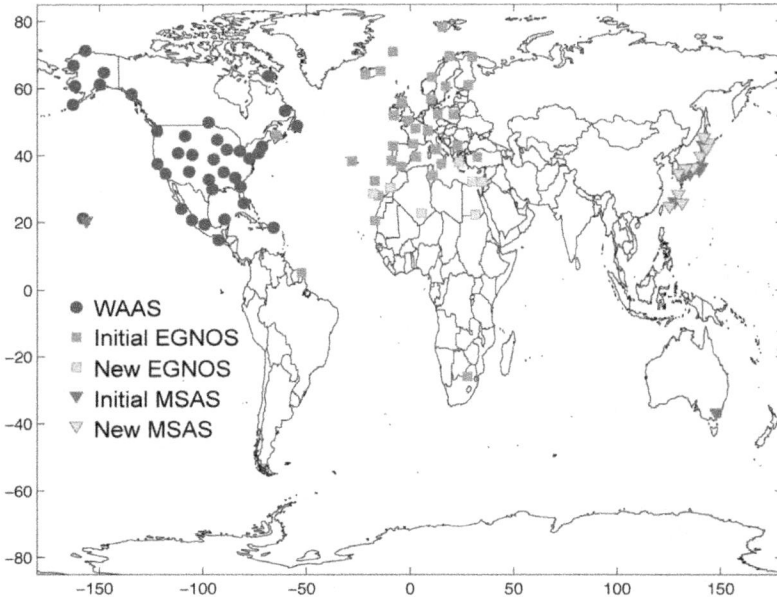

(a) Current Reference Networks

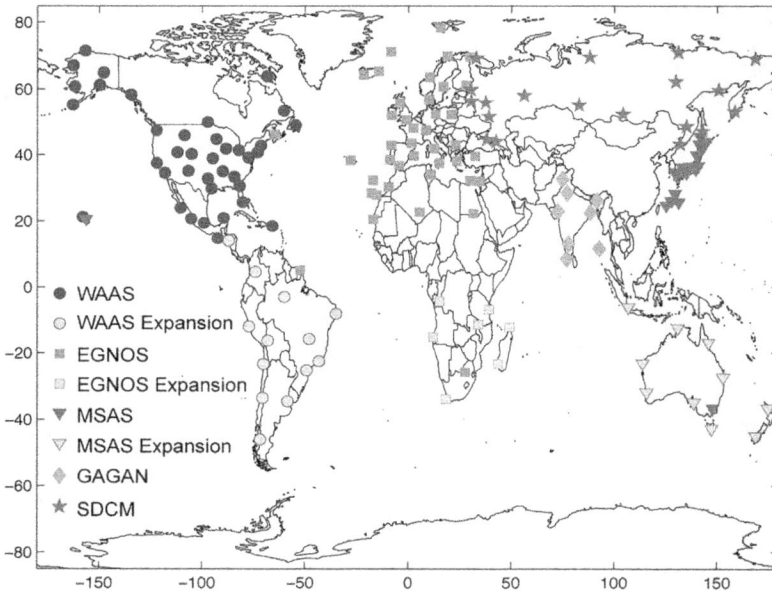

(b) Current Plans for Expanded Reference Networks

Figure 11:1 Existing and planned expansion of GNSS Space-based Augmentation Systems (SBAS) *(Credit: Todd Walter, WAAS Research, GPS Laboratory, Stanford University)*

and do not interfere with each other. An operator with an SBAS-capable receiver can benefit from the same level of service and performance of other augmentation systems, irrespective of the coverage area they are in. In order to ensure a smooth running of the SBAS infrastructure, the developers of the system have established *The SBAS (Satellite Based Augmentation System) Interoperability Working Group (IWG)*; it consists of more than 30 specialists that oversee the world's current five satellite navigation augmentation systems from the European Union, India, Japan, Russia and the United States of America.[64] A study of this document and an analysis of three of its images, in particular, namely:

- Current Reference Networks (Figure 11:1a);
- Current Plans for Expanded Reference Networks (Figure 11:1b); and
- Expanded Networks;

should shed light on the current coverage, immediate planned coverage and long-term expected coverage of the existing SBAS systems. Through these images, the reader can understand Africa's standing in today's world where PNT (position, navigation and timing) is critical to development. Accordingly, the African countries need to make the necessary collective and informed decisions to invest in a viable PNT network that can support development efforts in the continent, particularly in sub-Sahara Africa.

Africa and GNSS

In spite of the many application opportunities cited above, GNSS is yet to have any measurable impact on Africa's development. The reasons include the multiple and incompatible geodetic reference systems[65] that many African countries inherited from their colonial past and the slow process of evolving a uniform system that will

---

64 Report on '*Global SBAS Status*, Satellite Based Augmentation System (SBAS) Interoperability Working Group (IWG), June 2014'.

65 Precise positioning depends on a widely accepted coordinate system that is derived from a common datum. The Earth is not a perfect sphere; its rotation causes a slight 'bulge' at the Equator. Also none of the Earth's landmasses are fixed. Rather they 'float' on the magma beneath the crust—moving just a few millimetres each year, but moving nonetheless. This means that any datum and coordinate system that is selected must be capable of being updated. There is an International Terrestrial Reference Frame (ITRF) to which regional geodetic reference frames (GRFs), such as the African reference frame (GRF), are linked. GRFs form the basis for three-dimensional, time-dependent positioning that is essential for spatial applications such as the cadastre, engineering construction, precise navigation, geo-information acquisition, geodynamics, sea level, and other geo-scientific studies, such as the motion of tectonic plates.

facilitate regional and cross-border development programmes as articulated by the New Partnership for Africa's Development (NEPAD).[66] Also, to-date, there is no Africa-led study on the challenges, cost-benefit analysis, financial feasibility and funding for establishing and operating an African- GNSS system.

Geodetic Referencing in Africa using CORS, A Ground Based Augmentation System

But investment in the utilisation of a geodetic reference system yields a large stream of benefits for an agency or jurisdiction that typically requires accurate, compatible, spatial information, which GNSS can provide, for decision making.[67] Earlier efforts to attain the goal of developing such a uniform reference system, in Africa, through the Africa Doppler Survey (ADOS) project, did not materialize, in part because:[68]

- The logistics of carrying out the observations simultaneously proved exceptionally difficult and limited the amount of suitable data;
- The rationale was not fully understood by participating countries, resulting in a lack of motivation and enthusiasm; and
- The project was planned almost entirely by the International Association of Geodesy (IAG) and the international community, with little input from African countries.

African Reference Frame (AFREF)

The Africa Reference Framework project (AFREF), an African initiative, designed to unify the geodetic reference frames in Africa, using the Global Positioning System (GPS), as the primary positioning tool, is the successor to ADOS. AFREF was founded at a meeting of geo-spatial professionals in South Africa in October 2000;[69] it was formally established by the Windhoek Declaration of 2002.[70]

---

66 NEPAD is an economic development programme of the African Union; it was adopted at the 37th Session of the Assembly of Africa's Heads of State and Government in July 2001 in Lusaka, Zambia.

67 http://www.ngs.noaa.gov/FGCS/tech_pub/UseandValue.pdf

68 http://www.gim-international.com/content/article/an-africa-led-initiative-for-africa

69 Neilan, R. (2000). Summary of Meetings (at the margins of AFRICAGIS) on Unification of African Reference Frames, Cape Town and Pretoria, South Africa, 28 October—4 November 2005.

70 Windhoek Declaration on an African Geodetic Reference Frame (AFREF), United Nations Economic and Social Council, e/eca/disd/codi.3/11, 27 March 2003, Economic Commission for Africa, Addis Ababa, Ethiopia.

AFREF
African Reference Frame

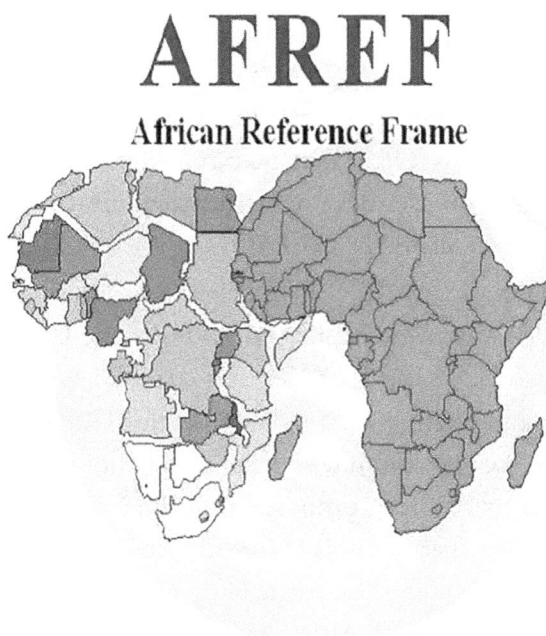

Figure 11:2 Logo of the African Reference Frame (AFREF)

All the above cited efforts are central to the success of Africa's regional and cross-border development programmes. By its design, AFREF should enhance the skills of African geodesists, surveyors and related researchers in geodesy and in the applications of Global Navigational Satellite Systems (GNSS).[71] While the left half of Figure 11:2, the Logo of AFREF, characterises the geodetic reference frames in Africa today, the right-half of the same figure depicts what is foreseen as the outcome of the AFREF project—a unified, uniform and consistent coordinate system, covering Africa, to be used as the fundamental reference system for all regional and continental geo-spatial information and planning and development projects across a wide spectrum of disciplines.

Because the *Continuously Operating Reference Station (CORS),* a GNSS Ground Based Augmentation System (GBAS),[72] addresses many applications of GPS tech-

---

71 http://www.afrefdata.org/

72 Continuously Operating Reference Station (CORS), provides Global Navigation Satellite System (GNSS) data consisting of carrier phase and code range measurements in support of three dimensional positioning, meteorology, space weather, and geophysical applications Agriculture, Construction and Civil Engineering, Mining, *Surveying & geo-spatial industry,* Science and Research programmes such as global climate change, sea level studies, modelling of the upper atmosphere in relation to storm and weather prediction, crustal strain and seismic deformation, and surface expression of hydrologic loading, Maritime, Transport system (Railway, Road transport).

nology, it is the GNSS tool that is at the heart of AFREF effort. '*CORS facilities collect and record, in an automated manner, the GPS data, at a known location that are required for relative positioning.*'[73] A completed AFREF network, made up of Continuously Operating Reference Stations (CORS) that are installed in each African country, will facilitate regional and cross-border development programmes and result in Africa's economic integration as articulated by NEPAD.

The national component of AFREF in Nigeria is the Nigeria GNSS Reference Network (NigNet). It was established in 2008 by the Office of the Surveyor General of Nigeria; as of 2011, 15 CORS stations had been established using AFREF's recommended guidelines of 500-1000 km for CORS station spacing. In order to "maximise the socio-economic benefits of satellite positioning and timing information for Nigeria, it has been proposed that the NigNet network be refined "with a spread of 10-15 km that is required for cadastral and large scale engineering applications."[74] Other African countries, particularly Algeria, Egypt, Ethiopia, Ghana, Kenya, Namibia, Tanzania and South Africa, are also establishing their own respective GNSS-CORS stations that will be integrated with AFREF.

Arrival of SBAS in Africa's Airspace

It is an established fact that CORS stations can provide a host of horizontal navigation services for application in the geodetic and geo-spatial fields, as shown under the Applications of GNSS section, in this chapter. Of particular interest to the aviation industry, however, are those GNSS augmentation systems which can provide vertical guidance services to an aircraft during final approach to a runway landing. The deployment of such technologies in an aircraft reduces the risk of accidents in which an airworthy aircraft, under pilot control, is unintentionally flown into the ground, a mountain, a body of water or into an obstacle, a situation known as Controlled Flight into Terrain (CFIT), in the aeronautical industry.[75]

---

73 http://www.ngs.noaa.gov/PUBS_LIB/GPS_CORS.html

74 Ojigi, L.M. (2014) Leveraging on GNSS Continuously Operating Reference Stations (CORS) Infrastructure for Network Real Time Kinematic Services in Nigeria In: Nkum R.K., Nani G., Atepor, L., Oppong, R.A., Awere E., and Bamfo-Agyei, E. (Eds) Procs 3rd Applied Research Conference in Africa (ARCA) Conference, 7-9 August 2014, Accra, Ghana. 24-37.

75 Controlled Flight into Terrain (CFIT) is a leading cause of airplane accidents involving the loss of life particularly since the introduction of commercial jets. There have been over 9,000 deaths due to CFIT since the introduction of commercial jets. Between 2000 and 2009, approximately 961 reportedly people died worldwide as a result of CFIT, the second largest cause of aircraft accident fatalities.

The urgent need to address the issue of CFIT led ICAO, at its 36th General Assembly in 2007, to resolve, in its Resolution A36-23, that "States and planning and implementation regional groups should complete a Performance Based Navigation (PBN) by 2009 to achieve, the implementation of [aircraft landing] approach procedures with vertical guidance (Baro-VNAV and/or SBAS) at all instrument runway ends, (with intermediate milestones as follows: 30% by 2010, 70% by 2014 and 100% by 2016).[76] Since most of the African countries, to-date, have no access to a GNSS Space Based Augmentation System (SBAS), it is not surprising that several of these countries, as reflected in their respective Performance Based Navigation (PBN) Implementation Plans, are leaning towards the implementation of Baro-VNAV.[77]

Figures 11:1a and 11:1b clearly show that Sub-Sahara Africa does not feature in the current and immediate future expansion SBAS coverage in Africa, a situation that has opened the door for outside interested parties and investors. In particular, both the European Union and the United Kingdom are taking a variety of steps that could result in Africa's participation in an SBAS extension programme that could provide a vertical guidance service for its aviation industry.[78, 79, 80, 81, 82, 83, 84] As of August 2016, China began to market its own GNSS system, Beidou, in Africa.[85] And Africa's own communication satellites could also be drafted into these plans.[86]

---

76 http://www.icao.int/Meetings/AMC/MA/Assembly%2036th%20Session/a36_res_prov_en.pdf

77 ICAO-PBN Implementation Tracking.html

78 Report of the Fourth EU-Africa Summit, April 2—3, 2014, Brussels—Roadmap 2014-2017

79 *https://www.ictp.it/trega.aspx;*

80 http://www.unoosa.org/pdf/sap/2015/RussiaGNSS/Presentations/47.pdf (2015 ICTP-TREGA project—Case Study).

81 http://www.gnss-africa.org/?page_id=23 (Algeria, Egypt, Libya, Morocco, and Tunisia);

82 The Agency for Aerial Navigation Safety in Africa and Madagascar (ASECNA) is an 18 member states organisation, with air navigation safety as its main mission; it cooperatively manages African airspace for the 18 African countries (mostly former French colonies in Africa) and France.

83 http://www.gmv.com/en/company/communication/news/2013/10/satsafinal.html

84 http://gpsworld.com/uk-space-agency-awards-sbas-africa-contract-to-avanti/

85 An International Training Workshop on Global Navigation Satellite Systems (GNSS) will be held August 8—13, 2016, at Ile-Ife, Nigeria. The workshop is being organised by the African Regional Centre for Space Science and Technology Education—English (ARCSSTEE), Nigeria in collaboration with Regional Centre for Space Science and Technology Education for Asia-Pacific (RCSSTEAP) China and Beihang University, Beijing, China.

86 http://aviation-africa.eu/sites/default/files/img/A4brochure_JPO_web_19.12.2015.pdf

Lessons for Africa

Meanwhile, in many countries including Australia, Canada, Sweden and the United States, industries such as land and maritime transport, agriculture, energy, solid mineral industry and construction, are using other augmentation technologies, particularly ground based augmentation systems (GBAS), such as CORS, for horizontal guidance. Unlike the aviation industry, several of these other industries do not believe, for now, that they need the added benefit of vertical guidance provided by SBAS, or the level of accuracy or high availability and integrity that SBAS offers. For aviation navigation, Australia plans to expand its use of Baro-VNAV while continuing to study the feasibility (possible benefits, costs and challenges) of implementing an SBAS nationally.[87] In Sweden, the aviation authority plans to expand its Baro-VNAV systems and cautiously apply SBAS where there are operational benefits.[88]

The countries in the Economic Commission for Latin America and the Caribbean region (ECLAC) decided to develop an independent and autonomous SBAS. They arrived at this decision after an ICAO-sponsored preliminary technical and operational study demonstrated that while it was possible to have an SBAS system in the region, severe ionospheric conditions in the geomagnetic Equator region (and +/- 20° around the Equator) and related degrading of SBAS signals, foreclosed the extension of the coverage of existing SBAS systems, particularly, WAAS and EGNOS.[89] As a way forward, in February 2015, the ECLAC member States that participated in the study subsequently recommended that "Each State or group of States [of the region] could consider the implementation of an SBAS system for the Region, after completing certain tasks required for validating and confirming its feasibility, namely: (a) Test bed operation (implementation of a platform of SBAS- type tests); (b) Cost-benefit analysis; and (c) Financial feasibility study." Meanwhile, to comply with ICAO's Resolution A36-23 on air navigation, many of these countries will be utilizing the Baro-VNAV to provide vertical guidance for their air navigation.

Africa's Approach to GNSS (SBAS)

Africa made a wise choice in selecting CORS, a GNSS-GBAS, for unifying the coordinate system in the continent. For the horizontal application of GNSS

---

87 https://infrastructure.gov.au/aviation/atmpolicy/sbas.aspx

88 Sweden's Performance Based Navigation (PBN) Implementation Plan—http://www.icao.int/safety/pbn/Lists/PBNImplementation/DispForm.aspx?ID=109

89 http://www.icao.int/SAM/Documents/2015-SAMIG15/SAMIG15_WP12%20SACCSA.pdf

augmentation, CORS, as detailed earlier in this script, is a proven technology that is satisfying various application needs, in many parts of the world. As each African country continues to establish its own COR stations, there is a need to recognise the value and importance of close spacing of those stations in order to ensure a greater accuracy. For example, Sweden began its own CORS programme since the early 1990s; it continued with the establishment of its second generation CORS stations in 2009, with a spacing of 70 km. As of June 2014, Sweden has CORS network of more with over than 300 permanent GNSS reference stations.[90]

To comply with ICAO's Resolution A36-23 on air navigation, member States have two options: Baro-VNAV and/or SBAS. In the absence of any current access to an SBAS system, several African countries are seriously aiming for the adoption of Baro-VNAV as their tool of vertical guidance for their respective air navigation operations. Even in Europe, where EGNOS-coverage is available, many countries in the region are enhancing their Baro-VNAV systems and are cautiously investing in EGNOS application in those areas where there are proven operational benefits. Others, such as Australia, are still carrying out extensive reviews, including justifications, cost implications, alternative technologies, and the future of today's SBAS technologies before taking an SBAS decision.

The Future of GNSS

It should be noted, however, that the GNSS we know today has become vulnerable and those that pioneered and invested heavily in it are trying and planning to develop alternatives to GNSS. The fact is that all is not well with the GNSS.[91, 92] At present, a very large number of different systems (telecommunication networks, financial services and a variety of transportation services) already have GPS as a shared dependency; a failure of the GPS signal could cause the simultaneous failure of many such services that are probably expected to be independent of each other. Known vulnerabilities of GNSS signals include system vulnerabilities, propagation channel vulnerabilities (The atmosphere and its deleterious effects, erroneous signals

90 Lilje, Mikael, Peter Wiklund and Gunnar Hedling (2014). *The Use of GNSS in Sweden and the National CORS Network SWEPOS.*FIG Congress 2014, Kuala Lumpur, Malaysia 16-21 June 2014.

91. *Global Navigation Space Systems: Reliance and Vulnerabilities,* Royal Academy of Engineering, London, United Kingdom, 2011 (up-dated in 2013).

92 Defense Fiscal Year 2016 Science and Technology Programs: Laying the Groundwork to Maintain Technological Superiority, Washington, DC, March 26, 2015.

picked up by receivers, etc.), accidental interference (electronic emissions from a variety of sources, such as high power transmitters, television, mobile satellite services and electronic devices), and deliberate interference (jamming, rebroadcasting and spoofing). The prevailing solution is anti-jamming.

Already, development of anti-jamming software is taking a big bite out of the SBAS investment and it is anticipated that the cost of GNSS anti-jamming programme would reach $4.082 billion by the year 2022.[93] Where is the point of no return for anti-jamming investment? Is there any anti-jamming software that can block any and all interference? If not, will there be an alternative PNT system. Indeed, the United States already began to seek ways to improve position, navigation and timing (PNT) without GPS. Part of that effort includes the proven capability of Locata, a private company in Australia that has successfully demonstrated the use of ground-based equipment, instead of satellites, "to project a radio signal over a localized area that is a million times stronger on arrival than GPS."[94] Hence, Africa's approach to investing in the GNSS (SBAS) infrastructure should be a very cautious one.

Africa's Resilient Position, Navigation and Timing (PNT) Tool

Presently, Africa is confronted with major challenges. The possibility of being yoked with the European Union and the United Kingdom on SBAS extension and coverage is one of them; China, with its own Beidou system, is also beckoning. What are the guidelines, including the pros and cons, of such a relationship for Africa since it will be one of unequal partners? We must also remind ourselves that, GNSS (SBAS), as we know it today, is vulnerable. Are there any other options beyond SBAS that will provide Africa a resilient position, navigation and timing (PNT) tool for its own use? These and other related questions can only be answered through an African-led study that takes into consideration the future of GNSS, the challenges ahead and the options that are available. Such a step should lead to informed-decisions by African member States on what should be the level of Africa's investment in the GNSS technology and how it would impact its development agenda.

The Africa-led study should focus on:

- GNSS today and its future direction, development and growth;
- The justifications for an African SBAS and the options that are opened to it;
- Africa's S&T contributions;

---

93 http://gpsworld.com/gps-anti-jam-increasingly-big-business/

94 www.locata.com/

- Ownership, operational and regulatory oversight structures;
- The scope of required infrastructure at national and regional levels;
- Estimated costs and benefits profile over the life of developing a GNSS system; and
- Possible funding options for establishing and operating the GNSS system.

The proposed Africa-led SBAS study should be initiated, sponsored and funded by the African Union Commission, and should be undertaken, jointly, by the African Union Space Working Group (AUSWG), the African Leadership Conference on Space Science and Technology for Sustainable Development (ALC) including its network of research scientists in Africa's relevant universities and research centres, and the African Civil Aviation Commission (AFCAC). In Africa's interest, it is a task that must be carried out diligently and speedily.

* * *

Chapter XII

NATIONAL SPACE PRIORITIES —PART TWO

This Chapter continues and concludes with the remaining recommended space priorities for Nigeria. These include the use of satellite services for national security; the constructive engagement of the talented and resourceful Nigerians who built NigeriaSat-X, as well those in the Diaspora, to accomplish similar goals at home. There is also a need for Nigeria to gain a better understanding of the equatorial plane and to use that knowledge for the nation's future space efforts. This chapter also proposes an alternative for carrying out space-based experiments of interest to the nation; it calls on Nigeria to build and nurture a viable private sector—an indispensable partner in the nation's space journey. It also addresses how Nigeria can use space tools to honour its regional and international obligations.

Threats to the Internal Security of Nigeria

Threatening the internal security of Nigeria today are ethno-religious conflicts that are being championed by a succession of war-lords in the north-east of the country, and by a variety of communal conflicts in many other parts of the country including the age-old oil-related crisis in the Niger Delta and its consequent environmental degradation of the region. While these conflicts are not unique to Nigeria, they all cause profound human, political, economic, psychological, social, cultural and environmental dislocation and weigh heavily on the people directly affected, and indirectly on the nation as a whole.

One of the nation's priorities today is how to nullify these security risks and threats so that Nigeria and its people can focus on being a nation that is working out its own development and growth strategies, rather than a nation that is forever struggling to end activities of terrorists in different parts of the country. To that end, the federal government established the Niger-Delta Development Commission (NDDC), in 2000, with the sole mandate to develop the oil-rich Niger Delta region of southern Nigeria. This act of government has only tempered the intensity and frequency of communal conflicts in the Niger-Delta; it has not ended the conflicts. Chapter II suggests a number of space-related solutions that can be adapted to address these and other conflicts, including the age-old communal conflicts between the nomadic herdsmen and the communities they encroach upon with their herds in Nigeria and in most of the West African States. Similarly, Chapter XIV recommends a number of solutions to space-based security threats, particularly from the impact of space debris, asteroids, comets and space weather.

Meanwhile, the ethno-religious conflict in the north-east of the country, the single most intractable crisis in Nigeria, today, lingers on. Part of the problem for Nigeria is that State terrorism has become international in nature and operation. It is facilitated by globalization and rapid advances in technology, with consequent weakened national borders and the use of enhanced information and communication technologies (ICTs). Given these circumstances, what are the counter-terrorism strategies and measures that the Nigerian security establishments can employ in response to the prevailing insecurity in the nation?

Security Planning and Strategies

First, Nigeria must realise that military hardware is only a means, and not an end in itself; "security planning and military strategies as well as the organisational infrastructures to implement the strategies are the key to success."[1] What the terrorists may lack in military fire-power they make up for in their resilience, mobility, technological savvy and hardy resurgence. They are also able to exploit the internet to plan, communicate and coordinate as well as feed on the global banking system to raise funds that support their operations.

To win the nation's war against terrorism, Nigeria's counter-terrorism strategies must be adaptive—it must be able to respond very rapidly to the changing circumstances in the conflict zones and beyond. Elements of the strategies should include Nigeria's comprehensive knowledge of the terrorists which can be gained

---

1 Mallik, Amitav (2004). *Technology and Security in the 21st Century: A Demand-side Perspective,* Stockholm International Peace Research Institute Report No. 20, Oxford University Press, United Kingdom.

by partnering with and mobilising local, regional and international communities as well as engaging science and technology (S&T) as a major component of the counter-terrorism tools. However, the effectiveness of these strategies depends on the commitment of the government and the dedication, skills and courage of the nation's security personnel.

Of paramount importance to the eventual success of Nigeria's counter-terrorism efforts is what the nation knows about the numerical strength of the terrorists, their geographical locations and spread; their adaptive methods within civilian populations; their sources of human, material and financial support; the cross-border inflow of such items as weapons, ammunitions, funds, paramilitary vehicles and fuel; and the information networks of the terrorists. Such knowledge should guide the country's subsequent plans-of-action and related operations against these terrorists.

Because Nigeria shares its land borders with four other countries, namely, Benin, Cameroon, Chad and Niger, it also shares with those same countries common security concerns. Thus the stalling of the proliferation of terrorists' cells and operations, beyond Nigeria's borders, demands greater concerted regional cooperation between Nigeria and these four countries. This is because, in addition to the sharing of borders, these territories serve as transport corridors for terrorists and their war machines to make in-roads into Nigeria. Nigeria also needs to partner with those countries that are more knowledgeable about counter-terrorism strategies and have established banks of intelligence information about terrorist operation. Over the years, the same countries have successfully foiled terrorist plans and operations locally, regionally and globally. The local inhabitants, who are the first victims of the terrorists' onslaughts, also have a partnering role to play in the war against terrorists. Ensuring that such local people are security conscious, remain alert and are supportive of the government offensives that could eventually protect them, is a key part of the nation's counter-terrorism efforts.

It is imperative that a special cadre among the personnel of the nation's security forces be well trained in the mastery, development and adaptation of science and technology (S&T) that is relevant to security. While S&T cannot eliminate the problem of terrorism, it can be deployed as a tool of surveillance and intelligence and as a deterrent against terrorists' attacks. With the aid of S&T, terrorists' plans can be detected and intercepted, thus leading to a reduction in the vulnerability of their intended targets, a limit to the damage they can do, and an enhanced opportunity for recovery. When fully imbibed, S&T will enhance the tempo of the nation's security operations and reduce the reaction time for making decisions that are critical to the nation's safety and security.

Satellite Services as Tools of Surveillance and Intelligence

A significant part of the S&T tools are satellite services such as satellite imagery and communications infrastructure, the internet, and the GPS (satellite global positioning system) tracker that can provide surveillance and intelligence information that will aid in the tracking of terrorists' movements and operations. To effectively utilise these services, there should be on the ground in Nigeria, a functional organisational structure, within the nation's security establishment, that understands how to task satellites and interpret the results of the imagery that is obtained. In addition to other image analysis and interpretation facilities that may be in place in the country, it is important that the Defence Space Command has its own independent and state-of-the art data analysis and interpretation facility that is able to communicate, without any interruption, with the security forces in action, in the field. And there are capable establishments (universities and image analysis entities) in Nigeria that can offer necessary assistance.

While the terrorists may not own any satellites of their own, it is safe to assume that they are using satellite-based services such as the internet and the GPS to plan and conduct their bombing raids and other incursions into Nigeria and the surrounding communities. A *GPS Tracker* should be able to provide requisite information that can be used by the security forces to intercept such plans and related actions. For example, a mobile phone is an indispensable operational tool for terrorists. Because a GPS tracking system is embedded in many mobile phones, law enforcement authorities are able to track the movements of known and suspected terrorists. A GPS tracking system is an invaluable asset in the investigation and prevention of bombings that are activated by a cell phone or text messages. By partnering with mobile telephone operators and internet service providers (ISPs) that are based in Nigeria, the nation can succeed through effective satellite enabled surveillance and interception measures to stem the bombing campaigns of the terrorists.

There are also other counter-terrorism space-based services. Among the few commercial providers of satellite images that can be used for counter-terrorism activities (such as surveillance and intelligence gathering) are DigitalGlobe with its WorldView-3 satellite,[2] Planet Inc with its SkySat series of satellites,[3] and Satellogic with its constellation of satellites that provide 1-metre resolution images as well as

---

2 https://www.digitalglobe.com/ (Accessed, November 5, 2015).

3 http://www.businessinsider.com/google-skybox-imaging-2014-6 (Accessed, November 5, 2015).

high-definition video.[4] Chapter VII provides more details on the capacities of these entities and on how to access their services and products.

The nation-wide knowledge needed to analyse, interpret and apply the data that Nigeria has been accumulating since 2003 by its Earth observation satellites, is the main problem, because it is still in its infancy; the same goes for radar data which requires greater understanding and proficiency. If it is absolutely necessary, radar data of Nigeria can be acquired, at minimal cost to Nigeria, using a radar data ground acquisition system and data processing facility installed in Nigeria, in partnership with radar satellite operators, who are knowledgeable about and are more familiar with radar technology, namely Canada, Germany, Italy, Japan, United Kingdom and the United States. For now and in the immediate future, DigitalGlobe, Planet Inc, and Satellogic offer very high resolution (1-metre resolution) imagery and video that Nigeria needs for its counter-terrorism efforts. By diligently following these steps, Nigeria should be able to accomplish its goal of stemming terrorism within its borders and beyond. In addition, Unmanned Aerial Vehicles, described below, have become invaluable, world-wide, in many security operations.

From Unmanned Aerial Vehicles (UAVs) to Experimental Nano-Satellites

Chapter XI alluded to the possible use of a UAV by an African farming community to track cattle and provide data on the health of each individual crop/economic tree in its community. Factually, UAV's, also known as drones, can do much more, including their use for security operations, border patrols, policing, parcel delivery, emergency services such as fire-fighting, policing, search and rescue, and surveying and mapping, and surveillance of pipelines. Today, UAVs also offer alternative options for acquiring Earth observation data of any country, and for executing search and rescue missions, at a cheaper cost, without going to a space altitude. Revolution in robotics and the invention of new light-weight materials, mean that all nations can afford to operate Earth observing sensing systems on all kinds of UAVs that they own and control, inside their national borders, in the service of their nation's interest and its people. These remotely operated systems are reliable and are becoming increasingly robust and inexpensive also. Payloads of UAVs for Earth observation vary from optical sensors, near-infrared cameras to synthetic aperture radar (SAR). With a UAV, it is also now possible to perform very low

---

4 *Satellogic to Offer Real-Time Earth Observation*, Earth Imaging Journal, ((Accessed, November 5, 2015).

altitude aerial photography, on a cloudy day. UAVs were originally developed for use in the military and in special operation applications, just as the aircraft and first generation of satellites were.

Today, UAVs are growing in civil applications in the USA, Europe, Asia, and in Latin America. Currently, South Korea is developing large UAVs for civilian use, such as in logistics and environmental monitoring.[5] In 2013, researchers at the University of Delft in the Netherlands developed and built what was then claimed to be the smallest UAV. It is not only smaller, but it can also carry more cameras and fly for a longer period of time than its predecessor. It also stood out for its use in safety and rescue operations.[6] Since that time, other UAVs that are as small as, and look like bees and flies, have taken to the airwaves.[7] As we look to the future, advances in the next-generation of robotics technology will accelerate the development of machines, such as drones, that can fly themselves, thus opening them up to a wider range of applications.[8]

In a similar manner, the highly specialised skills of the Nigerian engineers that successfully built NigeriaSat-X at Surrey, United Kingdom, should be productively engaged in a home-grown Unmanned Aerial Vehicles (UAVs) programme for Nigeria. These engineers should be deployed to frontline universities in Nigeria where they can use their unique talents and skills to nurture the development and growth of hands-on S&T activities. Their work could potentially include the development and building of UAVs and their payloads, such as sensor systems, on a competitive basis, on Nigerian soil. It is a programme that should be funded by the Federal Government of Nigeria, and it should be organised by the FMST, National University Commission and the National Academies of Engineering and Sciences.

With confidence gained in a successful UAV programme, the same UAV developers should commence a subsequent effort in the proof-of-concept, development and building of experimental nano-satellites and specific payloads for the collection of data needed to build up the nation's knowledge of the attributes and characteristics of the equatorial plane, in preparation for an informed future long-term space effort

5 *South Korea aims to become No. 3 UAV tech leader by 2023,* Yohanp News Agency, Sejong, South Korea, September 18, 2014, (Accessed, November 6, 2015).

6 *Dutch News In English—Netherlands, August 27, 2013,* (Accessed, November 7, 2015).

7 Piore, Adam (2014). *Rise of the insect drones,* Popular Science, January 29, 2014 (Accessed, November 7, 2015).

8 *Top 10 emerging technologies of 2015,* World Economic Forum, Davos, Switzerland, March 4, 2015 (Accessed, November 7, 2015).

of the country. From a space perspective, Nigeria's strategic geographical location near the equator calls for such a national engagement.

Sounding Rockets—Knowledge and Use of the Equatorial Orbit

Today, each nation or region strives to make its own mark in a number of ways.[9, 10] Nigeria can do the same in any aspect of the space enterprise it chooses, such as in star sensors, chemical processes, earth imaging sensors and cameras, navigation instruments, specialized software, and communication devices. Nigeria can also choose to invest in front-line research on the natural characteristics and attributes of its own geographical region and its endowments. The knowledge acquired in the process can be applied to meet the needs of the nation, the region and the international community. The Esrange Centre in Kiruna, Sweden, is such an establishment that was uniquely developed with Sweden's geographic location in mind.[11]

One of the distinct attributes of the Esrange Centre's location is its closeness to the North Pole, an attribute that the centre has transformed into a very buoyant space-related research enterprise that addresses a variety of the space needs of the global community in general and of Europe in particular. Because of its physical location, the centre, with its facilities, is able to receive data from most of the global polar orbiting satellites, including data that often ends up in a GPS device or in an atlas.

Equatorial Assets

The equator possesses comparable advantages that the countries within the equatorial and tropical belts need to study, master, develop and utilise as their contributions

---

9 Proponents of Brazil's participation in the ISS concluded that the return to the country in new technologies would energize its industries and would dwarf any amount Brazil invested in the ISS programme.

10 Similarly, Canada evaluated what was needed on ISS, assessed its own expertise and the possible commercial spin-offs that could accrue from its participation in the ISS. It subsequently came up with the development of the robotic "arm" that was used in the Space Shuttle; a more advanced version of the same arm is used in the International Space Station, called the "CANADARM", with a large Canadian logo on those arms. Today, CANADARM has given Canada national pride and respect as a leader in international cooperation within the space community; it also gave Canada an opportunity to showcase as well as enhance its capabilities which are having multiplier effects within the Canadian science and technology community as well as its economy.

11 *Sustainability Report 2014*, The Swedish Space Corporation (SSC), Stockholm, Sweden, (Accessed, November 7, 2015).

to their own social and economic development and to the advancement of human knowledge, exploration and use of outer space. To this end, in 2010, NASRDA, in collaboration with the International Academy of Astronautics, brought the international community together, in Abuja, to examine the characteristics and attributes of the equatorial plane and to explore how such knowledge and understanding can contribute to the region's development. What did Nigeria learn from this event? And what related programme(s) and project(s) emerged, thereafter, from this interaction? Because of their geographical locations, a number of other countries are also seeking knowledge of the equatorial plane. Nigeria should ally with these other potential partners, at the equator and within the tropics, such as Brazil, Indonesia, and Malaysia and Australia. These countries are undertaking a variety of experiments and studies about the equatorial plane, including the near equatorial orbit, because it is the most logical post in space for undertaking the Earth observation of near equatorial and tropical countries.

Among the endowments of the equatorial plane is the abundant energy that is wrapped in the intense rays the Sun beams, daily, over the equatorial regions of the world, including Nigeria, almost 12 hours daily. Thus, the countries within and adjacent to the equatorial belt, such as Nigeria, should be the ones that are in the fore-front of solar energy research for use in earth-bound and space-based human activities. We must end our dependence on fossil fuel, which, in addition to being a major contributor to climate change, will also run dry, sooner or later.

The equator also has other significant attributes. Because the surface of the earth has higher rotational speed at or near the equator, it is more advantageous to launch a space-bound object from a launch site that is closer to the equator than from any other location on Earth. While rotating on its axis, the Earth imparts some extra free energy to a departing rocket launched from an equatorial launching pad; the net benefit is a reduced fuel cost of about 20% or more in comparison with the cost of launching from other geographical locations on the surface of the Earth. By being close to the equator, rockets that take off from an equatorial launching base, such as the Brazilian launching pad in Alcantra, Brazil, the ESA launching pad at Kourou, French Guyana, and many others, listed in Chapter XI, can thus deliver 17 percent more payload into the high equatorial (geostationary orbit—GSO) orbit, which is often preferred for meteorological, communications and many other types of satellites. However, a rocket operating from either the Russian Baikonur Cosmodrome in Kazakhstan, at 45.6⁰ North latitude, or Cape Canaveral in the USA, at 28⁰ North, latitude, must expend more energy to spring from its launching pad and additional energy to transfer its payload to an equatorial orbit.

Individual satellites that are in equatorial low Earth orbits (LEO) complete a revolution every 90 minutes or so and are thus capable of re-visiting the same spot, on Earth, up to 16 times in a day. Such a high frequency of revisits can maximize the ability of a satellite that operates in the visual or near infra-red spectrum to exploit gaps in the ever-present cloud cover in the equatorial belt. For example, a constellation of three to four satellites, orbiting around the equator, can ensure a continuous re-visit of disaster events (such as forest fires, tropical storms and floods, and coastal oil spills), and should be able to monitor such strategic infrastructures as gas, oil and water pipe-lines, security installations, and border patrols, more than comparable polar orbiting satellites can. Thus, for Nigeria, a major focus of its immediate space technology development effort should be experimental research, and the build-up of essential knowledge on its future use of the equatorial orbit, instead of the current default polar orbit.

The space pioneers, Russia and the USA, and others that have since become space faring, namely Europe (through ESA), Japan, India and China, are all from the mid to high northern latitudes and they strategically chose to develop and locate their launching facilities within their own geographical territories which they politically control. Thus, the first set of equatorial or near-equatorial countries that acquired their satellites had to depend on the launchers of these northern latitude countries to place the satellites in the polar orbit. If instead, these satellites had been placed at the equatorial orbit, as shown and advocated in Chapters VII and VIII, such satellites would be serving their owners as well as other equatorial countries better, particularly for security purposes, for search and rescue operations and in times of emergency, such as natural disasters, largely because of their greater number of revisits per day to the same location, along or near the equator. The experiences of Brazil, Malaysia and Indonesia are worthy of study for Nigeria and other equatorial countries.

Brazil, in particular, has acquired some knowledge about the equatorial orbit that Nigeria and other equatorial countries need to learn from. Brazil's immediate plan is the on-going preparation for its EQUARS (Equatorial Atmosphere Research Satellite) Mission, to be launched as part of its LATTES satellite payload in 2017.[12] The main scientific objective of EQUARS Mission is the understanding of the coupling between the dynamic, electro-dynamic and photochemical processes in the low latitude neutral atmosphere and the ionosphere and the use of the

---

12 Villela, Thyrso (2011). *An Overview of the Brazilian Space Program*, Brazilian Space Agency, Ministry of Science, technology and Innovation, Brazilia (Accessed, November 8, 2015).

knowledge so acquired in the study of *Space Weather*. The LATTES satellite will also carry the MIRAX mission, a payload devoted to monitor the spectral and time variability of the X-ray sky using imaging telescopes. Italian scientists are using their experience with the X-ray astronomy satellite, as shown in Chapter XIV, to contribute to the development of the payloads of MIRAX. Also of high priority for the Government of Brazil is the development and use of its Alcantara launching centre. Moreover, since mid-2015, Brazil has been seeking a strategic space partner that will collaborate with it to turn Alcantara, an equatorial base, into a commercial satellite launching pad.[13]

Meanwhile, on July 14, 2009, Space X, a new private launch system developer, directly launched, with its Falcon-1 launcher, Malaysia's experimental satellite, RazakSAT-1 into the near equatorial orbit. Malaysia's Astronautic Technology (M) Sdn Bhd (ATSB), a local pioneer company in satellite development that developed RazakSAT-1 is also responsible for the development of RazakSAT-2 between 2016 and 2019.[14] Similarly, Indonesia's first indigenous satellite, LAPAN–A2, developed by LAPAN, is monitoring the Earth from its equatorial orbit. The satellite, launched on September 28, 2015, along with ASTROSAT, an Indian astronomical observatory and other satellites, by ISRO, from India's Satish Dhawan Space Centre. LAPAN-A2 carries a video camera along with an amateur radio communication transmitter for disaster management, and a digital camera for land use, natural resources and environment management. The satellite also carries an Automatic Identification System payload for tracking ships within the equatorial region.

But not all equatorial countries should launch satellites or anything else into space from their territories. Because the Earth rotates from west to east, Brazil is able to launch eastwards from Alcantra, over the Atlantic Ocean, with minimal danger to any populated area. The same goes for other potential equatorial launch sites such as San Marco launch platform in Kenya, as well as Sepik, Manus and New Ireland provinces in Papua New Guinea—these are well known favourable potential launch locations with open sea to the east for ease of recovery of first stage boosters. In the case of Nigeria, however, launching any satellite, eastwards, means that the spent rocket bodies and more could fall on Nigeria's neighbouring countries and their population centres, east of Nigeria. The crash-landing of such Nigeria-produced

13 http://www.reuters.com/article/us-brazil-space-idUSKBN0OV22C20150615 (Accessed, November 8, 2015).

14 http://tda.my/icp_programmes/development-of-razaksat-2-satellite/ (Accessed May 15, 2016).

space debris on territories, other than its own, will not bode well for Nigeria's future relationships with its neighbours.

Nigeria can partner with any other equatorial country on space matters; but it will be able to do so only by bringing a number of competences, such as its sound knowledge of the attributes and characteristics of the equatorial plane, to the negotiating table. The on-going national programme on sounding rockets, at Epe, offers such an opportunity. For the time being, this effort should focus on developing and using these rockets to acquire necessary knowledge about the Earth's atmosphere and radiation, particularly in our near-equatorial environment. The sounding rockets are also needed to test instruments that one day will end-up on satellites, either large or small, such as nano, pico and cube-satellites. Such knowledge development efforts are necessary preparations for Nigeria's future space activities and will distinguish Nigeria as a worthy partner in any future collaboration. For example, Nigeria might wish to take advantage of the opportunity provided by the United Nations/Japan Cooperation Programme on CubeSat Deployment, to launch its experimental satellites from the International Space Station (ISS),[15] or from any other launch service provider that will deploy such experimental satellites to the equatorial orbit.

Participation in Human Space Flight Initiative

An Astronaut programme has also been on Nigeria's space agenda since the publication, in 2005, of the original 25-Year Road Map for the Nigerian Space Mission document; it was retained in the 2007 review report. Early in 2014, the price tag for the Nigerian Astronaut programme was estimated to be close to ₦18.5billion or US$92.5 million, as of that time.

According to the 2007 review report, the nation's astronaut programme would: *"Promote Science Culture and a Sense of National Pride and also lead to the Acquisition and Development of allied infrastructure, facilities and skills."*

I had the privilege of being asked by NASRDA, in 2011, to comment on its 2007 review report. After analysing the brief on the astronaut programme at that time, I commented as follows:

"Kindly note that there is nothing convincingly stated in Appendix 3 nor in the entire document, to justify Nigeria's investment in an astronaut programme."

Today, the above quoted view, which I expressed in 2011 on the Nigerian Astronaut Programme, remains the same. The mitigating factors include cost implications

---

15 http://www.unoosa.org/oosa/en/ourwork/psa/hsti/kibocube.html (Accessed, November 10, 2015).

and the nation's inadequate preparedness to translate research and development results into practical goods and services for the people of Nigeria.

The five partners of the ISS, namely, Canada, Europe [ESA], Japan, Russia and the United States, collaborated and jointly developed and built the space platform; it was completed in 2011. They are using the space station to carry out a wide range of cutting-edge research activities which are expected to improve the efficiency of the production processes as well as the quality of goods produced on Earth. The attendant economic impact of these research efforts could be very significant and far-reaching.[16] Indications are that some of these goods are already impacting the way we live and work on Earth.[17] Nigeria's meaningful contribution to such efforts is of a higher priority.

But prospects for direct access to the ISS, particularly by non-ISS partners, in the immediate future, may be rare and limited, in part, because, on July 21, 2011, the United States phased out its Space Shuttle programme. For now, human access to the ISS is available only through the Russian *Soyuz* spacecraft. By 2019, commercial developers of space transportation systems such as *Space X* and the *Boeing Company*, two USA-based commercial companies, are expected to provide similar transportation services to the ISS.[18] Additional launch opportunities may become available when the Chinese *Tiangong* Space Station is launched in 2020. These and other similar developments could provide opportunities, in the future, for other countries, including Nigeria, to conduct experiments on board these space stations.

---

16 The areas of ISS research interest include, *inter-alia*:
- Bio-medicine (new antibiotics and antibodies and other treatments for cancer, malaria, HIV-AIDS and other diseases),
- The growth of collagen fibres to be used in the repair and replacement of human connective tissues,
- Organic and polymer chemistry compositions for use in such areas as advanced data processing, and
- New materials which include new alloys, the growth of crystals for semiconductor devices, the development of iron with high carbon content and lubricants—all in their purest forms because of the micro-gravity environment of outer space. The use of these new materials in combustion engineering is also expected to yield new stress-resistant and higher strength materials for Earth-based vehicles and machines as well as substitutes for non-renewable resources.
- Material processing in space is also expected to influence the development and production of alternative sources of energy on Earth.; and
- On-going research on the ISS also includes the study of microbial virulence in microgravity and telemedicine to waste recycling technology and water quality monitoring.

17 https://spinoff.nasa.gov/Spinoff2010/Iss%20Spinoffs.html(Accessed, November 11, 2015).

18 http://www.nasa.gov/press/2014/september/nasa-chooses-american-companies-to-transport-us-astronauts-to-international/#.VBiuBPldURo (Accessed, November 11, 2015).

International Cooperation

To-date, only a few countries have successfully participated directly in the utilisation of the facilities on board the ISS to carry out experiments that are in their own interest. Meanwhile, the participation of a limited number of countries in transition, such as Brazil, Malaysia and South Korea, in an astronaut programme, on board the International Space Station (ISS), is well known. For example, experiments conducted by the Brazilian astronaut, Marcos Cesar Pontes aboard the ISS, in March 2006, included DNA Repair, Seed Germination, Interacting Protein Clusters, Miniature Wire Heat Transfer Tube, and Chlorophyll Chromatography.[19] The observation and understanding of protein structure is instrumental to understanding the mechanisms of life. Aboard the ISS, in October 2007, Sheikh Muszaphar Shukor of Malaysia performed experiments on the characteristics and growth of liver cancer and leukemia cells, and the crystallisation of various proteins (lipases) and microbes in space. Lipases is a type of protein enzyme used in the manufacturing of a diverse range of products, from textiles to cosmetics.[20] And in April 2008, Yi So-yeon of South Korea carried out science experiments on board the ISS that included the monitoring of the way the changes in gravity and other environmental conditions could alter the behaviour of fruit flies, the growth of plants in space, the cultivation of microbes and how to reduce the noise level inside the space station.[21] However, as a pre-requisite to exploiting the results of their research undertakings in the ISS, such countries would need to pre-establish the needed research and development (R&D) and commercial infrastructure in their own territories. The five ISS partners have produced a compendium on the impact of ISS experiments on Earth, known as the *"International Space Station Benefits for Humanity."*[22]

But the cost of going to the ISS is very high. The Brazilian government spent or invested about US$20 million to send Marcos Cesar Pontes on the Russian *Soyuz* spacecraft to the ISS in 2006. Malaysia's bill for sending Muszaphar Shukor to the

---

19 http://www.spaceref.com/news/viewsr.html?pid=20081 (Accessed, November 12, 2015).

20 http://www.thestar.com.my/story/?file=%2f2007%2f10%2f11%2fnation%2f19136025&sec=n ation (Accessed, November 12, 2015).

21 http://english.yonhapnews.co.kr/techscience/2008/04/08/28/0601000000AEN20080408005600 0320F.HTML (Accessed, November 12, 2015).

22 International Space Station Benefits for Humanity (2015), NASA Publication #NP-2012-02-003-JSC, Washington, D.C., USA (Accessed, November 12, 2015).

ISS in 2007 amounted to US$26 million). Similarly the government of South Korea paid Russia approximately US$28 million to send Yi So-yeo to the ISS in 2008.

Given the above cost implications of an astronaut programme and its potential limitations, how other interested countries might be encouraged to participate in space-based experiments received attention at the 1999 UNISPACE III Conference. The conference recognised that global challenges can best be met by global dialogues, and that the conception and execution of joint projects between spacefaring and developing countries should be encouraged and facilitated, particularly within the framework of ISS. In the interest of international cooperation and the need to encourage the participation of those other countries that would want to participate in the ISS endeavour but could be hindered by cost limitations, the conference recommended, and the United Nations General Assembly approved, the enhancement of the use of the following mechanisms:

International, intergovernmental and non-governmental organisations and arrangements, ad hoc inter-agency mechanisms, bilateral and regional agreements, programme-specific agreements and transnational commercial activities.[23]

The Brazilian experience with the American Space Shuttle missions was such a mechanism. Prior to sending an astronaut into space in 2003, Brazil carried out a number of microgravity experiments, on board the space shuttle missions STS 83, 84, 94, 97 and 107, from 1997 through 2003, with the assistance of America's astronauts. The experiments focused on areas of critical importance to Brazil, particularly on hybrid seeds, and in the growth of proteins as critical input into the manufacture of pharmaceutical products. In recognition of the economic importance and value of the Space Shuttle-based collaboration between the two countries, in October 1998, President Fernando Henrique Cardosos of Brazil wrote a letter to President Bill Clinton of the USA in which he stated that "the research being carried out by the Brazilian scientific community on the Space Shuttle was extremely important to Brazilians, being an important advance in the research for both new drugs and agricultural research."[24]

---

23 A/CONF.184/6 (1999). Report of the Third United Nations Conference on the Exploration and Peaceful Uses of outer Space—UNISPACE III, Vienna, Austria, July 19-30, 1999.

24 Vaz, J. G. and J. A. Guimaraes (1999) *Brazilian participation in the International Space Station*, in the Proceedings of the International Symposium on International Space Station: The Next Space Marketplace, 26-28 May 1999, Strasbourg, France, Haskell, G and Michael J. Rycrof (Editors).(Accessed, November 14, 2015).

The success of the Brazilian experience and that of its experiments carried out on the Space Shuttle missions possibly influenced the recommendation of the conference on the participation of non-ISS partners in space research aboard the ISS and similar future space stations. In compliance with the GA-endorsed recommendations of UNISPACE III and the fore-knowledge of the Brazilian experiments, the United Nations Office for Outer Space Affairs (UN-OOSA) developed an initiative, known as *The Human Space Technology Initiative (HSTI)*.[25] The latter enables the developing countries to participate in and benefit from the scientific programmes of ISS without incurring the astronomical cost of sending an astronaut into space and maintaining him/her there for a specific period of time.

Human Space Technology Initiative (HSTI)

The HSTI offers all interested developing countries, including Nigeria, the prospect of participating in the utilisation of the ISS and similar facilities for their own benefit. The objectives of HSTI include:
- Expanding the use of a space station as a platform for:
  ◊ Space science and technology development;
  ◊ Education and outreach programmes; and for
  ◊ Discovering and utilising human space technology applications (spin-offs).

To accomplish these objectives, and on behalf of the developing countries, UN-OOSA is in an on-going dialogue with the ISS partners to contribute to HSTI activities by ensuring a wider use of all space stations. The focus of the negotiation is to secure the consent of those who own laboratory modules on a space station to provide physical space and instruments (for space station experiments), and time of their space station personnel (such as astronauts, cosmonauts, or taikonauts) in support of the experiments commissioned by developing countries. Through this arrangement, HSTI will work with space station owners, serve as an advocate for their use by developing countries and, in the process, bring the benefits of this unique facility to more people in the world.

Possible Focus of Nigeria's Space Experiments[26]

The focus of Nigeria's participation in space station activities would need to be guided by the key areas of each space station. The focus of the Chinese Space Station

---

25 *HSTI Guide Booklet*, United Nations Office for Outer Space Affairs, Vienna, Austria.

26 Details of how to participate in the HSTI, are available at the Office for Outer Space Affairs, United Nations Office at Vienna, Vienna, Austria.

will evolve in the coming years. In the case of the ISS, the focus includes: Scientific, Engineering, Utilisation and Education. Within the framework of HSTI, or similar but bi-lateral arrangement(s) with owners of current and future space stations, experiments from Nigeria may focus on the following:

- Healthcare—medicines and pharmaceuticals for a tropical environment;
- More efficient processes in industry;
- Advanced high-performance materials for automotive, medical and industrial applications;
- Agricultural experiments on hybrid seeds;
- A variety of experiments focusing on astronomy, biology, meteorology (climate change) and physics; and
- Education outreach.

On the whole, Nigeria would need to determine, enhance and sustain its local capacities and capabilities to support its participation in human space flight activities; these would require the development and sharing of fundamental and applied knowledge among the participating scientists and their institutions and input from the nation's private sector. Of critical importance are the priority needs of Nigeria, the areas where it would want to make its mark and make a difference, and the readiness of the nation's research and development (R&D) and commercial infrastructure to transform the results of the space-based experiments into practical goods and services for the people of Nigeria. The nation's assurance of success in carrying-out such space-based experiments will be higher if its private sector is a motivated partner.

The Private Sector

In most of the advanced countries, the contributions of governments and their institutions to the development of their societies are often complemented by those of their private establishments. When the UK government launched its Tactical Operational Satellite (TopSat) in 2005, right in its backyard, outside London, was the Surrey Satellite Technology Ltd (SSTL) that took on the assignment and got it done; it developed, tested and built the satellite. Because of its experience over the years, Britain asked the same company to develop its new radar satellite, NovaSar; the first in the NovaSar constellation is scheduled for launch in 2016.[27] Scientists and

---

27 On Nov. 29, 2011, U.K. Minister of State for Universities and Science, David Willetts announced an investment of £21M that will enable British small satellite pioneer Surrey Satellite Technology Ltd. (SSTL) to launch NovaSAR, an innovative and highly competitive new space-based radar remote

(Continued On Next Page)

engineers at universities and research institutions around the world are experiencing the same joy of being challenged by their governments and have become established as part of the development, testing and building of components that eventually make it to space. In the United States, it was the Jet Propulsion Laboratory at the California Institute of Technology that laid the foundation for America's road to space. It was also the way that SSTL, which built and sold such satellites to Nigeria and other countries, began its own space journey at the University of Surrey in 1985.

Chapter IX highlights how, today, the University of Birmingham, UK and Drexel University, USA, are helping their faculties and students to access research funds and to transform their research results into marketable products. One of the main goals of the National University of Singapore is how to cultivate and nurture, among its students, the global entrepreneurs of tomorrow. These and many other universities around the world are facilitating the building of industrial partnerships with both the public and private sectors by offering their own breakthroughs in medicine, engineering, energy and social science as the main entrepreneurial attractions. Today, universities have become enablers that are establishing successful technology ventures which are benefitting their communities and beyond, through new products and new job opportunities.

But, in the absence of government or university funding support, Nigerian universities are awash with unfulfilled dreams with many projects that reached their *terminus* at the research and development stage; they never got transformed nor matured into the expected useful products and related spin-offs that could have served the nation and the world market. A 1986 example presented in Chapter VIII is that of the effort of Prof. Balogun and his crew; they successfully developed, built and tested a functional satellite receiving station that was able to download data from operational meteorological satellites, but could not secure local funding support to commercialise their product. In this connection, the development and growth of SSTL should offer Nigeria an important lesson. The SSTL of today was conceived and born as a research and development project in 1985; it was subsequently nurtured by students, staff and radio amateurs at the University of Surrey in the United Kingdom. The opportunities are still there today and there are many justifications why Nigeria should make amends and support the many young minds that Prof. Balogun and other educators nurtured and continue to nurture.

---

sensing programme in the international market. http://eijournal.com/industry-insights-trends/united-kingdom-invests-in-radar-earth-observation-satellites. (Accessed, November 15, 2015).

For strategic and economic reasons, private sector participation is very critical to the success of a nation's development efforts just as SSTL, which is now a private company and an integral part of EADS Astrium, has been to the United Kingdom.[28] Such was also the case with SUNSAT, South Africa's first satellite, which was developed and built at Stellenbosch University in South Africa with significant guidance from South Africa's private sector. It was launched into space on a USA Delta II Launch Vehicle on February 23, 1999. South Africa's private sector also made a significant input into the country's indigenous Earth observation satellite, SumbandilaSat, which was launched on September 17, 2009 by Russia. Today, India, Japan and South Korea are in their ascendancy in science and technology because they grew and nurtured their respective space-capable private sectors. A nation that undertakes such steps, as these countries have done and continue to do, would, in the process, use that same indigenous knowledge or its derivatives to develop other advanced technologies such as nanotechnology, robotics, bio-engineering, material science, energy in various forms, communications, computer science, and data analysis and information technology. Harnessing such capabilities would eventually result in a wide array of direct and indirect social and economic benefits for the nation and its people; it would also minimize the dependence of the nation on technology transfer.

In contrast to what operates in other lands, when Nigeria decided to go into space in 1999, the government could not and did not call on any indigenous private sector establishment to back that decision because such a capability was non-existent in Nigeria. And instead of forging ahead to develop and nurture such a capability, Nigeria took its petro-naira abroad and bought the services of foreign entities to develop, build and launch its first satellite. Nigeria did the same for its four subsequent satellites, all without any contribution from Nigeria's private sector.

Nigeria cannot continue on its *"buy-and–purchase-from-abroad"* approach on programmes that are very germane to the nation's knowledge and skill development, without facing unfavourable and dire consequences in the immediate future. At the nation's decision making level, it is certainly well known that a robust indigenous industrial capability in Nigeria is of vital importance and would ensure self-reliance. It will enhance the ability and the productivity of the nation on all fronts; it will also

---

28 The European Aeronautic Defence and Space Company (EADS)/Astrium EADS Astrium is a private company that is based in Suresnes, France. It specialties are space transportation, satellite systems and satellite services for civil and military needs.

strengthen the nation's opportunities for international collaboration.[29] However, both the 2001 Nigeria's space policy document and the 25-year Road Map for Nigeria's Space Mission (2005-2030) contained no specific directives on how to develop and nurture such *indigenous* private participation in Nigeria's space activities.

Building and Nurturing the Private Sector

Private sector development and nurturing have been achieved in the industrialised countries because each of the countries concerned created opportunities, by law, that allowed individuals and private companies to bid for contract awards, to facilitate technology and know-how development or to provide technical consultancy in a given field of expertise in their own societies. A case in point can be found in Europe, where, under the *European Union Horizon 2020 Strategy*, the European Union is providing new funding to support and implement government-backed, early-stage, high-risk R&D innovations that are championing the participation of the region in the downstream market of Earth observation.[30] In the United States, it is done through open "*Calls for Proposals*" solicitations/invitations.

In the case of Brazil, the government took a number of steps to promote and stimulate the growth of its private sector, namely:

- It passed several laws which stimulate the application of industrial funds for research;
- In support of the nation's pharmaceutical industry, the government passed a law which regulates the production and commercialization of biotechnological and other products; and
- It put in place a beneficial patent law. [31]

As part of its Vision 2020, the Federal Government of Nigeria should fund the development and nurturing of such a private sector capability. The *European Union*

29 In negotiating for India's participation in ESA's Galileo programme, an Indian senior official is quoted as saying: "If we are putting in € 300 million, we must have a say in the control of the satellite." And in November 2003, the Indian prime Minister made it clear that India would participate in the project as an "equal partner" and not as a "mere customer"...... *South Asia Monitor, Sept. 16, 2004* (Accessed, November 15, 2015).

30 https://ec.europa.eu/programmes/horizon2020/(Accessed, November 15, 2015).

31 Vaz, J. G. and J. A. Guimaraes (1999) *Brazil's participation in the International Space Station*, in the Proceedings of the International Symposium on International Space Station: The Next Space Marketplace, 26-28 May 1999, Strasbourg, France, Haskell, G and Michael J. Rycrof (Editors). (Accessed, November 14, 2015).

Horizon 2020 Strategy and the Brazilian model cited above are worthy of study as lessons in development.

To accomplish the above, the federal government, with the assistance of *The Think Tank* that is proposed in Chapter XV, should:

- Formulate a national policy that will transform Nigeria into a *knowledge-rich society*, a society that invests in state-of-the-art research and development activities that are highly competitive by international standard. These efforts should be at the frontiers of knowledge, applied research and innovation;

- Dialogue with the private sector and establish a public-private partnership that can:

  ◊ Identify the priority needs of the nation in all fields, including space;

  ◊ Establish a process that would lead to the identification and selection of existing relevant indigenous private enterprises that can contribute to the public-private partnership undertaking;

  ◊ Set-up at least five (5) key private small to medium-sized enterprises that can operate in five (5) space-related specialty areas that address the needs of the nation; and

  ◊ Establish bench-marks and quality controls that will guide the activities of these enterprises.

- Provide all the incentives needed to encourage capable Nigerian scientists and engineers at home and in the Diaspora to contribute to the evolution of such enterprises;

- Support the activities of the small, medium-sized enterprises (SMEs) at the research centres and local universities in Nigeria by investing in their research, development and innovation efforts;[32] and

- Encourage and promote collaboration among all these three entities, namely, NASRDA and its centres, universities and other national research centres and the SMEs.

- The SMEs should:

  ◊ Be the final development and production centres for the research and intermediate-development output of the research institutions and the universities;

---

32 The optimal support level for the private sector must be found because it is very possible to over- or under-support it. If it is under-supported, the programme will fail; and if it is over-supported, then failure is becomes inevitable when the support is removed.

◊ Adhere strictly to a nationally approved minimum local content (*at least 30%, as recommended earlier*) in all product development and fabrication;

◊ Institute a value system that adheres to transparency and established periodic peer review; and

◊ Maintain a level of excellence that will attract national as well as regional and international cooperation and collaboration.

The Think Tank can play a critical role in assisting the National Space Council to develop the policy incentives and procedures needed to cultivate and nurture Nigeria's private sector.

Outcome

By developing and nurturing a private sector participation strategy across the nation, Nigeria will reinforce the *can-do* mentality in Nigerians. This will stimulate local investment and provide significant employment opportunities for Nigeria's graduates; it will also challenge the indigenous investors and their employees to give of their best. Such an approach, particularly in the development and manufacture of engineering components and software, would eventually result in Nigerian products that are not only acceptable but also attractive to investors, nationally and within the international market. It will also enhance the ability of Nigeria to engage, with confidence and authority, in international collaboration, since it is through such a venue that the country can reap the full benefits of partnerships.

A Model Contributor to the Solution of Global Problems

Through collaborations and partnerships, Nigeria can become a model contributor to the solutions, being forged within inter-governmental fora, to address global problems. Prominent among such problems, today, are (i) Global warming and the attendant *Climate Change* problems, (ii) The sustainability of outer space activities, and (iii) Counter-terrorism in our world. To understand why Nigeria should take these steps, we need to remind ourselves that Nigeria is a major oil and gas producing country. Available records presented in Chapter XIII also show that Nigeria is equally a major contributor to global warming that is fuelling climate change. "Extreme weather events," particularly severe storms and related floods, which are linked to climate change, are among the leading global risks today and they may be around for a while. Nigeria and other African and developing countries should contribute, in a tangible manner, to the solution of these problems by bringing requisite knowledge and skills to the collaboration table. Nigeria also has physical assets in space and its citizens on the surface of the Earth. As shown in Chapter XIV, the survival of humankind and machines in both environments continues to be threatened by space debris, Near Earth objects (NEOs) such as asteroids, comets and meteorites, and the impact of space weather. And as already deliberated upon in this chapter, Nigeria is not immune from terrorism, home-grown or externally inspired. It is thus incumbent on Nigeria and other developing countries to respond in kind by contributing, in practical and concrete terms, to a variety of mitigation measures as recommended in this chapter and in the next two chapters that follow, Chapters XIII and XIV.

* * *

CONTRIBUTIONS
TO THE RESOLUTION OF
GLOBAL CONCERNS

Chapter XIII

GLOBAL WARMING
AND CLIMATE CHANGE

Each year, on April 22, the global community celebrates *Earth Day*[1], a day when the global citizens *"demand that our world leaders stop delaying and begin protecting our planet now. We cannot wait to act; our Earth's future is at stake."*[2] But from whom or why do we need to protect our planet? From natural causes and human inhabitants of planet Earth, particularly, you and me! The physical interconnectedness of our world—the oceans, rivers and the atmosphere—transports dangerous pollutants, far from where they are first released, to different parts of the world. We now know that these pollutants, particularly the greenhouse gases,[3] are harming all living things in the global circulatory system. Figure 13:1 tells the consequent story of our over-heated planet, through 2015, and the reason why everyone, the industrialised and the developing countries alike, and all the inhabitants of this planet, including

---

1 Earth Day is an annual event, first celebrated on April 22, 1970, as a day when, worldwide, demonstrations are held to show support for environmental protection.

2 http://www.earthday.org/mobilize-the-earth (Accessed August 16, 2015).

3 Greenhouse gases—Many chemical compounds found in the Earth's atmosphere act as "greenhouse gases." These gases allow sunlight to enter the atmosphere freely. When sunlight strikes the Earth's surface, some of it is reflected back towards space as infrared radiation (heat). Greenhouse gases absorb this infrared radiation and trap the heat in the atmosphere. Over time, the amount of energy sent from the sun to the Earth's surface should be about the same as the amount of energy radiated back into space, leaving the temperature of the Earth's surface roughly constant. Many gases exhibit these "greenhouse" properties. Some of them occur in nature (water vapor, carbon dioxide, methane, and nitrous oxide), while others are exclusively human-made (like gases used for aerosols).

you and me, should not only be concerned but should also actively participate in finding a solution to global warming[4] and the associated climate change that may eventually engulf us all. In doing so, we need to fully digest the latest sombre information from the World Meteorological Organisation (WMO): "We have witnessed a prolonged period of extraordinary heat, which is set to become the new norm,…It is looking likely that 2016 will be the hottest year on record, surpassing the incredible temperatures witnessed in 2015."[5] In this same news bulleting, NASA, NOAA and the European Centre for Medium Range Weather Forecasting reported that "August [2016] was the hottest August on record for both land and oceans. The year [2016] to date has [also] smashed all existing temperature records."

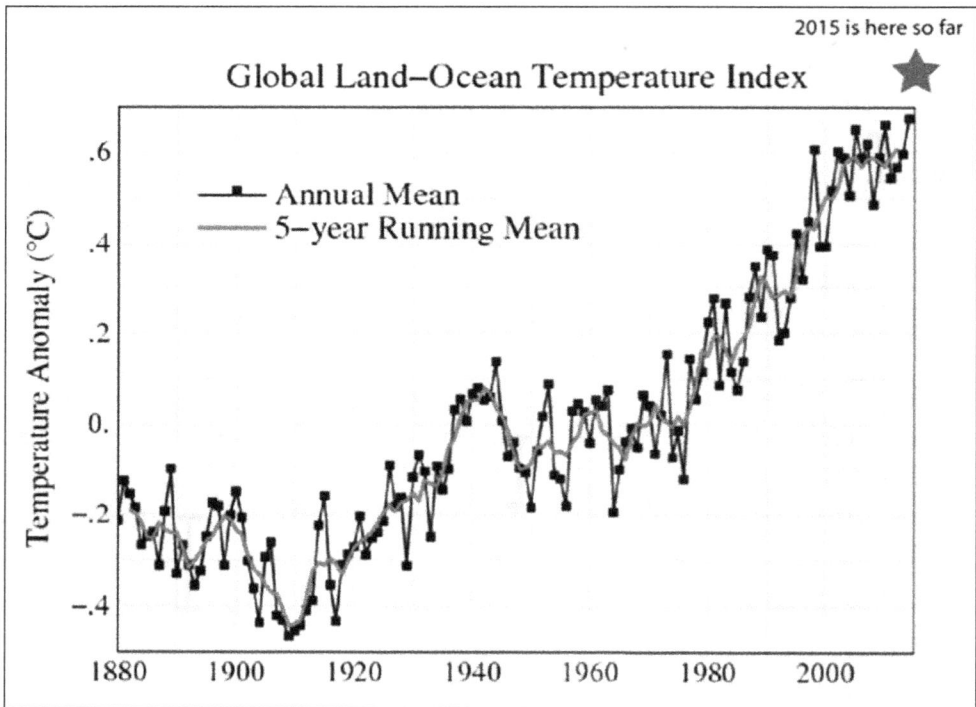

Figure 13:1 Global mean land-ocean temperature change from 1880–2015, relative to the 1951–1980 mean. *(Source: NASA/Goddard Institute for Space Studies).*

But, first, we need to understand the impact of nature and humans on our planet. Nature's impact on planet Earth includes (i) the energy from the sun

---

4 Global warming is the observed and projected increases in the average temperature of the Earth's atmosphere and oceans.

5 Extraordinary global heat continues, WMO News, Geneva, Switzerland, September 21, 2016.

which heats the earth's surface; (ii) large volcanic eruptions, such as the June 15,1991 *Mount Pinatubo* volcanic eruption in the Philippines (Figure 13:2 on next page) and the March 20 to June 23, 2010 multi-phased *Eyjafjallajökull* volcanic eruptions in Iceland,[6] both of which injected ashes and light-reflecting particles, known as aerosols[7, 8] into the atmosphere, as high as the stratosphere; and (iii) the heat-trapping gases (particularly water vapour (H_2O), carbon dioxide (CO_2), nitrous oxide (N_2O) and methane (CH_4), which exist naturally in the atmosphere (Figure 13:3 on next page). Because of the ability of these gases to trap some of the sunlight reflected as infrared radiation (heat) from the Earth's surface, and retain it as heat in a manner similar to the glass panels of a greenhouse, they are referred to as greenhouse gases. By trapping some of the sun's energy in the atmosphere, the greenhouse effect acts as a natural process needed to maintain life on Earth. In the absence of such a process, the resulting lower temperature may not sustain life forms on Earth as we know them today.

Indeed, there has been time-tested equilibrium between planet Earth and these natural forces; however, the greatest danger today is the impact of humans on our home planet. To some, our home planet is not only warming but it may become extinct.

The prevailing concern today is this: Given the daily human interactions, at every level, with planet Earth, what would our world look like in the next 100 years? According to Jim Hansen of Goddard Space Centre in New York,

---

6 http://news.bbc.co.uk/2/hi/europe/8634944.stm (How volcano chaos unfolded: in graphics).

7 The volcanic eruption of Mount Pinatubo on June 15, 1991 injected over 20 million tons of ashes in form of sulphur dioxide (SO_2) and large quantities of light–reflecting particles (aerosols,). Over the following months, the aerosols formed a global layer of sulphuric acid haze, global temperatures dropped by about 0.5 °C (0.900 °F), and ozone (O_3) depletion temporarily increased substantially. (See *Mt. Pinatubo 's cloud shades global climate, Science News, Free Library, July 18, 1992*). Similarly, the ash clouds that resulted from the *Eyjafjallajökull* volcanic eruptions caused large-scale disruption to air travel throughout most of Western Europe with smaller effects on farming in Iceland.

8 Aerosols are common specks of matter that are present in both very clear air and hazy skies, as solid particles and liquid droplets, and are inhaled, in millions, as each human takes a breath. Natural sources of aerosol range from mountains (volcanic ash), oceans (sea salt), deserts (sand storms), forests (volatile organic compounds from forest fires—smoke and black carbon) and many other ecosystems. Man-made sources of aerosol include forest fires, automobiles, incinerators, smelters, and power plants which are prolific producers of sulphates, nitrates, black carbon, and other particles.

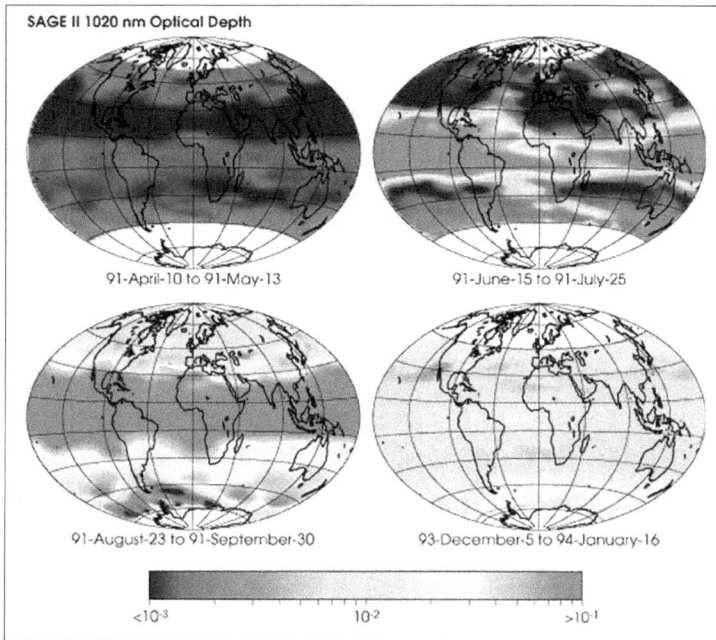

Figure 13:2 Global impact from June 15, 1991 to January 16, 1992 of Mt. Pinatubo's volcanic eruption *(Source: NASA)*

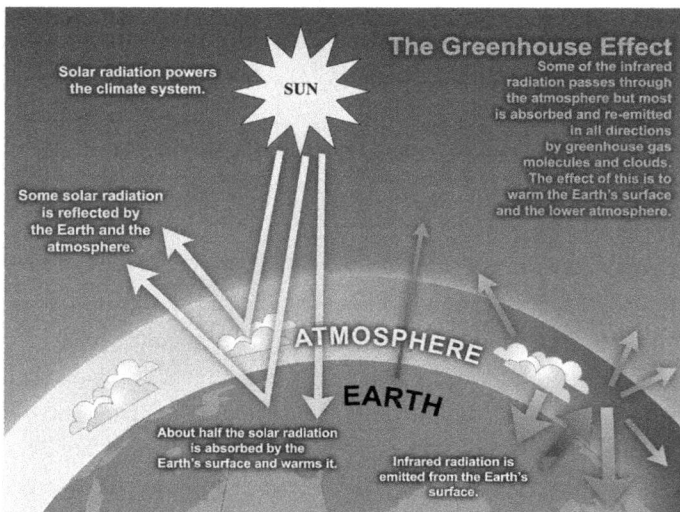

Figure 13:3 An idealised model of the natural greenhouse effect *(Source: IPCC)*[9]

---

9 FAQ 1.3, Figure 1 from *Climate Change 2007: The Physical Science Basis. Working Group I Contribution to the Fourth Assessment Report of the Intergovernmental Panel on Climate Change* [Solomon, S., D. Qin, M. Manning, Z. Chen, M. Marquis, K.B. Averyt, M. Tignor and H.L. Miller (eds.)]. Cambridge University Press, Cambridge, United Kingdom and New York, NY, USA.

Climate change is likely to be the predominant scientific, economic, political and moral issue of the 21st century. The fate of humanity and nature may depend upon early recognition and understanding of human-made effects on Earth's climate." [10]

Hansen is both alerting and warning us that, based on today's overall pattern of human consumption, worldwide, as well as human interaction with planet Earth, we should all expect a very bewildering and numbing future and the associated ecological consequences that await humankind.

To understand what we humans are doing, today, to the Earth's climate, it will be necessary to reflect on what we have done in the past, particularly, in the last one-hundred years and more. First, there was the 1896 warning by Svante Arrhenius, a Swedish scientist that fossil fuel (*coal, petroleum and natural gas*) combustion may eventually result in enhanced global warming." [11] The world paid little attention to Arrhenius' prediction until 1957 when the problems of global warming that he had predicted, six decades earlier, re-surfaced. That was when, in February 1957, two scientists, Roger Revelle and Hans Suess, at the Scripps Institution of Oceanography in California, corroborated the 1896 findings of Svante Arrhenius and also warned that the combustion of fossil fuel was resulting in the accumulation of carbon dioxide in the atmosphere. [12]

Subsequent climate-related data, most of it acquired by a variety of satellites, including the recent constellation of A-Train satellites, (Figure 13:4) [13] have confirmed a rapid increase in atmospheric concentrations of carbon dioxide and other long-lived greenhouse gases. These increases have led to mounting scientific concerns over possible future adverse effects on humans and other life-forms, especially since disruptions in the forces that influence climate will not be easily reversible. In-depth analysis of these copious space-acquired data has made us more appreciative of the

---

10 http://www.columbia.edu/~jeh1/mailings/2011/20110118_MilankovicPaper.pdf (Accessed, August 18, 2015).

11 Arrhenius, Svante (1896). *On the Influence of Carbonic Acid in the Air upon the Temperature of the Ground,* Philosophical Magazine and Journal of Science, Series 5, Volume 41, April 1896, pages 237-276, London, Edinburgh and Dublin.

12 Revelle, Roger, Hanss E. Suess (1957). *Carbon Dioxide Exchange Between Atmosphere and Ocean and the Question of an Increase of Atmospheric CO_2 during the Past Decades,* TELLUS, Volume 9, Issue 1, pp.18-27, February 27, 1957

13 The missions of the A-Train satellite constellation include: Carbon (Primarily carbon dioxide and other greenhouse gases such as methane) tracking, ocean circulation, atmospheric chemistry, aerosols and clouds, etc.

way we are stressing our life support systems and the subsequent impact of that stress on our home planet and its inhabitants.[14]

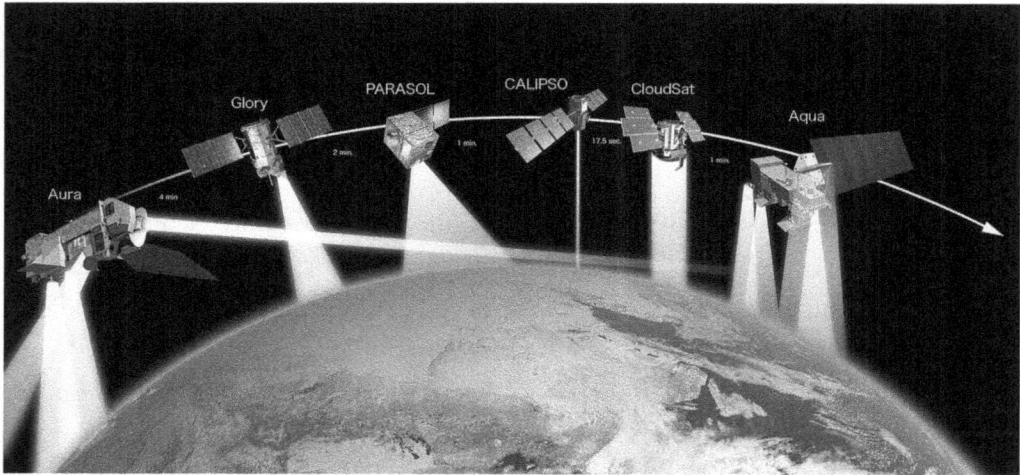

Figure 13:4 Satellites that are contributing to the monitoring of the Greenhouse gases (*Source: NASA*)

Among the human inflicted stresses on planet Earth is our insatiable desire for its raw materials which are fuelling the engines of our *perceived development* and the consequent unfamiliar environmental conditions of global warming and its attendant *climate change, sea level rise and acid rain, among other disasters.* Human activities that are contributing to the increasing greenhouse gases in the atmosphere and are impacting the climate system include (i) carbon emissions from industrial processes, particularly cement plants and oil refineries, (ii) agriculture—methane emissions from livestock and manure, and nitrous oxide emissions from chemical fertilizers, (iii) carbon emissions from transportation systems, and (iv) carbon emissions from the use of fossil fuel (coal, petroleum and gas) to generate energy (excluding transport). Figure 13:5 shows the contributions of fossil fuel combustion

---

14 Definitions of Life support systems—(i) "People have basic needs for food, water, health, and a place to live, and additionally have to produce energy and other products from natural resources to maintain standards of living that each culture considers adequate. Fulfilling all of these needs for all people is not possible in the absence of a healthy, well-functioning global ecosystem. The "global ecosystem" is basically the complex ways that all life forms on Earth—including us—interact with each other and with their physical environment (water, soil, air, and so on). The total of all those myriad interactions compose the planet's, and our, life support systems.http://www.newscientist.com/special/ocean-to-ozone-earths-nine-life-support-systems

through gas flaring, globally from 2013 through 2015. Because trees are essentially 'wet sticks of carbon,' when burnt, trees release CO_2 into the atmosphere and the process of photosynthesis—through which trees absorb CO_2 and combine it with sunlight, nutrients and water to convert it into energy to grow—is thus thwarted. Accordingly, changes in land-use, particularly through deforestation (wood harvesting and timber), forest fires and agriculture, contribute significantly to greenhouse gases.

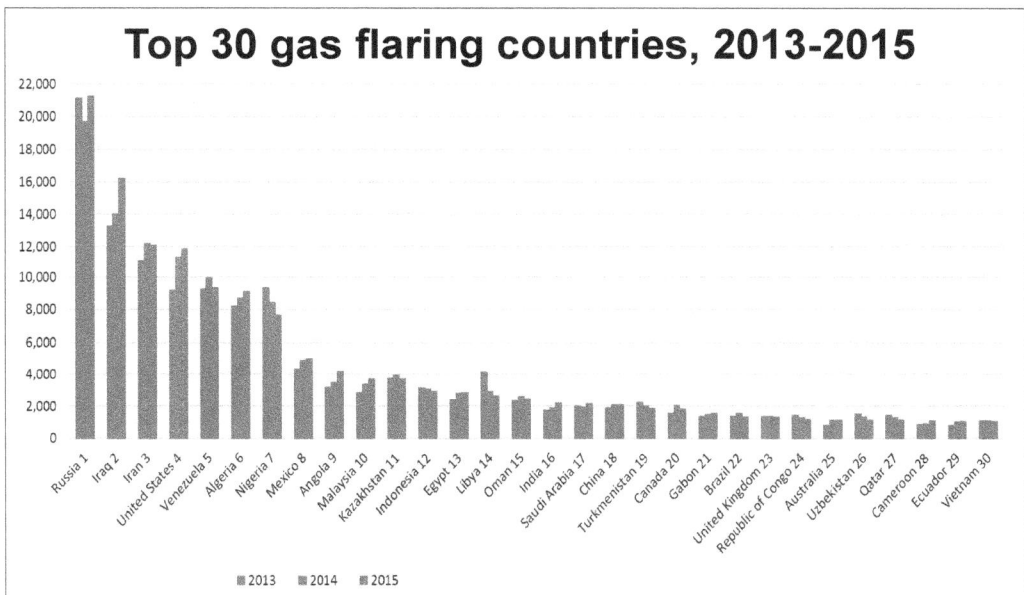

Figure 13:5 Top 30 gas flaring countries, 2013–2015 data showing volume in billion cubic metres *(Source: World Bank; Data Source: NOAA)*[15]

Global Actions

Convinced of the magnitude and variety of potential problems that could result from human impact on the Earth's environment, the global community continues to respond with a series of international actions, beginning with the convening of the 1972 United Nations Conference on the Human Environment. The conference established the United Nations Environment Programme (UNEP) with a 1997 expanded mandate that directed it *"to be the leading global environmental authority that sets the global environmental agenda, that promotes the coherent implementation of the environmental dimensions of sustainable development*

15 http://www.worldbank.org/en/programs/gasflaringreduction (Accessed, December 15, 2016).

within the United Nations system and that serves as an authoritative advocate for the global environment"[16]

One of the most challenging pre-occupations of humanity, today, is global warming with its attendant climate change. The heat waves of the late 1980s and in particular that of 1988, heightened public attention to the possibility of changing climate as the global community witnessed extreme storms over Europe, record heat waves and drought in North America, and weather anomalies elsewhere in the world. These and other concerns culminated in the establishment, in that year, of the Intergovernmental Panel on Climate Change (IPCC) by both the WMO and UNEP, with a mandate *"to provide the governments of* [the] *world with a clear scientific view on the current state of knowledge in climate change and its potential environmental and socio-economic impacts."*[17] The IPCC subsequently *"warned that only strong measures to stop greenhouse gas emissions would prevent serious global warming."*

Subsequent IPCC findings led to the 1992 UN Conference on Environment and Development (UNCED), also known as Rio'92, which convened in Rio de Janeiro and hammered out a convention on the reduction of greenhouse gas emissions that member States could agree on. The latter was the United Nations Framework Convention on Climate Change (UNFCCC); it was signed by 154 nations. The UNFCCC *"agreed to prevent 'dangerous' warming from greenhouse gases and set voluntary targets for reducing emissions."*[18] Unfortunately and predictably, only a very limited number of states met the targets they voluntarily set for themselves.

Since Rio'92, other global collective deliberations have also focused on how to stem the increasing rise in man-made greenhouse gases in the atmosphere in order to reduce the impact of global warming. Among the notable ones are: "The Third Session of the Conference of the Parties to The United Nations Framework Convention on Climate Change (COP-3)," held in Kyoto, Japan, December 1 -10, 1997 and known as *The 1997 Kyoto Climate Conference;* it resulted in the

---

16 http://www.unep.org/PDF/Natural_Allies_en/Natural_Allies_chapter1_eng.pdf (Accessed, August 18, 2015).

17 http://www.ipcc.ch/organisation/organisation.shtml#.T7K9rcWDr8k (Accessed, August 19, 2015).

18 Rio '92—The United Nations Framework Convention on Climate Change, United Nations, New York, 1992 (Accessed, January 10, 2015).

land-mark Kyoto Protocol[19] which committed the industrialised countries to emission reduction targets. Worthy of note is that countries such as China, India, Nigeria, Saudi Arabia and South Korea got classified as developing countries and were thus excluded from meeting the binding emission targets. The second notable conclave was *The Seventeenth Session of the Conference of the Parties to The United Nations Framework Convention on Climate Change (COP-17)*, known as *The 2011 United Nations Climate Change Conference*; it was held in Durban, South Africa. November 28—December 11, 2011. The Durban conference agreed to a *legally binding* emission mitigation regime for all countries which was to be negotiated and completed by 2015, and would take effect in 2020.[20]

Negotiations actually took place at the subsequent United Nations Climate Change Conference held in Paris, France, November 30 to *December 12, 2015. At the conference,* member States collectively acknowledged *"the importance of averting, minimizing and addressing loss and damage associated with the adverse effects of climate change."* Moving forward, the conference agreed, inter-alia, as follows:[21]

- To keep a global temperature rise this century well below 2 degrees Celsius and to drive efforts to limit the temperature increase even further to 1.5 degrees Celsius above pre-industrial levels;

- That all countries should submit updated plans that would ratchet up the stringency of emissions by 2020 and every five years thereafter;

- That all countries should monitor, verify and report their greenhouse gas emissions using the same global system;

---

19 The Kyoto Protocol is an international agreement linked to the United Nations Framework Convention on Climate Change. The major feature of the Kyoto Protocol is that it sets binding targets for only 37 industrialized countries and the European community for reducing greenhouse gas (GHG) emissions. This amounts to an average of five per cent against 1990 levels over the five-year period 2008-2012 The Kyoto Protocol was adopted in Kyoto, Japan, on 11 December 1997 and entered into force on 16 February 2005. The Kyoto Protocol is generally regarded as an important first step towards a truly global emission reduction regime that will stabilize GHG emissions, and provides the essential architecture for any future international agreement on climate change.

20 COP 17: The Durban Platform for Enhanced Action—In its final declaration, *The Durban Conference "Also decided to launch a process to develop a protocol, another legal instrument or an agreed outcome with legal force under the United Nations Framework Convention on Climate Change applicable to all Parties…"*

21 http://web.unep.org/climatechange/cop21/highlights (Accessed, January 5, 2016).

- To "reach global peaking of greenhouse gas emissions" as soon as possible; and
- Established a "Capacity-Building Initiative for Transparency" to help developing countries meet a new requirement that they regularly provide a national "inventory report" of human-caused emissions, by source, and track their progress in meeting their national goals.

To demonstrate their support for the aspirations of the conference, the Energy institutions present "*recognised the need to further embed energy efficiency investment principles into the way in which they engage with their clients.*" The Secretary-General of the United Nations, Mr. Ban Ki-moon, hailed the Paris climate accord as 'health insurance policy for the planet'[22]

The Bottle-necks To-date

Although each of these climate-focused global gatherings registered expressions of a universal obligation to reduce greenhouse gas emissions, achieving any progress before the Durban Conference was painfully slow. A number of countries that agreed to the decisions reached in Brazil in 1992 and/or signed the 1997 Kyoto Protocol reneged. Furthermore, there is too much at stake, and reconciling the different national interests became an uphill task. Here are some of the bottle-necks: greenhouse gas emissions have driven the industrial development and economic progress known to-date, resulting in mechanized agriculture; the construction of super highways and the vehicles that use them; a variety of industries including cement plants that are scattered over 94 countries, oil refineries and energy generation plants, all of which depend mostly on fossil fuels. It is the belief of the IPCC that increases in anthropogenic greenhouse gas concentrations could have caused most of the increases in global average temperatures since the mid-20th century.[23] And according to WMO, "the amount of greenhouse gases in the atmosphere reached a new record high in 2013, propelled by a surge in levels of carbon dioxide."[24]

To lower the emission of greenhouse gases would demand a radical departure from the current normal way of life as the world knows it. Standing in the way are major polluters and in this category are none other than major petroleum exporting

---

22 http://www.un.org/sustainabledevelopment/cop21/ (Accessed, January 5, 2016).

23 Climate Change 2007:Synthesis Report, IPCC Plenary XXVII (Valencia, Spain, 12-17 November 2007). (Accessed, January 10, 2015)

24 http://www.wmo.int/pages/mediacentre/press_releases/pr_1002_en.html (Accessed, January 10, 2015).

countries who continue to deny their contributions to greenhouse gases. There are also "governments *that* appeared unwilling to confront powerful industrial interests head-on *and thus refrained from* enacting sector-specific policy measures to limit use of fossil fuels, *particularly* transportation or utilities."[25] The huge size of delegations at these deliberations also constituted negotiating-bottlenecks while many countries, particularly the developing counties, perceived the cost of investments in new technologies that would reduce greenhouse gas emissions unaffordable.

Much of the blame for global warming and related climate change has been placed at the door of the industrial countries of the world; however, the emerging nations and the developing countries are not innocent either. For example, the rainforests of the world in Africa, Asia and Latin America are rich in biodiversity. However, today, wood remains, to a large extent, the source of energy in the rural communities of most developing countries. Floods, associated soil erosion and subsequent loss of the fertility of the land have continued to plague these areas, annu- ally, because of the denuding of these forests, particularly for timber (Cameroon), fuel-wood (Kenya), land cultivation (Brazil) and cultivation of mono-culture oil palm (Borneo, South East Asia).[26] In Nigeria, the United Nations Food and Agriculture Organisation (FAO) noted that between 1990 and 2010, the country lost 47.5% of its forest cover, or around 8,193,000 ha, resulting in one of the highest rates of deforestation in the world;[27,28] the net effect has been excessive soil erosion and unabated sand dust in the Nigerian air space, particularly during the harmattan season (December through March).

Most of these same developing countries are the destinations of second-hand imported vehicles and engines. To accommodate these vehicles in their respective host countries, new roads are being built to link already crowded cities. Exhausts from the running engines of these second-hand and new vehicles (motor-cycles, cars, trucks and buses, and the heavy-duty machines) consist, mostly of carbon dioxide (CO_2), water (H_2O), and nitrous oxide (N_2O), all of which are greenhouse gases. The emerging nations as well as the developing countries are also big producers of cement, the

25 Benedick, Richard E. (1999). Contrasting Approaches—The Ozone layer, Climate Change and resolving the Kyoto Dilemma, (Accessed, January 12, 2015).

26 See NASA/Goggle Earth images. (Accessed, January 12, 2015)

27 FAO 2010 Country Profile—Nigeria (Accessed, January 12, 2015)

28 FAO Country Programming Framework (CPF) Federal Republic of Nigeria 2013—2017, Food and Agricultural Organisation of the United Nations, Rome, Italy (Accessed, January 12, 2015).

third-largest industrial contributor of greenhouse gases in the world. Supplying more cement for buildings, roads and bridges makes big emission reductions impossible. Worse still, CO_2 emissions from small cement plants are two or more times higher than plants in industrialised nations due to poor efficiencies that require more fuel use.[29]

The Need to Act Collectively

According to the results of computer models, increasing greenhouse gases may do irreparable harm because of the effects of gradual changes in climate. "There is *also* no indication of the probability, timing, location or severity of the long list of potential negative impacts of *climate change* ranging from flood to tropical diseases and severe storms."[30] The ways in which planet Earth will react to these predictions are also alarming—"The model projections paint a portrait of increasing ecological change and stress in Earth's biosphere, with many plant and animal species facing increasing competition for survival, as well as significant species turnover, as some species invade areas occupied by other species." Such conditions are partly attributable to human transformation of the geographical landscape, particularly through growing global deforestation and the urbanisation of rural areas. Above all, no part of the world is escaping the fury of climate change.[31]

In Africa, for example, rain patterns are changing/shifting, thus putting stress on agricultural production; the long-term consequences in terms of crop failures, food security and human health are incalculable. Prolonged global warming will also result in the expansion of ocean water and in the melting of mountain glaciers and polar ice caps that will culminate in sea level rise for many low lying regions of the world. At its 2014 gathering in Davos, Switzerland, the World Economic Forum listed [human] "failure to mitigate or adapt to climate change" as one of the ten global crisis of the year. In Table1:3 of its report, the gathering went further and identified "*extreme weather events*" particularly severe storms and related floods, and "climate change" as two of the five 2014 "Top Global Risks in Terms of Likelihood."[32] Figure 1 of

---

29 Jung-Myung Cho and Suzanne Giannini-Spohn (2003). A China Environmental Health Research Brief—Environmental and Health Threats from Cement Production in China, Woodrow Wilson International Center for Scholars, Washington, DC 20004-3027 (Accessed, January 14, 2015).

30 Buis, Alan (2011). NASA: Climate Change May Bring Big Ecosystem Changes, Jet Propulsion Laboratory, Pasadena, CA, USA, December 14, 2011, (Accessed, January 14, 2015).

31 https://www.ncdc.noaa.gov/sotc/service/global/extremes/201507.gif (Accessed, September 3, 2015).

32 Global Risks 2014, 9th Edition, Published by the World Economic Forum, Geneva, Switzerland, (Accessed, January 21, 2015).

2015 Davos report also listed Extreme weather events[33] as second in the *Top 10 risks in terms of Likelihood*; *Interstate conflict* topped the ten highest risks.[34] A world-wide dissemination of these World Economic Forum (Davos) conclusions will go a long way to generate needed awareness as well as pave the way towards the mitigation of climate change.

Need for Informed Awareness and Action Particularly in the Developing Countries

For now, the core dilemma for the developing and emerging nations, particularly the large ones, including those that had perceived the cost of investments in new technologies needed to reduce greenhouse gas emissions as unaffordable, is how to grow without inflicting more damage on the environment. Globally, however, human fate is at risk and how to avert the dangers ahead by stemming the possibility of global warming and related climate change in the next few decades is a global responsibility to be shared by all. Building the needed public awareness that can mushroom into concrete action, at each national/regional level and globally, requires the assimilation of a general knowledge and appreciation of climate change and its impact for the political class and the general public in a language they can understand.

In 1936, President Franklin Roosevelt of the United States pondered on similar issues and conveyed his thoughts in these words:

> *The history of every nation is eventually written*
> *in the way in which it cares for its soil.*[35]

In ages past, human practices affected climate and the soil just as our interactions with planet Earth are doing today. But the inadequacy of human knowledge at that time, particularly its inability to connect human activities to climate change, resulted, to give just an example, in large-scale deforestation of northwest Syria in order to make room for olive groves. Subsequent lack of any corrective action, such as reforestation, that could stem large scale soil erosion led to the demise, centuries

---

33 Extreme weather events include, *inter-alia*: Storms (Hurricanes, Typhoons, Cyclones, Wind and Sand), heavy rainfall and associated floods, drought and related water crises and risks, extreme temperatures and associated heat and cold waves wild fires, major biodiversity loss and ecosystem collapse.

34 Global Risks 2015, 10th Edition, Published by the World Economic Forum, Geneva, Switzerland, (Accessed, February 12, 2015).

35 Statement of President Franklin Roosevelt on *"Signing the Soil Conservation and Domestic Allotment Act,"* Washington, DC., *USA, March 1, 1936.* (Accessed, October 11, 2015).

ago, of Syria's 'dead cities' such as *Qal'at Sim'an, Serjilla* and *al Bara*.[36] In the absence of needed soil that could absorb the rain at that time, springs dried up and the sustenance of the people collapsed, leaving behind in Syria, today, a semblance of the surface of Mars.

Fortunately, as a result of advances in science and technology, of which space science and technology is a major component, today's generation and future ones are at a greater advantage. Unlike in the ancient times, today, we are able to enhance our knowledge of our environment through human analysis of copious information being acquired by ground-based data collecting stations and space tools such as the A-Train constellation of satellites in Figure 13:4. But the global nature of today's problems, such as climate change, demands that the developing countries should also complement these initiatives of the industrialised countries and make comparable and significant contributions in support of climate change mitigation. In the process, we can all come together for the love of our planet and our own self-preservation. The issue is one of survival of this and future generations. Should we decide to ignore the overwhelming warnings, such as potential *Big Ecosystem Changes, including Health and Environmental threats,* we do so at our own peril.

There are many approaches to accomplishing this objective, beginning with the development of an appropriate national/regional policy as called for by the UNFCCC.[37] Such policies require adequate information that is underpinned by essential knowledge which needs to be developed and acquired through learning and research. The focus of a sound climate change policy should include the actions needed to (i) reduce greenhouse gas emissions, (ii) adapt to impacts of climate change, particularly the unavoidable ones, and (iii) support the challenge of finding a global climate change solution. Each of these actions will succeed only through each country's/region's capability to continually assess where it stands and where it needs to go next on climate change.

Because climate change affects all of us, world-wide, how to equip the developing countries to be active participants in its mitigation is receiving attention in some quarters. Specifically, the United Nations, particularly through its UNFCCC-Climate

36 Mann, Charles C. (2008). *Our Good Earth*, The National Geographic Magazine, pp. 80-107, September 2008, New York, NY 10022 (Accessed, January 14, 2015).

37 Climate Change: Impacts, Vulnerabilities and Adaptation in Developing Countries: http://unfccc. int/resource/docs/publications/impacts.pdf (Accessed, January 14, 2015).

Technology Centre and Network,[38] the World Bank,[39] and a number of other governmental and non-governmental organizations are providing a variety of assistance to several developing countries, including those in Africa, that will enable them to cope with climate change, promote gas flaring reduction and low carbon growth, and facilitate the preparation and implementation of related technology projects and strategies. In Tanzania, the University of Dar-es-Salaam is developing and implementing innovative research on climate change,[40] while the Climate Systems Analysis Group (CSAG) at the University of Cape Town in South Africa *"seeks to apply* [its] *core research to meet the knowledge needs of responding to climate variability and change."*[41] The mission of the African Climate Change Fellowship Program (ACCFP), sponsored by START International Ltd, is "To increase opportunities for research and training that strengthen scientific capacities in developing countries to understand, communicate and motivate action on critical global environmental change challenges."[42] All of the above notwithstanding, the World Bank continues to emphasise that "African countries should focus on investing in research and advisory services to develop and disseminate adaptation options, and scaling-up investments that build resiliency."

In the absence of needed adequate research facilities and local funding support, Africa's capacity to undertake climate change studies, today, is miniscule. Consequently, the continent continues to depend, almost exclusively, on the scientific knowledge as well as on data and information, including those on its natural resources and environment, that are acquired and generated by and in the northern hemisphere where most research is carried out. But such foreign knowledge will not and cannot provide all the critical information needed for decision making in Africa and other developing areas of the world. Each nation and region needs to understand those aspects of climate change that could impact its future, a knowledge that will shape an effective policy response based on national/regional needs.

Africa needs to develop critical indigenous capacity, particularly the essential knowledge of climate system science, through rigorous research, in order to acquire

---

38 unfccc.int/ttclear/templates/render_cms_page?TEM_ctcn (Accessed, January 20, 2015).

39 http://web.worldbank.org/WBSITE/EXTERNAL/COUNTRIES/AFRICAEXT/0,,contentMDK: 22410211~pagePK:146736~piPK:146830~theSitePK:258644,00.html (Accessed, January 20, 2015).

40 http://start.org/programs/cc-research-education-outreach (Accessed, January 20, 2015).

41 http://www.csag.uct.ac.za/ (Accessed, January 20, 2015).

42 http://start.org/programs/accfp1 (Accessed, January 21, 2015).

a greater understanding of climate change, its mitigation, needed adaptation strategies, and how to quantify its current trend and predict its future impact; climate change modelling is one of such capacities which should be built by and operated in Africa's institutions. According to the WMO, "simulating climate change at the regional and national levels is essential for policymaking. Only by assessing what the real impact will be on different countries will it be possible to justify difficult social and economic policies to avert a dangerous deterioration in the global climate."[43] Climate change modelling and other capabilities, built at the institutional and individual levels, are critical to Africa's effective contribution to major international scientific programmes, such as: Global Atmosphere Watch (GAW), World Climate Research Programme (WCRP), International Geosphere-Biosphere Programme (IGBP), Global Climate Observing System (GCOS) and Inter-governmental Panel on Climate Change (IPCC).

The core capacities in question are those that will advance climate system science in the developing countries and the essential knowledge needed to track and account for their own local greenhouse gas emissions and shape their input into a global climate change solution. Among the critical research tools needed for information and data acquisition are ground-based data collecting stations on land and on the oceans, and air and space components such as Unmanned Aerial Vehicles (UAVs) and constellations of nano and cube satellites as well as appropriate data analysis systems. While a few world-class climate-dedicated research stations, such as the Baseline Surface Radiation Networks (BSRN) stations, are located at Tamanrasset, Algeria; Ilorin, Nigeria and at De Aar, South Africa, there is no indigenous African satellite in orbit or an African UAV that is currently dedicated to climate research. However, students at Sri Ramaswamy Memorial University, in India, developed and built such an experimental nano-satellite, named *SRMSAT*, with a secondary mission of monitoring greenhouse gases; it was launched by the Indian Space Research Organisation in October 2011.

The knowledge is there and such an effort is not beyond the capability of proven tertiary institutions in Africa if they will collaborate to do so. For example, the first satellite built on African soil was *Sunsat*; it was developed, and built at Stellenbosch University in South Africa with a significant guidance from South Africa's private sector and launched into space on a USA's Delta II Launch Vehicle on February 23, 1999. Ten years later, South Africa built an indigenous Earth observation satellite,

---

43 http://www.wmo.int/pages/themes/climate/climate_models.php (Accessed, January 21, 2015).

SumbandilaSat, which was launched by Russia on September 17, 2009. To accomplish such a regional collaboration, interactions with experienced international institutions will be an advantage while individual national political support will be indispensable. It would also be to the advantage of the fossil fuel and cement producing countries and companies of Africa that face common climate change challenges and are responsible for the preponderance of Africa's greenhouse gas emissions, to fund such research efforts. The commitment of all parties to such a regional collaboration effort should result in knowledge generation and sharing, skill development and joint execution of meaningful plans of action that can address national and regional climate change issues as well as contribute to the international response to climate change.

As a potential determinant of our collective future that has been acknowledged by several international fora, including the 2015 United Nations Climate Conference that was held in Paris, climate change has become a major global concern. Today, its abatement is a recognised shared global responsibility which can only be attained through a faithful collective compliance to the 2015 Paris Climate Summit Agreement. We are obliged to manifest our contributions, through knowledge development and sharing as well as concrete climate research programmes; Africa and the rest of the developing world should not be found wanting on this life-saving mission.

* * *

Chapter XIV

THE OUTER SPACE ENVIRONMENT

Today, the security and safety of humankind on Earth and in space, and those of space-based assets, have taken centre stage in the exploration and utilisation of outer space. When any disaster strikes, be it on land (floods, sandstorms or earthquake), on the ocean (cyclone/hurricane), from air space (air crash) or from outer space (space debris, meteorites, asteroids and solar storms), providing alert and associated location information is critical to the mounting of any successful rescue operation.[1, 2] Certainly, all societies can benefit from such systems. The escalating number of disaster victims and the associated colossal economic losses dictate the need for such services. For example, from 1991-2000, a total of 665,598 people lost their lives to natural disasters alone; in that same period, natural catastrophes reportedly cost an estimated US$78.7 billion per annum (2000 prices).[3] The international community and the United Nations Committee on the Peaceful Uses of Outer Space (COPUOS) are working to address these global concerns. As a country with aviation and space assets, how is Nigeria and other developing countries responding and contributing to these global efforts on land, in the air, and in space?

---

1 Space debris refers to the remnants of rockets, satellites and other objects that were left in space by human activities because they no longer serve any useful purpose. See also footnotes 3 and 4 on meteorites and asteroids.

2 United Nations Treaties and Principles on Outer Space (ST/SPACE/11 or A/AC.105/722)—According to the *Liability Convention* (Convention on International Liability for Damage Caused by Space Objects), all member states, including Nigeria, that own satellites or space vehicles in space (functional or not) are liable for damages caused to other satellites if there is a collision or to foreign territory if there is uncontrolled re-entry in a foreign territory.

3 Red Cross Red Crescent—World Disasters Report 2001—Chapter 8 (Accessed, September 1, 2015).

Search and Rescue System

The global clamour for a solution, particularly to aviation and maritime accidents and related tragedies, triggered the October 5, 1984 formal agreement between Canada, France, the former USSR, and the USA that resulted in the establishment of the International Search and Rescue Satellite System (COSPAS-SARSAT). By this agreement, the signatories decided, in accordance with international law, to support, using their space assets, the implementation of a global search and rescue programme and to provide access to the system, to all States, free of charge and on a non-discriminatory basis, for the end-user in distress. From its inception in 1982 until August 2010, the COSPAS-SARSAT System provided assistance in rescuing more than 28,000 persons in over 7,500 incidents (e.g. land, fishing, boating, shipping, snowmobile, snow-avalanche, and aviation accidents) world-wide.[4] Before February 1, 2009, the COSPAS-SARSAT space-based system was able to process distress signals from both analogue and digital beacons. Thereafter, COSPAS-SARSAT stopped processing analogue 121.5 MHz distress signals and switched completely to the processing of digital signals from *406 MHz Beacons*. The COSPAS-SARSAT operational architecture is shown in Figure 1:2 of Chapter I. The latest beacons incorporate the Global Positioning System (GPS) receivers that can transmit highly accurate positions (within about 20 metres) of the object in distress.[5]

---

4 http://www.sarsat.noaa.gov/ (Accessed, September 1, 2015).

5 The Global Positioning System (GPS), operated by the USA, and operational since 1994, is a space-based satellite navigation system—with a constellation of 24 satellites—that provides free location and time information in all weather, anywhere, on or near the Earth, where there is an unobstructed line of sight to four or more GPS satellites. The Russian equivalent of the GPS is the Global Navigation Satellite System (GLONASS) also with a constellation of 24 satellites; GLONASS was made available for civilian use in 2007. Other satellite navigation systems in the horizon include: The European Union planned Galileo positioning system; when completed, it will consist of a constellation of 30 satellites (24 operational satellites plus six in-orbit spares); initial services will be made available by the end of 2016. As the constellation is built-up beyond that period, new services will be tested and made available, with system completion scheduled for 2020. The Indian Regional Navigational Satellite System, consisting of a constellation of seven satellites, is expected to be operational from 2016. The Chinese BeiDou Satellite Navigation System, which is operational regionally, is being expanded; the new system will be a constellation of 35 satellites (5 geostationary orbit (GEO) satellites and 30 medium Earth orbit (MEO) satellites), that will offer global coverage by 2020. Japan's regional Quasi-Zenith Satellite System (a regional time transfer system and Satellite Based Augmentation System for the Global Positioning System, that would be receivable within Japan) will consist of four satellites that are scheduled for launch before the end of 2017.

Until January 2004, Nigeria was far removed from any structured search and rescue programme that was comparable to the COSPAS-SARSAT of today. Only the north-west corner of the country (Sokoto State) was within the footprint of the nearest COSPAS-SARSAT station, i.e., the Spanish Mission Control Centre in Maspalomas, Gran Canaria. From Nigeria's record of aviation disasters, it is apparent that, for a long time, Nigeria had been in need of such a rescue system. For example, between November 20, 1969, when Nigeria experienced its first recorded air crash, and May 2010, the loss of lives due to aircraft–related disasters has been staggering.[6]

Specifically, "*between 1965 and 2002 (37 years) when the Nigeria Airways was in operation, it recorded six crashes in which 219 persons died. Between 1988 and March 15, 2008, a period of 20 years, the private airlines recorded 16 crashes, with a total of 866 lost lives.*"[7] All these figures do not include the death tolls from the military airplane crash of September 26, 1992 at Ejigbo, Lagos that killed 157 Nigerian military officers.

The absence of COSPAS-SARSAT emergency beacons in the country's civil and military aircraft when these accidents occurred contributed to the difficulties experienced in locating the sites of these accidents prior to mounting rescue operations; an effective rescue effort, in most cases, could have resulted in a number of lives being saved. At that time, Nigeria and most other African countries could not organise any meaningful search and rescue programme in times of distress.

The mounting casualties from recurring aircraft crashes led the United Nations Office for Outer Space Affairs (UN-OOSA) and COSPAS-SARSAT to explore how to provide such a space-related service to interested African countries. In collaboration with the European Space Agency and the Spanish National Institute for Aerospace Technology (INTA), UN-OOSA organised *The First UN/INTA/ESA Workshop on Space Technology for Emergency Aid/Search and Rescue Satellite-Aided Tracking Systems*, in September 1998. The workshop was held specifically for the benefit of the African countries that were within the footprints of the Maspalomas Station in order to

---

6 Nigeria recorded its first plane crash on November 20, 1969, when a government-owned DC-10 aircraft on a flight from London crash-landed in Lagos and killed all 87 passengers and crew on board. Between November 1969 and November 2008, over 1000 persons were killed in plane crashes on Nigerian soil. http://allafrica.com/stories/200610301057.html; (Accessed October 15, 2015).

7 Tolar, Dagga (2005). *Another fatal air crash, the dividend of privatization,* Democratic Socialist Movement, December 15, 2005, (Accessed October 15, 2015).

encourage these countries to take advantage of such a life-saving programme. A second course followed in 1999. The participation of both the National Emergency Management Agency (NEMA) and the Nigerian Airspace Management Agency (NAMA), at the management level, at each of these two workshops, and the related exposures gained in the process, paved the way for President Olusegun Obasanjo to commission the national COSPAS-SARSAT Mission Control Centre at NEMA's Headquarters in Abuja, on January 13, 2004. This action made Nigeria the fifth country in Africa with COSPAS-SARSAT facilities; the others are Algeria, Madagascar, South Africa and Tunisia.

The COSPAS-SARSAT facility in Nigeria should facilitate Nigeria's efforts to locate, distressed ships, land vehicles, and aircraft and their passengers as well as individuals on foot, such as members of the armed forces who may be in danger or in distress. The most critical requirement is that those who could be in need of rescue, at any time in their mission, should have on them the necessary beacon(s) that would transmit their co-ordinates (i.e., geographical location(s)), to the monitoring station via the COSPAS-SARSAT satellites on sight at the time of need.[8] However, the inability of the nation's COSPAS-SARSAT system to rapidly and successfully aid in search and rescue operations, as expected, in times of such tragedies, has called into question, in a number of quarters in Nigeria, the merits of the investment Nigeria made in the COSPAS-SARSAT station in Abuja.[9] Most of the time, the problem is not one of machine; it is a human problem, either with the airline management and operators or with the COSPAS-SARSAT station and its operators. What is needed is how to shed some light on the operation of the system. A mandatory semi-annual report of NEMA to the nation on the detailed operations of Nigeria's COSPAS-SARSAT System, including information on incidents the organisation (NEMA) responded to using the facilities of the system, should enable government authorities to determine its effectiveness and that of its management.

---

8 This pre-supposes that there is a corresponding and funded programme to purchase and equip those who need them with the appropriate beacons.

9 The Nigerian COSPAS-SARSAT system did not perform as expected on October 22, 2005 when a Bellview Boeing 737 plane, with 117 people on board, crashed soon after take-off from Lagos, killing everyone on board. Similarly, what role did Nigeria's COSPAS-SARSAT station play when, on September 17, 2006, the Nigerian Air Force Dornier 228-212 aircraft (with 18 people on board including 10 generals and 4 colonels) crashed into the hills of Ushongo village in Benue State, on its way to Obudu in Cross River State?

Sustainability of the Outer Space Environment

The threat to human and personal securities in all parts of the world is not limited to falling aircraft from air space; we all are also at the mercy of falling meteorites, asteroids and space-debris. From time to time, they do happen, just as the Mars meteorite, addressed in Chapter III, landed in Zagami, Katsina State, on October 3, 1962. The Zagami community, along with the entire nation, escaped a major catastrophe, in part, because of the very limited size of the meteorite (only 18,000 kg of the Mars' rock survived re-entry) and the scanty population at its landing site.

Near Earth Objects—Asteroids and Meteorites

To gain a better appreciation of what could have happened in Nigeria if *Zagami* had been a very large rock of about 0.5 km in diameter or more, we should recall the *Chicxulub crater* that was credited to an asteroid that impacted the *Yucatán Peninsula* in Mexico about 66 million years ago.[10] The crater is more than 180 kilometres (110 mi) in diameter and 20 km (12 mi) in depth, making the feature one of the largest confirmed impact structures on Earth. South Africa is home to the largest impact crater on the African continent. Our next stop is Lake Bosumtwi in Ghana, about 30 km south-east of Kumasi, where, 1.07 million years ago, the inhabitants were not as lucky as the Zagami people were in 1962. That was the time a meteorite landed in today's Ghana and created the 10.5 km wide (in diameter) impact crater that today is filled with water and is known as Lake Bosumtwi. The crater, shown in Figure 14:1, is ranked amongst the ten most impressive impact craters in the world; two other impact craters in Africa that are also in that rank are the Aorounga crater in Chad and the largest and oldest known impact crater in the world, the Vredefort crater in South Africa. The latter is 250 km in diameter, and resulted from a meteorite impact about two billion years ago.[11]

Because of the massive size of the meteorite that landed in Ghana and the cosmic encounter-velocity with which it impacted the Earth surface, the shock waves that resulted from the impact deposited sufficient thermal energy to completely and instantaneously melt the target rocks at its impact site in Bosumtwi. The rapid distribution of the impact-melt throughout the resulting crater followed, and it produced a variety of unusual crystalline and glassy igneous rocks. The melted and vaporized

10 http://www.passc.net/EarthImpactDatabase/chicxulub.html (Accessed, October 16, 2015).

11 http://www.universetoday.com/19616/earths-10-most-impressive-impact-craters/ (Accessed, October 17, 2015).

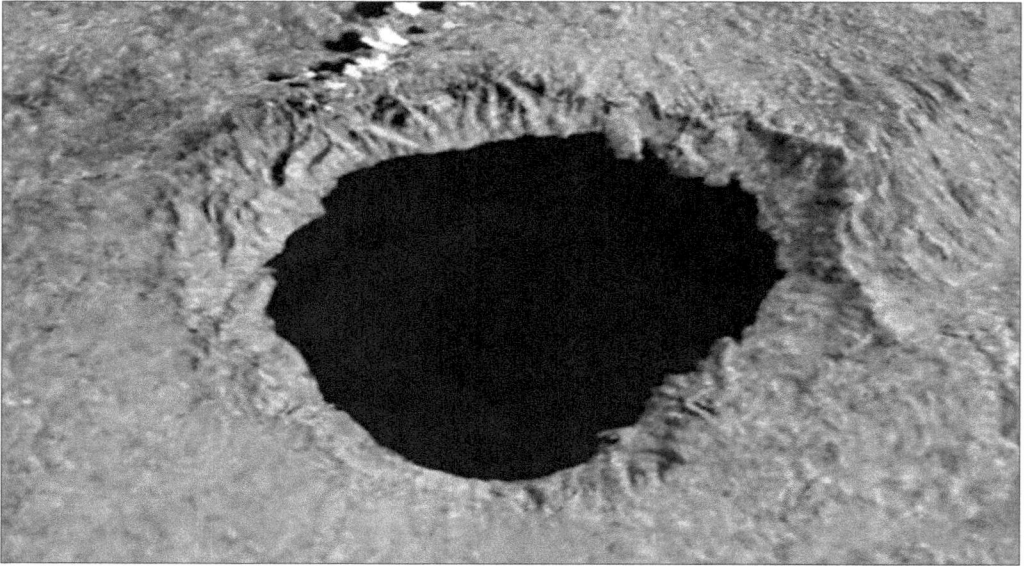

Figure 14:1 Image of Lake Bosumtwi, Ghana *(Image Credit: Created with NASA WorldWind by Vesta using Landsat 7 (false color) satellite image.)*

materials were then ejected, at a high velocity, to a distance of 300 km – 350 km away, inland and along the coast of today's Ivory Coast, where they formed deposits of small and large glassy bodies (tektites), typically a few millimeters to about 20 cm, in sizes, as shown in Figure 14:2.[12]

Figure 14:2 Samples of Ivory Coast tektites
(Credit: Aubrey Whymark, tektites.co.uk)

---

12 Melosh, H. J. (1989). *Impact cratering: A geologic process*, Oxford University Press (Oxford Monographs on Geology and Geophysics, No. 11), 1989, 253 p. (Accessed, October 17, 2015).

How could one be sure that there is a great connection between the crater in Bosumtwi, Ghana and the tektites-strewn field in Ivory Coast? Three of the factors that led to such a conclusion were the geographical proximity of the crater to the tektite-strewn field in Ivory Coast, the very similar ages of the crater in Bosumtwi and of the tektites found in Ivory Coast and the comparable chemical compositions of the tektites in Ivory Coast and the gross physical characteristics of the rocks and the rock formations around the rim of Bosumtwi Lake.[13]

Over the thousands of years since this cosmic event wiped away all signs of life in an area close to 10 km radius from today's site of Lake Bosumtwi, the region went through several transformations and much rejuvenation. Today, life is back to normal, and tourism is being promoted at a site that was, for ages, noted for its lifelessness.

A revisit to the most recent devastating meteorite impact on Earth, known as the Tunguska Explosion, can also serve as an illustration of the havoc a bigger *Zagami* could have wrought in Nigeria. That explosion took place at 7:17 am, on June 30, 1908 in Siberia, Russia. The Tunguska Explosion resulted from what is now believed to be an airburst of a large meteoroid at an altitude estimated to be about 5-10 km, (3.1—6.2 miles); the energy of the blast has been estimated at about 1,000 times as powerful as the atomic bomb the USA dropped on Hiroshima, Japan, on August 6, 1945. It has also been estimated that the shock wave from the blast would have measured about 5.0 on the *Richter* scale.[14] The explosion levelled 800 square miles of trees, and much of the area has not fully recovered.[15, 16]

Mitigation of Near Earth Objects (NEOs)

Because of the grave danger posed to our planet Earth by such tragic realities as *The Tunguska Explosion*, the global community has begun to pay increasing attention to the threats of NEOs.[17] The on-going efforts on *NEOs science* and management

---

13 http://www.scribd.com/doc/38710580/Asteroids-Comets-and-Meteorites (Accessed, October 16, 2015).

14 Verma, Surendra (2005). *The Tunguska Fireball: Solving One of the Great Mysteries of the 20th Century*, Icon Books Ltd, London, UK (Accessed October 27, 2015).

15 The Tunguska Impact—100 years later (http://science.nasa.gov/science-news/science-at-nasa/2008/30jun_tunguska/) (Accessed October 27, 2015).

16 http://www-th.bo.infn.it/tunguska/ (Accessed October 27, 2015).

17 NEOs include near-Earth asteroids (NEAs), near-Earth comets (NECs), and meteorites, large enough to be tracked in space before striking the Earth. Asteroids and meteorites are extra-terrestrial rocks that are mostly composed of carbonaceous, stony and metallic (mainly iron) materials. Comets are fragile, irregularly shaped cosmic bodies composed of frozen gases, rock and dust.

began in the 1994-1995 period, and have since focused and coalesced on the discovery, tracking and the observation as well as possible deflection of NEOs in order to be able to mitigate their impact on Earth.

Two recent events in the outer space environment, in 1994, triggered the on-going NEOs impact global awareness and have kept the mitigation efforts in focus. The first was the collision of the Comet Shoemaker—Levy 9 (SL9) with Jupiter, the largest planet in the Solar System. Eugene and Carolyn Shoemaker and David Levy discovered the comet on March 24, 1993. It was torn into pieces as a result of its close approach to Jupiter, in July 1992. Scientists monitored the train of the comet and concluded that it would again pass, on July 19, 1994, within 25,000 kilometres of the centre of Jupiter. Because the passing distance between the comet and Jupiter was less than the radius of Jupiter (69,911 km), it was predicted that parts of the comet's train would collide with Jupiter; and they did, between July 16 and 22. The second NEO event took place on December 9, 1994; it was a near-miss as the asteroid XL1, about 300 m in diameter, and known as *1994 XL1*, passed within 105,000 km of our planet Earth.[18] The asteroid, which was about the size of the Empire State Building in New York City, was discovered, just outside the Earth's orbit, by Robert H. McNaught of Siding Spring, Australia. But SL9 collision with Jupiter got most of the global attention at that time.

Several space-based and terrestrial telescopes, including the German ROSAT X-ray observing satellite, the USA-ESA Ulysses Space Probe and three USA's space-craft (the Galileo, which was on its way to rendezvous with Jupiter in 1995, Voyager 2 and The Hubble Space Telescope) and the JPL's Deep Space Network facility at Goldstone, California, captured the Jupiter-SL9 collision live, and/or recorded various measurements from the collision. Scientific analysis of the observations and data recorded provided the following results and observations:

- The comet consisted of about 20 fragments of stony and icy matter ranging up to 2 km (1.2 miles) in diameter.

- These fragments collided with Jupiter's southern hemisphere between July 16 and July 22, 1994, at a speed of approximately 60 km/s (37 mi/s) or 216,000 km/h (134,000 mph).[19]

---

18 http://www2.jpl.nasa.gov/sl9/news47.html (Accessed October 18, 2015).

19 http://en.wikipedia.org/wiki/Comet_Shoemaker%E2%80%93Levy_9 (Accessed October 19, 2015).

- Contrary to pre-impact predictions, astronomers had not expected to see the fireballs from the impacts[20] and did not have any idea in advance how visible the other atmospheric effects of the impacts would be from Earth. Observers soon saw a huge dark spot after the first impact. The first impact represented the most energetic event ever witnessed in the Solar System besides the daily output from the Sun; it produced huge plasma flares and sharp increases in radio wave emission.

- The impact of the fragments caused Jupiter to exhibit a number of highly visible big dark scars. The scar from the first impact was visible even in very small telescopes, and was about 6,000 km (3,700 miles) (one Earth radius) across.[21]

- Global stratospheric temperatures rose immediately after the impacts, and then fell to below pre-impact temperatures 2–3 weeks afterwards, before rising slowly to normal temperatures.[22]

- Perhaps the largest impact was produced on July 18; it was estimated to pack the explosive power of 6 million megatons of TNT (600 times the world's nuclear arsenal).[23]

- Spectroscopic observers found that ammonia and carbon disulphide persisted in the atmosphere for at least fourteen months after the collisions, with a considerable amount of ammonia being present in the stratosphere as opposed to its normal location in the troposphere.[24]

On the whole, the multiple impacts of comet SL9 fragments with Jupiter in July 1994 provided a historic opportunity to directly observe the phenomena resulting

20 Weissman, Paul (July 14, 1994). "The Big Fizzle is coming." *Nature* 370: 94–95. Bibcode:1994Natur.370...94W. doi:10.1038/370094a0 (Accessed October 19, 2015).

21 Hammel, H.B. (December 1994). *The Spectacular Swan Song of Shoemaker–Levy 9*. 185th AAS Meeting 26. American Astronomical Society. p. 1425. (Accessed October 19, 2015).

22 Moreno, R.; Marten, A; Biraud, Y; Bézard, B; Lellouch, E; Paubert, G; Wild, W (June 2001). "Jovian Stratospheric Temperature during the Two Months Following the Impacts of Comet Shoemaker–Levy 9". *Planetary and Space Science* 49 (5): 473–486. Bibcode:2001P&SS...49..473M. doi:10.1016/S0032-0633(00)00139-2. (Accessed October 21, 2015).

23 Bruton, Dan (February 1996). "What were some of the effects of the collisions?" *Frequently Asked Questions about the Collision of Comet Shoemaker–Levy 9 with Jupiter*. Stephen F. Austin State University. Retrieved 2014-01-27. (Accessed October 21, 2014).

24 McGrath, M.A.; Yelle, R. V.; Betremieux, Y. (September 1996). "Long-term Chemical Evolution of the Jupiter Stratosphere Following the SL9 Impacts." *Bulletin of the American Astronomical Society* 28: 1149. Bibcode:1996DPS....28.2241M. (Accessed October 25, 2014).

from hypervelocity collisions on a planet. Detailed analysis of the data, including related information, obtained thereof, has advanced human understanding of comets, of Jupiter, and of the collisional processes that shaped the solar system.[25] The analysis of the data, collected during the collision by the international scientific community, has helped in the development of better models that are now being used to predict and assess the consequences of a comet or an asteroid impact on Earth.

These two 1994 NEOs events jolted the global community into action, and in the years that followed, the danger from and probability of a collision of a Near-Earth Object with Earth was discussed with increasing frequency at intergovernmental scientific meetings and at newly founded conferences of high-velocity impact physicists, astrophysicists and experts on the detection of NEOs. Under the banner of the United Nations, the world leading scientists met and deliberated on NEOs threat to planet Earth at the First United Nations International Conference on Near Earth Objects, held in New York, April 24-26, 1995. Shortly, thereafter, unlikely allies—the USA, Russia and China—co-sponsored and co-organised "The Planetary Defense Workshop," held at the Lawrence Livermore National Laboratory in Livermore, California, May 22–26, 1995; the latter focused on the issues that surfaced at the United Nations conference and on needed actions for mitigating apparent threat to collective safety and security at each national level as well as collectively, at the international level.

A number of countries subsequently embarked on national and regional efforts to build up their NEOs impact mitigation capabilities. These included:

- The Hayabusa mission: challenge to near-Earth asteroid sample return and new insights into solar system origin by Japan;
- Dawn, Deep Impact, Stardust and the Wide-field Infrared Survey Explorer (WISE) spacecraft missions of the United States;
- The Near Earth Object Surveillance Satellite mission of Canada;
- The Marco Polo near-Earth object sample return mission of ESA and JAXA of JAPAN;
- The prospective Asteroid-Finder spacecraft mission of Germany;
- Arecibo (Puerto Rico) and Goldstone (USA) Radio Telescope facilities;

---

25 Boslough, M. B. and D. A. Crawford (1995). The Shoemaker-Levy 9 Impact Plumes on Jupiter: Implications for Threat to Satellites in Low-Earth Orbit Sandia National Laboratories, Proceedings of the Planetary Defence Workshop held at the Lawrence Livermore National Laboratory, Livermore, California, May 22-26, 1995 (Accessed October 25, 2014).

- The USA's Panoramic Survey Telescope and Rapid Response System (Pan-STARRS);
- The Skalnaté Pleso Observatory in Slovakia;
- The Asia-Pacific Ground-based Optical Satellite Observation System in APSCO Member States;[26] and
- The International Scientific Optical Network (ISON) that is managed by the Keldysh Institute of Applied Mathematics, part of the Russian Academy of Sciences.

The NASA video on 100 Tons Space Junks[27] provides a better appreciation of the importance of building up the above NEOs impact mitigation capabilities. The first part of the video shows the October 7, 2008 crash-landing of meteorite 2008 TC3, in the Nubian desert of Sudan. It was the first time that the actual crash-landing of NEOs had been observed, live. And there are still many more similar objects in space.

As of May 30, 2015, 12,744 Near-Earth objects have been discovered with the aid of these space tools. Some 870 of these NEOs are asteroids with a diameter of approximately 1 kilometer or larger, while 1,589 of them have been classified as *Potentially Hazardous Asteroids (PHAs)*. The nation-States with these capabilities also recognised and understood the need for an international collaboration on NEOs impact mitigation. The net outcome of this recognition and understanding was a major UNISPACE III recommendation which called for an improved international coordination of activities related to near-Earth objects.[28]

The issues involved in protecting the Earth from an asteroid impact can be classified into four categories: (a) Finding NEOs, (b) Determining the risk of NEOs impact, (c) Deciding on a course of action in cases where the risk is relatively high and if a deflection is necessary, and (d) Implementing a space mission campaign to deflect the asteroid. At UNISPACE III, there was consensus that '*the international community should 'improve the international coordination of activities related to near-Earth objects.*' Following the approval of the General Assembly in November 2007, the S&T of COPUOS began work on the development of a global decision-making process that could prevent an Earth impact by a near-Earth object. In support of this effort, the Association of Space Explorers (ASE) offered its own contribution. It subsequently

---

26 APSCO Member States include: Bangladesh, China, Indonesia, Iran, Mongolia, Pakistan, Peru, Thailand and Turkey.

27 https://www.youtube.com/watch?v=lplsonmuiVQ

28 http://www.unoosa.org/pdf/sap/hsti/UNISPACE_report.pdf (Accessed October 27, 2015).

identified and invited individuals to serve, in its NEOs Committee, as *principals* on its "Panel on Asteroid Threat Mitigation." The report of the ASE Panel on NEOs was submitted to COPUOS at its 2009 Session;[29] it was a major contribution to the proposal that the Committee's Working Group on NEOs finalized and presented on NEO mitigation to COPUOS in June 2013. The proposals, accepted and adopted by COPUOS in 2013, focused on NEOs mitigation, and included "*the development of an overall international governance model that can respond to a threatening Near Earth Object,*"[30] and "*an international interagency plan to develop joint technologies that could deflect a rogue asteroid including a space demonstration of how to change the course of such an asteroid.*"[31] As shown in the footnotes below, the two entities responsible for these activities began their work in 2014.

Because of its commitment to NEOs impact mitigation and its belief that this can only be achieved through a concerted and collective global action, the ASE Committee on NEOs succeeded in energizing more than 100 leading scientists, astronauts and business leaders around the world to sign a declaration, *The Asteroid Day Declaration*, which called for a hundredfold increase in the detection and monitoring of asteroids.[32] *The Asteroid Day*, marked worldwide for the first time on June 30, 2015, "is a global awareness movement where people from around the world come together to learn about asteroids and what we can do to *protect our planet*, our families, communities, and future generations." Why June 30? To remind us of the impact of the Tunguska Explosion that occurred on June 30, 1908.

Space Debris

The outer space environment within which the International Space Station and over 3,000 active satellites are operating today, alongside over 7,000 pieces of debris that are orbiting the Earth, is also getting crowded, and safety of space assets within

---

29 http://www.space-explorers.org/ATACGR.pdf (Accessed October 27, 2015).

30 An international asteroid warning network (IAWN) which would link together the institutions that are already performing many of the proposed (asteroid warning) functions.

31 A space mission planning advisory group (SMPAG) made up of member States with space agencies, SMPAG's responsibilities include (i) Lay out the framework, timeline and options for initiating and executing response activities; Inform the civil-defence community about the nature of impact disasters; and (ii) Incorporate the civil-defence community into the overall mitigation planning process through an impact disaster planning advisory group.

32 http://www.asteroidday.org/ (Accessed October 20, 2015).

it is now of major concern to all.[33] Most of these pieces of debris, otherwise known as *space debris,* include dead satellites, broken parts of satellites as well as broken parts of rockets used in launching these satellites into space; they are much, much smaller than large meteors. But space debris may cause injury, even death, and property damage. In general, they do not survive re-entry into the Earth's atmosphere because of the *air resistance* that opposes their motion. Space debris made of materials that have low melting temperatures, such as aluminium, are very susceptible to overheating as a result of atmospheric drag; hence they burn up before re-entry. Others with components that are made of large pieces of materials of moderate melting temperatures may survive re-entry, since they can easily radiate heat over their large surface areas. The components that do survive re-entry are usually made of stainless steel, titanium, and other materials with high melting temperatures. Large and heavy pieces of space components often survive re-entry, and when they do, they often pose danger to everything—living or non-living—in their paths. That was the case on April 27, 2000 when parts of the Delta rocket, used in the launch of a USA GPS satellite, re-entered the Earth's atmosphere, with its debris littering the landscapes of Brazil, South Africa and the United States (Texas).[34]

It is doubtful that space debris was ever a subject of national conversation or interest in Nigeria until the errant Italian X-ray astronomy satellite, Bepposax, made the front page news in the Nigerian daily newspapers of April 29, 2003 with the head-line: *Bepposax Satellite crash-landing over Nigeria's territory today or tomorrow.* The head-line and associated details, both credited to NAMA, raised the blood pressure levels in many high places in Nigeria, where, at that time, the country had no plan for handling such a fiasco if the satellite had actually crash-landed there. Luckily for Nigeria and its citizens, the errant satellite crash-landed in the equatorial Pacific, about 186 miles northwest of Galapagos Island (Ecuador), on April 29, 2003 at 10:57 pm, Nigerian time. Again, as it was in the case of Zagami, Nigeria was spared of what could have amounted to a major national disaster.

The 1,400 kg Bepposax satellite was a joint Italian-Dutch project for X-ray observation. It was launched by American ATLAS I orbital launch vehicle into a circular orbit on April 30, 1996. By September 21, 2002, its flight altitude was about 435 km, a far cry from 600 km at its deployment in 1996. Its operators predicted its uncontrolled re-entry because its control mechanism had totally failed. As expected,

---

33 http://www.nasa.gov/worldbook/artificial_satellites_worldbook.html (Accessed October 20, 2015).

34 http://www.aerospace.org/cords/spacecraft-reentry/ Accessed October 20, 2015).

because of its high mass, parts of Bepposax survived re-entry into the Earth's atmosphere. Thus, towards the end of its own life, the satellite became a major risk to life and property on Earth, particularly along its path, within latitudes 4^0 North and 4^0 South of the equator, as shown in Figure 14:3. On April 30, 2003, I submitted to the President of Nigeria, my advisory on "How Nigeria should prepare its citizens for such an eventuality, as the crash-landing of Bepposax, in Nigeria, in the future."

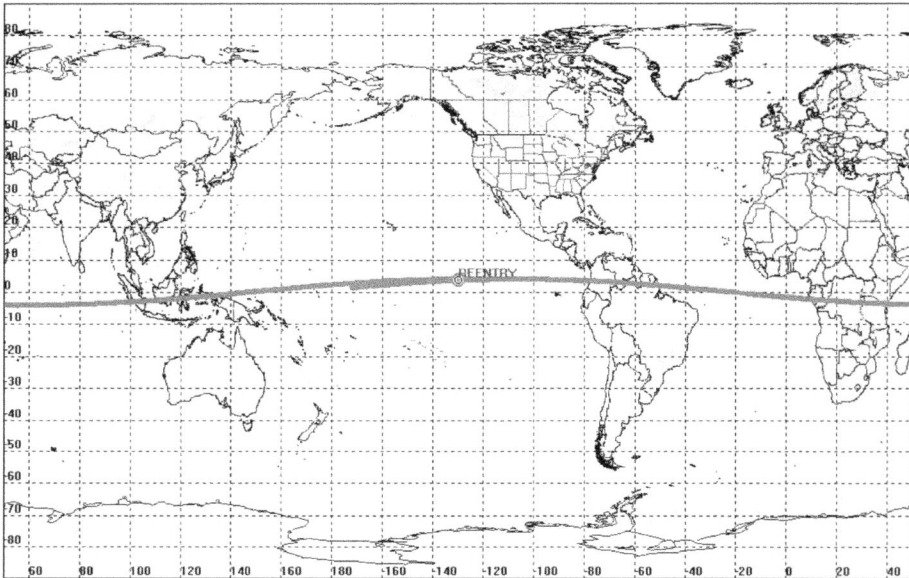

Figure 14: 3 Bepposax Reentry Path, April 30, 2003, 09:30 UTC
(Credit: Luciano Anselmo & Pardini Carmen, CNR di Pisa, Italy)

Unlike the demise of Bepposax, the last phase of which was left to natural elements in space, at least two satellites were deliberately destroyed in recent years. The first was the 2007 anti-satellite missile test which China conducted on January 11, 2007. At that time, China launched a multistage solid-fuel missile to destroy its polar-orbiting weather satellite, *Fengyun-1C*, in a head-on engagement, at an altitude of 865 kilometres (537 mi). Just over a year later, on February 20, 2008, the USA military shot down, with a missile, its highly classified errant spacecraft, USA-193 because it was in a deteriorating orbit and was expected to crash into the Earth within weeks. The USA satellite, launched on December 14, 2006, malfunctioned shortly after its deployment. These two intentional acts have increased, by tens of thousands, the space debris population.

Rare accidents also do happen in space just as on highways on Earth—the latter, albeit, more frequently. That was the case in space, on February 12, 2009, when

Iridium 33 and Cosmos 2251 satellites collided, at a closing speed of well over 15,000 mph, at an altitude of 790 km (490 miles) above the Earth's surface over Siberia, Russia. Before the accident, the Russian communications satellite, Cosmos 2251, launched in 1993, was inactive and could no longer respond to any maneuvering. The American Iridium 33 communication satellite was active and operational, but there was no warning that it was in any immediate danger. That was the first time ever, in the history of the space age that two whole satellites would slam into each other. The global space community described the collision as *"extremely unusual."* The collision resulted in a cloud of over 60,000 pieces of broken objects moving about in space, uncontrolled.

What are the implications of an intentional destruction and in-orbit collision of satellites in space? Only one thing—every man-made object in space, between altitude 300 km and 1300 km above the Earth's surface, including any space object(s) launched into that highly populated belt, is at risk of being hit by space junk.

Because of the growing congestion and increasing chances of collision amongst the various assets within the outer space environment, the United States Joint Space Operations Center (JSpOC), a part of the United States Strategic Command, is now charged with the responsibility of protecting all USA's space systems, a responsibility that requires a constant awareness of the outer space environment, including space weather. JSpOC also provides a new "service" which alerts a satellite operator to the risks its satellite is exposed to in space, albeit, at no cost to the operator. The JSpOC service, which began since the Siberian space collision, operates around-the-clock, every day.[35] Thus, whenever an operator's satellite is in a collision path or is approaching a close encounter with other satellites or space junk, chances are high that JSpOC will alert the operator. Following the exchange of necessary information between JSpOC and the satellite operator, the operational satellite that is at risk can be steered away from the collision course and given a new lease on life. Today, JSpOC is the only entity in the world with this capability. Indeed, without any knowledge of the fate of its space asset, *NigeriaSat-1* satellite, Nigeria was a beneficiary of the USA space-related alert system in both January and March of 2010. Those were the times when Nigeria's Earth observation satellite, NigeriaSat-1, was in a collision path with Space Junk 28955 on January 3, 2010 and Space Junk

---

35 A number of the leading space countries have already launched their projects to track down potentially dangerous asteroids, including NASA's Near Earth Asteroid Tracking (NEAT), Deep Space 1, Deep Impact, Dawn, and Stardust, as well as ESA's Rosetta and JAXA's Hayabusa.

01716 on March 8, 2010. Fortunately, because of the JSpOC's assistance, a collision was avoided on both occasions.

At the inter-governmental level, the issue of space debris first came up for discussion in 1994 at the thirty-first session of the S&T Sub-Committee of COPUOS. Between that period, and 2000, the Sub-committee progressively focused on scientific research on space debris, with emphasis on mathematical modelling and characterization of the space debris environment, and on measures to mitigate the risks of space debris, including spacecraft design measures that could provide protection against space debris.[36,37] Similar to the case of NEOs, at UNISPACE III in 1999, there was consensus that the international community *should improve the protection of the near–Earth space and outer space environments through mitigation of space debris.* After several years of technical meetings amongst its members, the Inter-Agency Space Debris Coordination Committee *(IADC) developed* "a set of high-level qualitative guidelines that gained a wider acceptance among the global space community."[38] On its part, COPUOS, upon the recommendations of its S&T Sub-Committee, took into consideration the United Nations treaties and principles on outer space and developed a set of recommended guidelines based on the technical content and the basic definitions of the IADC space debris mitigation guidelines. In its resolution 62/217 of 22 December 2007, the General Assembly endorsed the Space Debris Mitigation Guidelines proposed by COPUOS. It also agreed that the voluntary guidelines for the mitigation of space debris reflected the existing practices as developed by a number of national and international organisations, and it invited Member States to implement those guidelines through relevant national mechanisms.

From its own experience with Bepposax and NigeriaSat-1, the issue of space debris should be of fundamental interest not only to all Nigerians, but in particular to the nation's leadership at the highest level, including its security and safety-related authorities. Following the crash-landing of Bepposax, this author submitted a set of recommendations, to the then Nigerian Administration in 2003, on the critical steps the nation must take in case NEO or space debris should crash-land on its territory.

---

36 United Nations Document A/AC.105/571, paras. 63-74.

37 United Nations Document A/48/20, para. 87.

38 The Inter-Agency Space Debris Coordination Committee (IADC), made up of twelve of the leading space agencies, is an international governmental forum for the worldwide coordination of activities related to the issues of man-made and natural debris in space.

And as a nation with space assets, Nigerian authorities should be fully conversant with the international regulations that govern human exploration and use of outer space, including the ownership of responsibility for any damage caused in space or on Earth by an errant satellite.

These and many other concerns were addressed by the *1967 Treaty on Principles Governing the Activities of States in the Exploration and Use of Outer Space, including the Moon and Other Celestial Bodies (The Space Treaty)*. As of July 6, 1967, Nigeria became a co-signatory to *The Space Treaty*. Article VI of this treaty stipulates that "all member States are internationally responsible for all national (public and private) activities in outer space that are conducted from their respective territories." As of 2005, Nigeria also became a co-signatory to *The Liability Convention of 1972* and *The Registration Convention of 1975*. These two conventions are elaborated in Articles VII and VIII of *The Space Treaty*. By the articles of these two conventions, it is incumbent on all states to regulate public and private space activities within their territories since each concerned State is liable for damage(s) caused by the space objects that are launched from its territory or that it owns. It is thus incumbent on Nigeria to enact and adhere to a requisite national space law for the regulation of outer space activities in and by Nigeria; the same applies to all other countries in a similar situation.

Space Weather[39]

Thunder and lightning often accompany tropical storms such as is frequently witnessed in Nigeria and other tropical countries of the world. In a number of such moments, electrical transformers explode because they are struck by the powerful electromagnetic pulse of lightning. Similarly, from time to time, almost at 11- to 12- year intervals, the sun erupts, ejecting, in the process, enormous solar flares that are accompanied by an intense burst of radio waves as shown in Figure 14:4 on the next page.

---

39 Space Weather is the term scientists use to describe the ever changing conditions in space. Explosions on the Sun create storms of radiation, fluctuating magnetic fields, and swarms of energetic particles. These phenomena travel outward through the Solar System with the solar wind. Upon arrival at Earth, they interact in complex ways with Earth's magnetic field, creating Earth's radiation belts and the Aurora. Some space weather storms can damage satellites, disable electric power grids, and disrupt cell phone communications systems. http://www.windows2universe.org/space_weather/space_weather.html (Accessed October 5, 2015).

The website: https://youtu.be/C1Kact6QHG0 contains a video clip of the solar flare shown in the above Figure 14:4. Solar flares produce high energy particles and radiation (X-rays); they are dangerous to living organisms and man-made machines, and are not visible to the naked eye—thus they can only be observed using special instruments such as telescopes. The sum total of these Sun-induced activities in space is referred to as Solar Weather.[40]

Although the Earth's magnetic field and atmosphere shield the Earth's surface from the effects of solar flares and associated activities, such radiations from the Sun, however, do disturb the Earth's ionosphere, which in turn disturbs some radio communications.

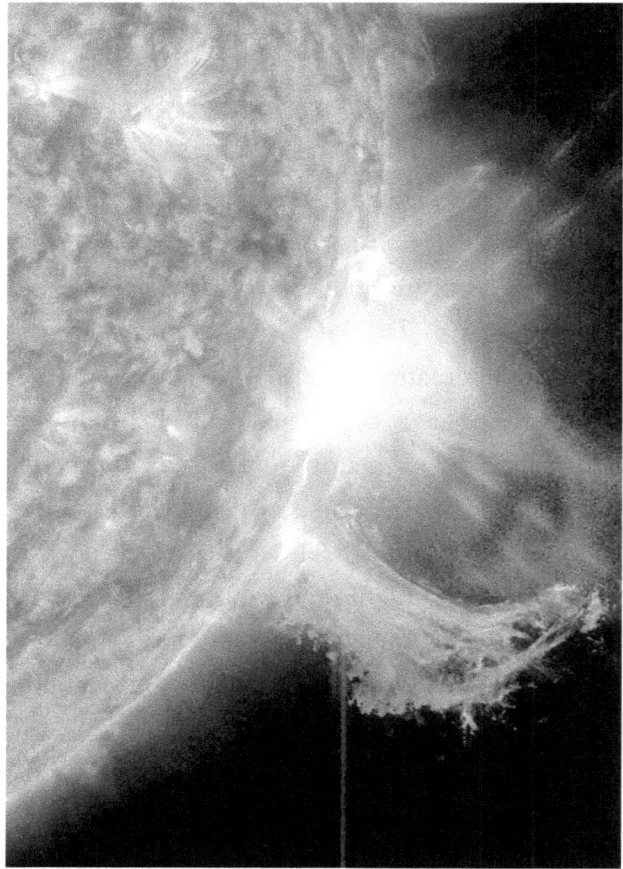

Figure 14:4 Image of an actual solar flare, in action, captured by NASA's Solar Dynamics Observatory on October 2, 2014. *(Credit: NASA)*

Along with energetic ultraviolet radiation, solar flares also heat the Earth's outer atmosphere, causing it to expand. The latter increases the drag on Earth-orbiting satellites, and in the process reduces their lifetime in orbit.[41] Both the intense radio emission from solar flares and the corresponding impact on the Earth's atmosphere can also degrade the precision of Global Positioning System (GPS) measurements, including those in cars, aircraft and mobile phones. Solar storms can have destabilising effects on earth and in the space environment against aircraft, electric power

40 The website: https://www.youtube.com/watch?v=HFT7ATLQQx8 contains videos of additional solar flares (Accessed February 5, 2015).

41 http://hesperia.gsfc.nasa.gov/rhessi2/home/mission/science/the-impact-of-flares/ (February 15, 2015).

grids, and spacecraft. Indeed, solar storms caused the electric power blackout in eastern Canada, Britain and in the north east of USA on March 13, 1989.

From time-to-time, NASA issues warning of impending solar activities as it did in 2010 when it warned that: "If forecasters are correct, the solar cycle will peak during the years around 2013. And while it probably won't be the biggest peak on record, human society has never been more vulnerable. The basics of daily life—from communications to weather forecasting to financial services—depend on satellites and high-tech electronics. A 2008 report by the National Academy of Sciences warned that a century-class solar storm could cause billions in economic damage."[42] According to scientists, such a scenario could pose a danger to the world's nuclear power plants and consequently to the world population because of the tidal wave of solar emissions from such intense solar activity. Such a fear is being fuelled by the fact that *"the largest active region (AR 1339) seen on the Sun since 2005 and visible to the naked eye, has rotated to the centre of the Sun's face, as seen from* (the) *Earth—which means any eruptions it produces will be aimed right at us."*[43] Based on the images of AR 1339 taken November 1 to 10, 2011 by NASA's Solar Dynamic Observatory (SDO), the belief is that the global community may experience a repeat (or worse) of the 1859 solar storm—called the *Carrington Event* or the *Solar Superstorm*. "No one knew flares existed until that September 1, 1859 morning when one super-flare produced enough light to rival the brightness of the sun itself."[44] In 1859, there were also few electrical devices in operation; the most prominent one was the telegraph.[45]

---

42 http://science.nasa.gov/science-news/science-at-nasa/2010/16jul_ilws/ (February 15, 2015).

43 Major, Jason (2011). National Geographic News, USA, November 10, 2011.

44 http://science.nasa.gov/science-news/science-at-nasa/2008/06may_carringtonflare/(February 7, 2015).

45 Magnetic observatories world-wide recorded disturbances in Earth's field so extreme that magnetometer traces were driven off scale, and telegraph networks experienced major disruptions and outages. The electricity which attended this beautiful phenomenon (*Solar Superstorm*) took possession of the magnetic wires throughout the country, the *Philadelphia Evening Bulletin* reported, and there were numerous side displays in the telegraph offices where fantastical and unreadable messages came through the instruments, and where the atmospheric fireworks assumed shape and substance in brilliant sparks. In several locations, operators disconnected their systems from the batteries and sent messages using only the current induced by the aurora. In Havana, Cuba, the sky that night appeared "stained with blood and in a state of general conflagration" and auroras were observed as far south as Hawaii and northern Venezuela.... (http://www.wunderground.com/blog/JeffMasters/comment.html?entrynum=1206) (Accessed, February 7, 2015).

Today, the story is different. The global community is interconnected and networked by an array of satellites, telephone lines, and power grid, all of which can be victims of another Solar Superstorm.

What are the options available today to humankind when threatened by solar flares? According to the report of a 2008 NASA-sponsored workshop on solar flares, the effective mitigation requirements include *inter-alia*:

Enhanced forecasting and monitoring capabilities including advance notice of higher solar activity is needed for launch vehicles and space operations, including manned operation on the Moon and in transit to Mars; and

Improved infrastructures designed to better withstand geomagnetic disturbances with emphasis on new technologies (GPS modernization with new signals and codes, new-generation radiation-hardened electronics), and improved operational procedures.[46]

If the forecasting of solar storms is more reliable, utility and satellite operators with such forewarned knowledge can take necessary measures to reduce damage by disconnecting wires, shielding vulnerable electronics and by powering down critical hardware.

In order to provide such up-to-date solar storm awareness as well as understand the Sun-Earth connectedness, a variety of space tools are currently in use, and others are proposed, globally, to gather necessary information on space weather and related solar activities, including solar flares (See Figure 14:4). Amongst these tools are:

- The National Oceanic and Atmospheric Administration's (NOAA) Geostationary Operational Environmental Satellites (GOES);
- NASA's Solar Terrestrial Relations Observatory (STEREO);
- The joint European Space Agency and NASA mission Solar and Heliospheric Observatory (SOHO);
- NASA's Advanced Composition Explorer (ACE);
- NASA's Solar Dynamics Observatory (SDO) with its the Extreme Ultraviolet Variability Experiment;
- SOLAR-B Satellite of Japan, otherwise known as "*HINODE*;" and
- ESA's Solar orbiter scheduled for launch in 2017.

Because of these and other threats to the safety and security of humankind, other inhabitants and machines here on planet Earth and within the outer space environment, the GA, in its *Resolution 63/90 of December 5, 2008*, stated:

---

46 *Severe Space Weather Events—Understanding Societal and Economic Impacts*—Workshop Report, National Research Council, Washington, D.C., 2008 (Accessed, February 9, 2015).

…considers that it is essential that Member States pay more attention to the problem of collisions of space debris, including those with nuclear power sources, with space debris, and other aspects of space debris; calls for the continuation of national research on this question, for the development of improved technology, for the monitoring of space debris, and for the compilation and dissemination of data on space debris; also considers that, to the extent possible, information thereon should be provided to the Scientific and Technical Subcommittee; and agrees that international cooperation is needed to expand appropriate and affordable strategies to minimize the impact of space debris on future space missions.

At its 52nd Session in June 2009, COPUOS responded to these GA concerns and agreed to and has continued to pursue two complimentary goals. The first is the *"Long-term sustainability of outer space activities,"* which began in 2010; the second is *"the collection, processing and availability of relevant data and information on objects in outer space for the promotion of a safe and sustainable development and peaceful uses of outer space."* Because the safety of the outer space environment is a necessary condition for the long-term sustainability of all human activities in outer space, the international community is now preoccupied with discerning the necessary collective global steps needed to ensure that the outer space environment is safe and secure for all.[47]

Collective Responsibilities and Minimum Contributions

To ensure an outer-space environment that is safe and secure for the benefit of all humankind, a number of space-faring and space-capable countries are already far ahead in their respective responses to the above *call-for-action* by the GA. They are taking these steps because of their realisation that continuing success in the exploration and utilisation of outer space that is safe and secure would require long-term international cooperation and public-private partnerships. The tens of thousands of space debris scattered within the outer space environment by deliberate acts or by accidents and the subsequent close encounters of space assets with errant space junk are clear proofs that the outer space environment is not safe and secure today. Certainly, the safe operation of all space assets is not only critical but should also be a major priority for the social and economic development of the countries that have made such investments. But, today, no global entity is legally responsible for the management of international space traffic. There is also no global legal mechanism

---

47 Relevant documents are available at the United Nations Office for Outer Space Affairs, Vienna, Austria http://www.unoosa.org/ (Accessed, February 10, 2015).

for sharing space awareness information among all member States with space assets. In the absence of such arrangements, the mitigation of asteroid impact and space debris is thus a collective global responsibility that requires collective global preparedness and response. In addition to embarking on well publicised local emergency preparedness, Chapter XI identifies the scope of practical action-oriented contributions that Nigeria and other developing countries should make to these global mitigation measures. There is no alternative.

* * *

MOVING AHEAD

Chapter XV

THE WAY FORWARD

The nature of Nigeria's space journey, to-date, prompts the questions: What are we aiming for? What path do we want to take? Most nations have designed their space programmes to address these basic questions. However, a space programme can be one of the most challenging and demanding national undertakings because it has the potential to revolutionise any nation's development and growth. With this in mind, Nigeria, with reflection and refocus, should properly redirect its space journey. This will require vision; talented people with inquiring minds, knowledge and skills; the essential financial resources that will come from a committed diversified economy that is not overwhelmingly dependent on the extraction of carbon deposits; and the unalloyed commitment of the government and people of the country to reshape and rebuild the programme. It is a national assignment that should be undertaken in the interest of Nigeria and for the benefit of all Nigerians. Thus, the refocused space journey requires the full participation of all, beginning with the unwavering commitment of the nation's leadership.

Chapter V chronicles how the journey took off in 1999 with misleading pieces of advice to the government and without a number of essential preparatory steps. Similarly, Chapter VII presents some of the unjustifiable justifications used to seek government support for a number of the activities proposed for the 25-year Road Map for Nigeria's Space Mission (2005-2030), hereafter referred to as 25-year Road Map. Today, the nation's space journey is in need of rescue. Its entire focus, including the 25-year Road Map should be redefined and redesigned by the Nigerian government to bring it into alignment with the needs of the people. In support of such an action, I have offered, in this chapter, a number of recommendations for consideration and action by the National Space Council (NSC).[1] The conscientious pursuit of these recommendations should move the nation forward in its space endeavours and in its attainment of the attendant rewards and benefits. To begin with, the recommended national actions include the following:

- The NSC should embody the vision of the nation's space programme;
- The NSC should, constitute a sound technical advisory body, hereby proposed as *The Think-Tank*, in order to redefine and redesign the nation's space agenda, as well as monitor its implementation and progress in order to ensure its success;
- *The Think-Tank* should offer meritorious advice to the NSC on the pursuit of the immediate and mid-term priorities of the nation's space programme as detailed in Chapters XI and XII;
- *The Think-Tank*, on behalf of the NSC, should conscientiously partner with all the stakeholders[2], for a united national space effort;
- The NSC should rally the resources that will ensure the sustainability of the nation's space programme; and
- The NSC should ensure that the nation takes necessary measures to learn vital lessons from the comparable steps of other countries. Specifically, we should consistently be asking: how did other nations approach this and what can we learn from their mistakes and achievements?

At this juncture, I will elaborate, in more detail, on these outlined recommendations.

Empowering the Nation

Looking ahead with a clear vision of the nation's future, Nigeria needs to undertake a number of critical steps that can transform all aspects of the nation's development and growth. Priority areas include: infrastructural and industrial

---

1 The National Space Council is the apex body charged with the responsibility of developing policy guidelines for Nigeria's activities in space and of overseeing the implementation of the National Space Programme. The NSC was established as part of the National Space Policy and Programme, adopted by the Nigerian Federal Executive Council, at Abuja, on July 4, 2001.

2 As used in this this book, the word stakeholders refers to the representatives *at the decision making levels* of government, academia and private establishments that have responsibilities for the Nigerian space programme and vested interest in the use of its products and services to meet the nation's needs. These stakeholders are identified in the Section on "Partnering with all the stakeholders for a united space effort," in this chapter.

development, particularly in transportation (roads, railways, and air-travel); energy generation and distribution; water resources development and management; delivery of communication and health-care services, both to the urban and the rural areas; sustainable agriculture that can ensure food security; and the management of the marine ecosystems and the Earth's environment. Other steps to ensure the nation's empowerment include the building and nurturing of the entrepreneurial spirit and the nation's commitment to first-class fundamental and applied research and development activities that can fuel technological innovations, as detailed in Chapters IX and XI. The nation's endowments that can support such efforts are noted in Chapter X. Such national undertakings will foster the emergence of science and technology enterprises that are critical to the nation's development and should sustain the growth of highly-skilled jobs in a variety of industries as has been the case in Brazil, China, India, Malaysia, Saudi Arabia, Singapore and South Africa. The sense of accomplishment derived from undertaking these transformative steps will naturally flow into the building of a viable national space programme, under the leadership of the National Space Council.

The National Space Council

Given the state of the nation's space programme as illustrated in Chapters VII, and VIII as well as the other concerns noted in this book, the National Space Council (NSC) urgently needs to redefine the nation's space aspirations. In order to translate such a vision into realistic, practical and achievable goals, the NSC will need a helping-hand to move the nation forward on a new redefined space agenda. That helping hand is *The Think-Tank*; it should consist of trusted Nigerian patriots that are very knowledgeable and represent the different areas of the nation's economy that are related to or affected by space science and technology. *The Think-Tank* shall be accountable to the Council, and shall carry-out Council-mandated duties. In order to enhance the effectiveness of the Council and the nation's space programme, the initial focus of *The Think-Tank* should include, among other things, the following:

- Design a new redefined national space programme for the Council;
- Ensure, through regular monitoring and evaluative measures, that the nation's space journey is on track, thereafter;
- Periodically communicate to the Nigerian people the steps taken and the goals attained as well as the ways in which the space efforts are yielding useful products and services that meet the needs of the nation and the Nigerian citizenry; and

- Ensure that the activities of the national space programme shall gradually and eventually reinforce a *can-do* attitude in Nigerians.

The Think-Tank, with input from all the stakeholders, should also ensure: (i) *Increased public awareness of space, and* (ii) *A broad knowledge of the role of space in development for political leaders.*

Increased Public Awareness of Space

National political will and broad public support are critical for the sustainability of the nation's space activities. Accordingly, the *Think-Tank* should develop and engage all Nigerians in a variety of space-awareness programmes that will enable them: to understand and appreciate the role of space in their daily lives;[3] to increase their awareness of the world they live in and of their own immediate environment; and to understand the role of outer space in that environment, including our inter-connectedness with the rest of humankind and the universe. Another approach would be to have a functional planetarium (not only in Abuja, but also at each state level) that is funded by respective governments, and by the private sector, and which would provide unique theatre-opportunities that can inform and educate the general population about the stars, the planets—including our Home Planet, Earth, other celestial bodies and the outer space environment, and how these impact our daily lives. Chapter XI provides examples of such facilities in other African countries.

A Broad Knowledge of the Role of Space in Development for Political Leaders

With input from the stakeholders, *The Think-Tank* should organise, for the benefit of Nigeria's political leaders, regular, broad and continuing orientation on the scope of the nation's space activities, on their specific impact on the different aspects of Nigeria's economy and on how the outer space environment is being used, globally and locally, to improve the human condition. These measures will enable the nation's decision makers to understand and value the importance of space as an indispensable tool of national development. The knowledge gained in such a process should pave the way for an appreciation of what it takes to achieve a meaningful and successful space programme including the choice of appropriate priorities to focus on.

---

3 A public campaign, that includes the print and radio journalists, educators, and marketing gurus, could use print media, cell phone technology, add-on segments to the news hour on TV, and other digital media.

Immediate and Mid-term Priorities of the Nation's Space Programme

The design of a space programme draws from a palette of many choices, and each nation must identify those priorities that are relevant to its own aspirations and peculiar condition(s). Chapters XI and XII contain recommendations on key national space priorities for the attention and consideration of the NSC on behalf of the People and the Federal Government of Nigeria.

Partnering with all the Stakeholders for a United Space Effort

Because space affects all facets of human society, there is the need to partner, nationally, with all the stakeholders, for a united space effort, by purposefully encouraging all of them to actively participate in and contribute to the nation's space programme. A component of the partnering effort is an annual national space dialogue among all the stakeholders on the state and impact of the nation's space journey; the latter includes the '*use of satellite data to power the nation's economy and development*' as outlined in Chapter XI. Such a continuing annual dialogue will ensure the currency of the nation's space efforts, enhance its competitiveness and provide necessary opportunities for all the stakeholders to discuss and reposition the nation's focus, as appropriate. That should be the unambiguous message of *The Think-Tank* to all Nigerians. To be successful, the Nigerian Space Programme also requires appropriate Nigerian partners and mutually beneficial regional and international collaboration to enrich and drive Nigeria's agenda on space.

In Nigeria, the collaborating partners, who in essence are the stakeholders, should include, *among others*, the following:

- Representative(s) of (a) the Federal Government at the Executive Level, and (b) the Appropriate Committees of the National Parliament (for policy formulation and financial support);
- The user-ministries and their parastatals at the federal, state and local levels—examples include *Nigerian Meteorological Agency (NIMET)*, Office of the Surveyor General of the Federation, the *Nigerian Geological Survey Agency (NGSA), Nigeria's Raw Materials Research and Development Council (RMRDC), the National Emergency Management Agency (NEMA)* as well as the *Ministries of Transport and Aviation, Mines and Steel Development, Health, Education, Petroleum, Agriculture, Water Resources*, among others;
- National Population Commission;
- The Presidential Technical Committee on Land Reform;

- The national security establishments as both drivers and beneficiaries of space research and applications;
- The space organisation, NASRDA, and its centres (for research and development);
- The research and academic institutions and the Nigerian National Universities Commission (NUC) for their scientific and technological competences and input at the R&D and applications levels;
- The relevant national academies (such as the Nigerian Academy of Engineers and the Nigerian Academy of Sciences), and the non-governmental professional organisations, such as the Nigerian Mathematical Society, Nigerian Institute of Physics and Nigerian Society of Engineers (for their scientific and technological competences and input at the R&D and applications levels);
- The private sector (for technical and industrial know-how);
- Cognate regional and international research-based institutions for collaboration; and
- Representatives from the commercial end-user communities for input into the application processes that can address the interests of the citizenry as well as the end-user markets.

Annual National Space Dialogue

The Annual National Space Dialogue should be sponsored by the National Space Council and organised by *The Think-Tank*. Participants should include all the representatives, at decision-making levels, of all the stakeholders identified above and other appropriate entities that may be invited by *The Think-Tank*. Because of its manifest importance, experience shows that such an annual national in-house dialogue is a standard practice in many countries.[4] *It is not a media dialogue.* Through such a forum, which should encourage the exchange of ideas amongst and between all the interested parties on short-term, mid-term and long-term space-related issues, including the state of science and technology in the country and associated research and development (R&D) efforts, details of the nation's specific space activities should be aired, discussed and repositioned as appropriate. The role and contributions of most of the stakeholders, such as the nation's universities, research institutions, the nation's security establishments, the academies of engineering and sciences, the appropriate professional bodies and the private sector, are essential to the dialogue.

---

4 http://www.thespacereview.com (Accessed, November 19, 2015).

Specifically, the annual national space dialogue would also provide NASRDA, as the nation's space research and development agency, the National Space Command, the nation's universities and other relevant tertiary institutions, as well as the nation's user-agencies the opportunity to showcase their activities, their different stages of implementation, their products and services, their contributions to the national development agenda, and the successes and bottlenecks they have experienced in the process. The forum should also be an occasion for all the stakeholder-institutions to connect with the nation, as a whole, and articulate the current and proposed future activities of Nigeria's space programme, including their potential impact on the Nigerian people and the Nigerian economy.

To win both the confidence and the support of the Nigerian people for the nation's space programme, the annual national space dialogue should always devote a special session to *Space and the Nigerian citizen*, as well as shed light on the input of space into the other areas of the nation's development, including our life-support systems, personal safety and national security. Such a focus would give the political leaders, the scientific community and the general public an opportunity to collectively reflect on Nigeria's continuing presence in space for its own benefit and for the benefit of the internal community as addressed below.

The National Space Council should ensure that such an annual national dialogue, as proposed herein, becomes an established agenda of the nation's space programme.

Sustainability of the Nigerian Space Programme

To successfully apply space tools and services to address the nation's needs as detailed in this book requires a sound research and professional organisation. The long-term soundness and sustainability of the nation's space programme is inseparable from the sustainability of Nigeria's activities in outer space and the returns that accrue to the nation from its space investments. To be sustainable, Nigeria's space programme would need to focus on those activities that (i) address the needs of the nation, and (ii) that Nigeria's political leaders will support and commit to financially, and for the long-term, without any equivocation. The programme must reflect the input from and enjoy the confidence of the nation's scientific[5] and the user communities; and it must also resonate with the aspirations of the Nigerian citizens. Its activities

---

5 The specific scientific communities referred to include: Cognate Nigerian universities, the Nigerian Academy of Engineering, the Nigerian Academy of Science, the National Mathematical Centre, the Nigerian Mathematical Society, and the Nigerian Institute of Physics.

should also attract the best and the brightest Nigerian minds and should be ones that encourage meaningful contributions and collaborations from national and international scientific communities. Confidences and commitments of this nature are the main drivers of successful space programmes all over the world.

Financial Sustainability of NASRDA

To be financially sustainable, the national parliament should express and reflect its confidence in the nation's space programme through adequate budget allocations. Today, this is an up-hill *battle*. Part of the problem could be the absence of space advocates amongst Nigeria's elected officials; another is the need to include a budget line for space applications in the budget of the user-agencies.

Space is a long-term commitment that requires long-term investment. The concept of budgeting annually for a national space programme, as is the case in Nigeria today, makes planning difficult for an organisation such as NASRDA. To execute its capital projects, NASRDA needs to enter into long-term partnerships with contractors and scientific partners both at home and abroad. Research and development undertakings equally require long-term planning and financial commitments. Nigeria would also need to welcome intellectual collaboration and cooperation at the regional and international levels in order to develop and enhance its own space capabilities. We need to be cognizant of the value of multilateral cooperation: it can *"nurture space-related partnerships,"*[6] and *"secure sustained access to specific space-based services."*[7] Potential international partnerships cannot be sustained if one is unable to make a commitment beyond twelve months at a time. Simply put, Nigeria's decision makers must know that a space programme is a long-term enterprise that requires long-term commitments. Accordingly, the concept of an annual budget for a space programme defeats all the opportunities and benefits that can accrue to a nation from long-term investments. A four-to-five year budget cycle for the nation's space programme is a demonstration of the nation's commitment; it is in Nigeria's interest to do so. Thus, Nigeria's politicians, who normally operate on short- to mid-term agendas, would need to change gears and *think long-term* when the issue of budgeting for the nation's space programme is placed before them for legislative approval.

Table 15:1 is a brief national story of how NASRDA's budget, often the highest among the S&T establishments of government, suffered within the reporting period

---

6 South African National Space Agency (SANSA), Dept. of Trade and Industry, Pretoria, South Africa.

7 *Australia's Satellite Utilisation Policy*, Space Coordination Office, Canberra, Australia, 2013.

of 2007 to 2013. An analysis of the table shows that if NASRDA had any advocate in government or in the National Assembly, this advocate would need to work harder. In comparison with the National Assembly (NASS) whose 2010 budget rose astronomically by 1,442% above its 2007 value, NASRDA's 2010 budget increased only to 6.28% of its 2007 value. By 2013, that disparity reached an astonishing level of 1,710% as compared to NASRDA's 2013 budget that went down to -52.94% of its 2007 budget. Hard to believe? These are figures from the nation's budget office. While it is difficult to justify this extreme budget disparity, it is very apparent that science and technology and its space component have not reached an appreciable level of both understanding and acceptance in Nigeria, particularly at the decision making level.

Table 15:1 2007 to 2013 Budget Allocations to the National Assembly and the key R&D establishments of government, in Billions of Naira *(Source: Extracted from Nigeria's Federal Government Annual Budget records)*

Year	NASS[8]	NASRDA[9]	NASENI[10]	BIOTECH[11]	NMATHC[12]	NASRDA+ NASENI+ BIOTECH+ NMATHC
2007	8.284	4.907	0.614	0.646	0.336	6.503
2008	67.141	2.938	0.552	0.620	0.555	4.665
2009	106.642	2.844	1.016	1.212	0.393	5.465
2010	127.800	5.215	0.921	0.140	0.442	6.718
2011	150.000	3.488	0.879	1.715	0.892	6.974
2012	150.000	3.710	1.227	1.580	0.856	7.373
2013	150.000	2.309	1.322	1.266	0.917	5.814

8 NASS—Nigerian National Assembly

9 NASRDA—National Space Research and Development Agency

10 NASENI—National Agency for Science and Engineering Infrastructure

11 BIOTECH—Biotechnology Institute

12 NMATHC—National Mathematical Centre

To gain a better perspective on NASRDA's financial woes, it might be instructive to undertake a comparison of NASRDA's budget with those of other space aspiring countries of the world such as Brazil, South Korea and South Africa. Figure 15:1 provides an instructive comparison of 2009-2013 budgetary allocations for the respective space programmes of these three countries relative to similar funding in Nigeria. This figure is a confirmation of what was described in Chapter VI—that is, science, technology, engineering and mathematics, including space science and technology (SST), are often talked about in Nigeria, but are not appreciated as drivers needed to power the nation's development efforts by the nation's decision makers.

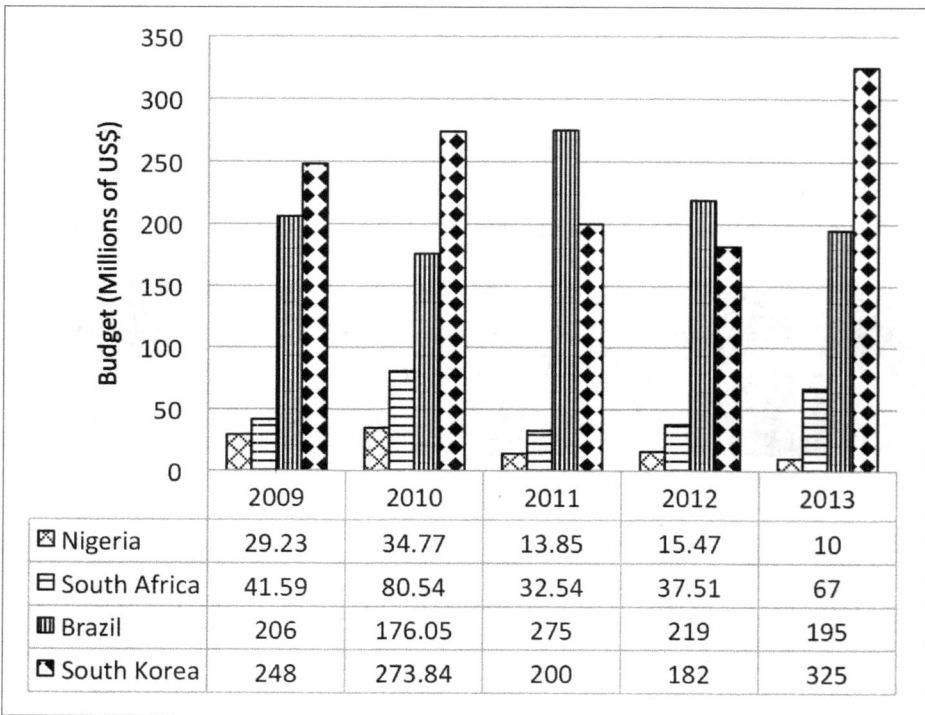

	2009	2010	2011	2012	2013
⊠ Nigeria	29.23	34.77	13.85	15.47	10
⊟ South Africa	41.59	80.54	32.54	37.51	67
▥ Brazil	206	176.05	275	219	195
◪ South Korea	248	273.84	200	182	325

Figure 15:1 Space Budgets of Nigeria, South Africa, Brazil and South Korea, 2009—2013, in Millions of US$ *(Sources: NASRDA (Nigeria); Brazil's Ministry of Planning & Budget; South Africa's Treasury; and OECD—The Space Economy at a Glance 2014)*

The under-funding of the nation's space programme and the inconsistent manner, over the years, in which the nation's parliament has parcelled out the wilted annual allocations to NASRDA have resulted in the following:

• The stifling of the nation's space initiatives and programmes;
• The inability of NASRDA to energetically carry out its statutory mandate;

- The under-utilisation of NASRDA's professional staff, their subsequent demoralization and the eventual exit of a number of them to other more reliable sectors of the economy, at home and abroad; and
- The inability of NASRDA to enter into long-term commitments as well as comply with the agreements and Memoranda of Understanding (MOUs) it has entered into with local and foreign partners. The attendant loss of respect for NASRDA's professed words of honour/agreements and the country, Nigeria, which it represents, has been incalculable.

A number of tangible steps, such as transparency and on-line availability of the reports of its activities, may positively reverse the fortunes of NASRDA on the national funding issue. The websites of the National Communication Commission, the National Population Commission, the Nigerian Airspace Management Agency, among others, should provide a guide. The Nigerian public has the right to know the accomplishments and the growing pains of NASRDA, a publicly funded entity. The public should be informed at the end of each year, on how NASDRA has spent the public funds allocated to it annually to achieve the nation's space objectives. This is one sure way through which NASRDA can effectively engage the Nigerian public to serve as its advocate at the national parliament.

To be a viable space-nation that can both face the challenges and subsequently reap the benefits of its space activities connotes that the nation will always provide the resources needed to sustain such an enterprise. Nigeria's current involvement in space is at the rudimentary stage. Data acquisition, transmission, processing, analysis and interpretation, to obtain the information that can be used in decision making, are yet to have a significant impact, nationally. Nigeria is yet to undertake, at home, significant space technology activities that involve proof of concepts, development and fabrication—a feat that should produce spinoffs that can contribute to the improvement of the nation's security, economy, employment, productivity and lifestyle. Figure 15:2 (next page) shows the growth in the global space economy and the collective benefits that are accruing to those countries that have made such a leap forward, through corresponding commitments in technology development and the input of a vibrant private sector.

As shown in Figure 15:3, the current funding pattern of the 25-Year Road Map will not attain the intended goals of the nation. However, through a greater sense of appreciation, by the nation's decision makers, of what space can do for Nigeria and its people, and which is manifested in tangible committed investment and nurturing, Nigeria's redefined and redesigned space efforts can be part of the future space economy statistics, similar to the ones reflected in Figure 15:2.

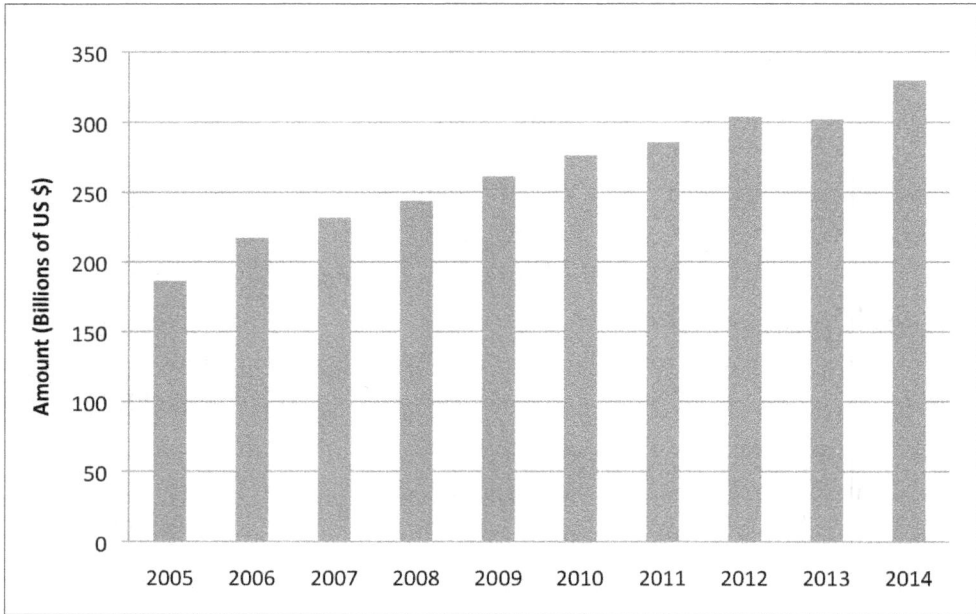

Figure 15:2 The Global Space Economy *(Source: The Space Reports (2006-2015), The Space Foundation, Washington DC, USA)*

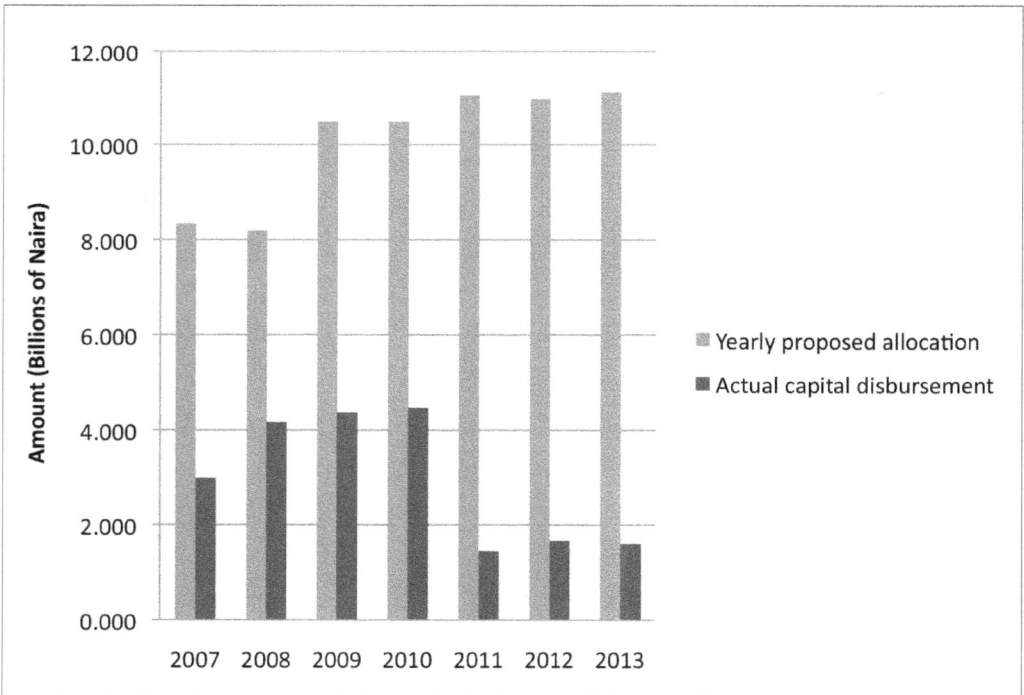

Figure 15:3—Funding of Phase 1 of the 25-year Road Map for Nigeria's Space Mission (2005-2030) *(Source: NASRDA)*

Learning from the Experience of Others

In this chapter, as well as in other sections of this book, I have proposed a fresh start for Nigeria's space journey. As shown in Chapter VIII, such a new beginning will enable the nation to benefit from what it should have learned from its past actions as well as from the experience of others. To reach that next stage and beyond, it would be expedient for a strategic small group, made up of Nigeria's decision makers, representatives of the nation's scientific and engineering communities including NASRDA, and the science and engineering students from the nation's universities, all acting as *fact-finders*, to undertake genuine working visits to the space and scientific research and development institutions of countries such as Australia, Brazil, India, South Africa and South Korea. Such strategic fact-finding missions will provide the *fact-finders* and the nation with a better perspective and appreciation of what these countries are doing nationally, regionally and globally, to sustain their active role in science and technology and in the space enterprise of today and tomorrow. *The knowledge gained from these missions can provide Nigeria with a deeper understanding of what works in the different settings visited and how to avoid the mistakes experienced by others.*

In each space establishment visited, the Nigerian *fact-finders* should both look at and examine the setup of the respective space organisations, assess the calibre of their employees especially that of their space experts and the leadership of each organisation; objectively evaluate the organisational culture and the kinds of discipline that permeate each organisational setting; observe and study the level of dedication to duty of these employees, the creative and nourishing environments in which they work and, last but certainly not the least, appraise their sense of duty and commitment to the goals set by their leaders. It could be illuminating if our *fact-finders* would also ask their hosts one or two specific questions: (a) If you were to start all over again, what would you do differently? (b) What lessons have you learned from your mistakes? If Nigeria is committed to becoming a veritable space-capable nation, such an exchange of views should provide the Nigerian fact-finders with a better sense of what it will take to do the same in Nigeria.

As the Nigerian visitors depart from each of their destinations, and head for home, they should not forget to note that in India, "*Our space application programme is user driven,*" that is, it reflects the focus and the impact of the nation's space programme on the lives of the Indian people. In case the fact-finding mission comes to Washington, D. C. and Moscow, the Nigerian visitors would note that the era of an unbridled space race, as advocated in the 25-year Road Map, is over. In fact, today, the Russians have stated that "*Space exploration is part of our national life.*" In the United States, the

"*Mars belongs to us*" euphoria of 1997, when the *Pathfinder* spacecraft landed on Mars, has now given way to "*Space enhances our lives.*" During its sojourn in South Africa, the mission should also assess the claim of the South Africans that they are "*Africa's eye on the universe*"—a claim made possible by HartRAO, SALT and the Square Kilometre Array (SKA) projects. At the end of the mission, it should be apparent to the Nigerian *fact-finders* that the space enterprise has become one of the critical and fundamental foundations of industrialization and will be much more so in the foreseeable future. They would also note that the space activities they just witnessed were all built on a solid foundation of science and technology.

Nigeria and Nigerians, at home and abroad, are anxiously waiting for the recommendations of these Nigerian fact-finders to the people and government of Nigeria. They want to know how Nigeria, as a nation, should redefine and redesign its road-map for its space journey, and the commitments it must make to attain the aspirations of that journey in order that Nigerians can also proclaim in the near future that "*Space is our operational tool of development.*"

* * *

Chapter XVI

CONCLUSION

Embarking on a journey implies an appreciation and an understanding of where the journey is heading and the preparations needed to get there. Thus, as a commitment to Nigerians of today and tomorrow, we should give our collective and undivided attention to our justifications and ultimate purpose for embarking on a new redefined and redesigned space journey as recommended in Chapter XV.

Nigeria started this journey because of its desire to employ space tools, products and services to meet the needs of the Nigerian people. Space will help us manage the nation's resource endowments and environment, provide for human safety and national security within the Nigerian society, and we can and should effectively utilise Nigeria's human resources in the pursuit of these challenges. No one can stop Nigeria from achieving these goals, except Nigerians. Similarly, no one will help Nigeria to accomplish them as they should, except Nigerians. If we apply our inquiring minds to address these critical human needs, as illustrated with a number of examples throughout this book, space tools and services can contribute to making the journey a more rewarding one for all of us.

One of the non-negotiable pre-requisites of this new refocused journey is Nigeria's determination to eschew the unbridled purchasing and importation of technologies from abroad at the expense of indigenous development of the technologies we need at home. Such predilections, which have taken root on the Nigerian soil, at the federal, state and local levels, over a period close to six decades, have only succeeded in robbing Nigerians of the opportunities to be more creative and inventive in trying to solve their own problems. Although we cannot make everything we need, we certainly must not buy everything we need. As shown in Chapter X, Nigeria is endowed with much natural wealth and human capability. Nigeria's private sector should be given every incentive and be nurtured to produce consumer goods with a minimum of 30% local content. Nigeria must pin its hopes and its future, including the future of its unborn generations, not on the sweat and labour that Nigeria's petro-naira can purchase and import from abroad, but on the backbone of the inventions and creativity of its own people.

The attainment of these goals requires two commitments. First, we must re-examine ourselves and our attitudes toward our own nation, and we should ask ourselves, whether, indeed, we carry out our respective daily duties *in the interest of Nigeria. If what we want and do for Nigeria is greater than any and all our individual striving, we will collectively reap the benefits.* Second, we must embrace science and technology as tools of development. A solid science and technology foundation is an indispensable prerequisite for Nigeria's development and growth and also for the success of its space journey. By taking such a path, Nigeria will develop, grow and nurture its own experts as it is the practice in those societies that have been the source of expertise on Nigeria's problems and needs for the past five decades and more. The Nigerian post-independence S&T foundation, which Nigeria's inquiring minds of that era proudly projected to the world as an intellectual giant, is dwarfed in the face of today's technological advances. Chapters IX through XII provide a number of practical, realistic and achievable remedies that can re-invigorate the nation's S&T foundation which should enable Nigeria to move on, confidently, with its newly re-charted (redefined and redesigned) space journey as proposed in Chapters XI, XII and XV.

The global future, of which Nigeria is a part, is now here; its tools of development are, in part, the convergence of standard industrial technologies and the evolving new ones. As described in Chapters IX and XI, Nigeria must be in step with the knowledge requirements of the new and advanced technologies that are powering the computer industry, renewable energy development, the space enterprise, communication products and services, transportation systems, health-care service delivery, and rural and urban development, just to name a few. Such advances will enable Nigeria to successfully diversify its economy and foster economic growth that can contribute to the creation of new goods, new services, new jobs, and new capital and can also attract unparalleled partnerships across the globe.

In furtherance of Nigeria's contribution to these advances, the federal and state governments should enact necessary regulations and provide incentives, including financial ones, that will not only encourage and facilitate, but will also promote and nurture working partnerships between (i) the nation's research institutes including the research and development centres at the nation's universities, (ii) the private sector—both at home and from abroad, and (iii) the public sector. The primary objective of such collaborations is to bring researchers and people with practical guidance and real-world thinking together to work for the nation's good. In the process, the private and public sectors benefit from the knowledge and resources available through the expertise at the nation's universities and the research institutes, while the partnerships cultivate and nurture global entrepreneurs of tomorrow in the students and their educators. The partnerships also offer the research institutions the opportunities to access funding from the participating public and the private sectors. By so doing, research efforts are advanced, and new ideas and technologies are developed. The latter not only respond to the needs and requirements of the society, but can also fuel the development of new products for the Nigerian consumers and the creation of new jobs at home. Chapters IX and XI contain a number of examples of such partnerships from different parts of the world. It is a proven approach that can facilitate industrial advancement of Nigeria as well as prepare the nation for the industrial competitiveness of the future.

Nigerians should also note that in the space age of today, countries that are able to help themselves, that invest in the generation of knowledge and can contribute to innovations, are often enthusiastically welcomed to interact, at the research and development levels, with their counterparts within the space-faring and space–capable communities of the world. Nigeria must cultivate and nurture such a character if it is to be highly regarded in both regional and international circles. Today, *subject to the principle of give and take*, the doors of new scientific and technological cooperation remain open to several of '*long ago perceived but no more contemporaries*' of Nigeria, namely, Brazil, China, India, Mexico, South Africa and South Korea, but not to Nigeria. Why? Because these countries are committed to long-term science and technology, investment and development programmes. Indeed, they all now belong to one or more of the international associations, such as BRICS,[1] MIKTA[2] and OECD[3] which were established with the goal of fostering collaboration in S&T,

---

1 BRICS—An association of five major emerging national economies, consisting of Brazil, Russia, India, China and South Africa, It had its first Heads of State Summit in 2009.

2 MIKTA—An association of five medium level powers, consisting of Mexico, Indonesia, South Korea, Turkey and Australia; they first met in 2013.

3 OECD—The Organisation for Economic Co-operation and Development (OECD), with a membership of 34 countries today. It was originally established as a European organisation in 1960. It subsequently expanded, and its membership now spans the globe, from North and South America to Europe and Asia-Pacific.

economic development and trade, as well as tackling the global challenges of our time, particularly, climate change and global security. Nigeria can accomplish the same feat as its former contemporaries have done, through its resolute commitment to a sound S&T foundation and its effective participation in the space enterprise. For example, Nigeria should put priority on collaborating with other like-minded countries within the equatorial and tropical belts to study, master, develop and utilise the equatorial plane for their own social and economic development and for the advancement of human exploration of outer space. Subsequent efforts should include indigenous development and deployment of experimental payloads to advance this knowledge.

We should not forget that Nigeria was once a very successful agricultural country: it was a large net exporter of food and the world's leading producer and exporter of palm oil, groundnuts and cocoa. The discovery and production of crude oil in Nigeria, and our subsequent singular devotion to petroleum economy, brought an end to the nation's agricultural productivity and ushered in its attendant economic woes. As shown in Chapter X, the down-hill trends can be reversed with Nigeria's acceptance of the reality that its future hinges on economic diversification. In concert with such a diversification agenda, Nigeria must also be steadfast in its on-going effort to enhance local food production; it must feed itself and regain its lustre as a food exporter, through sustainable agriculture and the mastery of new technologies needed to attain these goals. An effective national space programme is a critical contributor to the revitalisation of the agricultural sector of the economy.

As Nigeria proceeds on this re-invigorated space journey, it would be paramount for its drivers and the nation to focus on how space will advance the development and growth of Nigeria and enhance the wellbeing of its citizens. To attain these goals requires targeted investments in appropriate ground infrastructure, a sufficiently educated population that understands and appreciates the significance of space in development (including politicians and policy makers), and a trained workforce that will effectively use the data being acquired by the nation's satellites. Part of this effort should include how prospective researchers can gain access to satellite acquired data at little or no cost. Chapters VII and XI provide a guide. But if we continue to focus on what the journey will do for Nigeria's prowess and its continental leadership, particularly through continued and unsustainable acquisition of space hardware, we will be going nowhere.

As its name implies, the long-term focus of NASRDA should be research and development. And, in partnership with Nigeria's universities, NASRDA should devote all its attention, and commit its human and financial resources, to fundamental and applied research and the development of specific technologies that can mature into a variety of products and services that can benefit the Nigerian people. A number of such technologies could serve as research payloads and components of future Nigerian space efforts, while others could find application possibilities, as spin-offs, in various aspects of the nation's economic development programmes. Meanwhile the government should encourage the private sector, with adequate incentives, to focus on bringing into the market, the research results from the nation's universities and allied institutions.

Furthermore, countries that are space-bound practise long-term, multi-year financial commitments; they do not parcel out their national space funds on an annual basis. A multi-year financial plan, with built-in rigorous accountability procedure, will allow the national space organisation to engage in a set of goal-oriented national space activities; it would also enable Nigeria to respond, with commensurate contributions, to its regional and international obligations. Nigeria, as a potential competitor in the global market, should also not expect any other country or international organisation to subsidise its space aspirations. Specifically, Nigeria must put its own money where its mouth is.

In this book, I have reflected on Nigeria's space journey, including its origin and subsequent developments, as I know it, and as I believe it is today. I have also recommended that the Federal Government of Nigeria commit itself not only to a new refocused space journey, but also to the building of a buoyant S&T capability which can support that journey and revitalise all segments of Nigeria's economy. That commitment should be championed by the National Space Council, and should be manifested, not only in funds, but also through appropriate government oversight, incentives, and enabling legislations. Such legislations should result in tangible contributions of Nigeria's inquiring minds, at home and in the Diaspora, to frontline research and knowledge generation and development, in Nigeria. In hindsight, the status of S&T in Nigeria would have been better secured if there had been an enabling legislation to back up the 2006 initiative to establish the National Science and Technology Foundation with an initial endowment of US$5 billion.

If Nigeria is to embrace space as a tool of development, that process must begin with appropriate education. Educating Nigerians to appreciate and understand the role of S&T and space in national development, in their daily lives, and in the management of their environment, should be a paramount national undertaking, not only in the nation's classrooms and research institutions, but also at all the legislatures, through the media and in public fora. Equally imperative is a *continuing annual national space dialogue* that would foster long-term collaboration between and among the stakeholders, including, the government and its research institutes and user agencies, the nation's universities, the private sector, the user communities and the public in general.

If we stay the course and navigate our way with realistic planning and achievable goals for the refocused space journey, Nigeria's space effort should, among other things:

- Have a multiplier effect on Nigeria's economy and industrial production;
- Enhance the quality of life of Nigerians;
- Enhance the science and technology capability of the nation;
- Ingrain a can-do mentality in Nigerians;
- Reduce Nigeria's scientific and technological dependence on others;
- Strengthen the meaning of and give substance to technological collaboration and knowledge sharing between Nigeria and other countries;
- Strengthen the meaning of "independence" for Nigerians and Nigeria and serve as a significant symbol of national achievement; and
- Uplift Nigeria's image at home, in Africa and within the international community.

Through its long-term investment in the development and growth of science and technology, Nigeria will have the opportunity to shape its own space and economic future as well as become a leading partner and voice in determining Africa's future. While space-faring and space-capable nations of the world are charging ahead, could Nigeria be close behind? Our actions, as a nation and a people, will tell. The world is watching and hoping that the political wind of change that arrived on Nigeria's shores in early 2015, will also clear a path to progress for the country and all its citizens—progress that taps the full potential of the country's human and natural resources, explores all its frontiers, and builds a country with ever-expanding opportunities, where the sky is, indeed, the limit.

* * *

ABOUT THE AUTHOR

Born in the Village of Araromi Abiodun, Ogun State, Nigeria, Adigun Ade Abiodun was educated in Nigeria, the USA and Canada. He obtained his Ph.D. in 1971, in civil engineering from the University of Washington, Seattle, Washington, USA, and was a post-doctoral research fellow at the Canada Centre for Remote Sensing in Ottawa, Canada. From his academic post at the University of Ife (UNIFE), since re-named Obafemi Awolowo University (OAU), he was seconded in 1977, by the Nigerian government, to the United Nations Outer Space Affairs Division (OSAD), since renamed Office for Outer Space Affairs (OOSA). He was appointed in 1981 as the United Nations Expert on Space Applications and served the organization in that post until his retirement in 1999. With the goal of building indigenous capacity and capability in space science and technology in the developing countries, he led the United Nations efforts in the establishment of the United Nations-Affiliated Regional Centres for Space Science and Technology Education in India, Brasil/Mexico, Morocco and Nigeria. He also led the United Nations effort in the development of "A Space-Based Cooperative Information Network (COPINE)—Linking Scientists, Educators, Professionals and Decision Makers in Africa."

After his retirement from the United Nations, he was appointed a Member of the College of Commissioners of the United Nations Monitoring, Verification and Inspection Commission on Iraq (UNMOVIC), (2000-2007), and as the Senior Special Assistant (to the President of Nigeria) on Space Science and Technology (March 2000–May 2003). He was elected and served as the Chairman of the United Nations Committee on the Peaceful Uses of outer Space (June 2004–June 2006). He also served as one of the 20 global panelists that explored, in June 2006, *Humans and Space: The Next Thousand Years,* at the Foundation for the Future, Bellevue, State of Washington, USA; and as a Member of the Association of Space Explorers' (ASE) Panel on the Mitigation of Asteroids Impact (2007-2008). The ASE panel concluded its work with a published report titled: *Asteroid Threats: A Call for Global Response.* He was a contributor to the *First Asteroid Day Event,* June 30, 2015, with the signing of *The Asteroid Day 100x Declaration.*

He initiated and co-founded the African Association of Remote Sensing of the Environment (AARSE), and the African Leadership Conference on Space Science and Technology for Sustainable Development (ALC). He was a Board Member of the Space Generation Advisory Council and a Trustee of the International Society for Photogrammetry and Remote Sensing. He is currently a member of the Board of Directors, World Space Week Association (WSWA), which he once chaired. He is also a member of the African Union Space Working Group.

Ade Abiodun has lectured and published extensively on different aspects of space science and technology. He is the Founder of the African Space Foundation.

* * *

INDEX

Note: Page references followed by *f* or *t* indicate material in figures or tables, respectively. Page references containing *n* indicate material in notes. All references to programmes, agencies and governmental entities are Nigerian unless otherwise specified.

Australia
    ancestral knowledge in, 229
    economic impact of satellite data in, 265
    equatorial assets of, 294
    fact-finding in, 365
    navigation system in, 282, 283, 284
    Square Metre Array, 210–211, 261
automobile emissions, 321
automobile industry, 136, 221–222, 225
autonomy, loss in universities, 128–129

B

Baalke, Ron, 59
bagasse, 207
Baiada Center for Entrepreneurship (USA), 203
Balewa, Abubakar Tafawa
    18-Article Agreement (Nigeria–USA), 51
    satellite-based telephone conversation, 4, 55–56, 65
Balogun, E. Ekundayo, 66, 182–183, 303
Ban ki-Moon, 320
barite, 236t
Baro-VNAV system, 282, 283
Baseline Surface Radiation Network (BSRN), 74–75, 74f, 326
basic science education, 217–219
bauxite, 116, 236, 239
Bayero University, 149
BeiDou Navigation Satellite System, 26n1, 271n47, 274, 274n58, 281, 284, 330n5
Belgium, communication satellite of, 163
Bello, Sir Ahmadu, 56n29
Benin
    cultural artifacts of, 231–232
    marine ecosystem, 30–34, 30f
    as Niger Basin Authority member, 80, 80n42
    shared security concerns with, 289
Bepposax satellite, 174, 263, 341–342, 342f, 344
Berlin Wall, fall of, 3, 21, 21n14
Bhattasali, B. N., 242–243
Biafran War (Nigerian Civil War, 1967–70), 61–62, 87, 197–198, 213, 259
big data analytics, 253, 253n4
Biotechnology Institute, 361t
Birmingham Investment Fund (United Kingdom), 202–203
bitumen, 237t
Blue Print for Scientific and Technological Development, 94–97, 103–104, 105
Boeing Company, 171, 298

Borno Kingdom, 229–230
Bosumtwi Lake impact crater (Ghana), 333–335, 334f
Botswana, Square Kilometre Array, 211, 261
Brazil
    astronauts/ISS participation of, 171, 293n9, 299–301
    automobile industry in, 136
    deforestation in, 321
    equatorial assets of, 294, 295–296
    fact-finding in, 365
    foreign S&T talent in, 200–201
    healthcare initiatives in, 28, 200
    launch facility of, 172n54, 294, 296
    R&D in, 111, 125, 200–201, 225
    remote sensing satellite, 154, 184
    space budget of, 362, 362f
    space-capable private sector in, 305–306
    space shuttle projects of, 300–301
    state-level development in, 215
    State of São Paulo Research Foundation, 137–138, 200–201
    S&T enterprises in, 253, 355
    S&T success in (ethanol), 204–207
BRICS, 133, 133n53, 369, 369n1
broadband subscribers, number of, 112, 132, 143f
broadcasting systems, 65, 165–166
bronze artifacts, 231–232
"Brown Bag Journalism," 96–97
Buckley, David, 210
budget allocations, to space programme, 145, 360–363, 361t, 362f, 371
Buhari, Muhammadu, 78, 124, 130,
Burkina-Faso, as Niger Basin Authority member, 80, 80n42
buy and purchase
    communication satellites from China, 160–162, 168, 187–188
    eschewing approach of, 367–368
    indigenous development *vs.,* 5–6, 89–103, 106–110, 180–182, 304–305, 367–368
    lessons learned, 179–183
    Nigerian Satellite Project, 5–6, 89–103, 106–110, 146, 179–180

C

Cadbury Nigeria Plc, 134–135
California Institute of Technology (USA), 303
Calls for Proposals (USA), 305
Cameroon
    deforestation in, 321

www.ingramcontent.com/pod-product-compliance
Lightning Source LLC
Chambersburg PA
CBHW082126210326
41599CB00031B/5889